NEWS
Around the
WORLD

Practitioners, and the Public

NEWS
Around the
WORLD

Pamela J. Shoemaker
and
Akiba A. Cohen

Routledge
Taylor & Francis Group
New York   London

Published in 2006 by
Routledge
Taylor & Francis Group
270 Madison Avenue
New York, NY 10016

Published in Great Britain by
Routledge
Taylor & Francis Group
2 Park Square
Milton Park, Abingdon
Oxon OX14 4RN

© 2006 by Taylor & Francis Group, LLC
Routledge is an imprint of Taylor & Francis Group

Printed in the United States of America on acid-free paper
10 9 8 7 6 5 4 3 2 1

International Standard Book Number-10: 0-415-97505-0 (Hardcover) 0-415-97506-9 (Softcover)
International Standard Book Number-13: 978-0-415-97505-6 (Hardcover) 978-0-415-97506-3 (Softcover)
Library of Congress Card Number 2005026749

Library of Congress Cataloging-in-Publication Data

Shoemaker, Pamela J.
    News around the world : content, practitioners, and the public / Pamela J. Shoemaker and Akiba A. Cohen.
        p. cm.
    Includes bibliographical references and index.
    ISBN 0-415-97505-0 (hardback : alk. paper) -- ISBN 0-415-97506-9 (pbk. : alk. paper)
    1. Journalism. 2. Reporters and reporting. I. Cohen, Akiba A. II. Title.

PN4731.S48 2005
070.4--dc22   2005026749

Taylor & Francis Group
is the Academic Division of Informa plc.

**Visit the Taylor & Francis Web site at
http://www.taylorandfrancis.com**

**and the Routledge Web site at
http://www.routledge-ny.com**

Contents

Acknowledgments

We did this study for all of the usual reasons—the importance of studying news, especially in multiple countries, and testing a theory based on data previously collected in the United States. But to be truthful, the study became possible for a very practical reason: We came into some money.

"What are you going to do to make us famous?" That's what David Rubin asked Pam Shoemaker when he told her about the one-time accounting windfall her research chair account was receiving on July 1, 1999. They were both in the S.I. Newhouse School of Public Communications at Syracuse University in New York, with Rubin as dean and Shoemaker as the John Ben Snow Professor.

She had a few general ideas but wanted to think about it for a while. After a day or two she tossed preliminary ideas at her German research assistant, Martin Eichholz, a student in the Newhouse doctoral program. Eichholz was eager to do research and was not afraid to take on a big project.

The big idea was to take Shoemaker's theory about the nature of news from application in the United States to application around the world: Would her U.S. findings showing that deviance and social significance are predictors of newsworthiness generalize to other countries? The study would definitely involve content analysis of news media in several countries—as many as possible—and maybe a survey to measure what people think is newsworthy.

Eichholz and Shoemaker discussed the idea with Rubin, and he was enthusiastic. Shoemaker met with Jonathan Snow and Ann Scanlon of the John Ben Snow Foundation, which funded Shoemaker's research. They thought an international study was a good idea, and so the study was born.

Shoemaker wanted a research partner and coauthor with more experience in international research, both in methodological issues and in the number of international scholars known. She immediately thought about Akiba Cohen, chair of the Department of Communication at Tel Aviv University in Israel. Cohen (who was then at the Hebrew University) had invited Shoemaker to an international news conference in Jerusalem in 1988. He had done numerous international research projects, and he had contacts around the world as a result of serving as past president of the International Communication Association (ICA). But would he help with Shoemaker's project?

Shoemaker and Eichholz asked Cohen to meet with them at the 2000 International Communication Association meeting in San Francisco. Shoemaker outlined her theory, described her plan, and asked Cohen if he thought she was crazy.

No, not crazy, Cohen replied, but it was an ambitious plan, and maybe some changes should be made. Shoemaker listened and became more confident that she needed Cohen to help her do the study. Would he agree to help her get started? What was the first step?

Gather information and advice from people who knew about such studies, Cohen said. Since Shoemaker and Eichholz were planning to give a paper at the September 2000 World Association for Public Opinion Research (WAPOR) meeting in Paris, it was agreed that an advisory team should be assembled to meet there.

So Shoemaker and Eichholz invited several people to hear out her ideas and to talk about the sort of study that should be done:

- Akiba Cohen, Tel Aviv University, who has experience with several cross-country research projects
- Julianne Newton, Oregon University, Corvallis, who does research in visual communication
- Robert Stevenson, University of North Carolina, Chapel Hill, who has lots of experience in working with scholars in other countries
- Gideon Kouts, Sorbonne University, Paris, who teaches journalism and continues to be a freelance reporter for Israel's Channel 2 television news

The team met for two days. After hearing Shoemaker's objectives, various research options were considered, with much spirited discussion. How many countries? Which countries? Did anyone know mass communication researchers in these countries? Should research companies be hired instead of university scholars? How many news media? Which ones? Over which period of time? For how long? Should a survey be done? Could it be done in some countries? Was survey research too expensive? Could focus groups be done instead? How many cities in each country should be studied? Should visual content (images) be studied separately from verbal content (text)? And so on.

Shoemaker and Eichholz left Paris with the directive to write a study proposal, which the advisory team then reviewed. After the first proposal review, Shoemaker was even more convinced that any study with a good chance of achieving her goals and of being completed would have to include Cohen as a partner. She invited him to be co-investigator—a full partner—on the project. Fortunately, he agreed.

The partnership proved to be crucial. Both Shoemaker and Cohen made important changes at every stage of the project, which caused delays from time to time. Nothing seemed easy, but the study moved forward, and both knew that it would be good: Something valuable would be learned.

The plan evolved to include a quantitative content analysis of newspapers, television news, and radio news programs in two cities in each of ten countries. Shoemaker and Eichholz worked on the study protocols and visited Cohen in Tel Aviv later in the year to work on operationalizing the concepts. Cohen's research assistant, Noa Loffler-Elefant, joined the team and helped tremendously with planning the content analysis coding scheme and in planning and collecting data in Israel.

We were fortunate to work with many talented research assistants. In January 2000, Elizabeth Skewes joined the research team as a second research assistant for Shoemaker. Cohen met with the American team in May at the 2000 ICA meeting in Acapulco, Mexico, to finalize the countries. We decided to hold a meeting in New York City that September to train the directors who would represent each country. Eichholz and Skewes were in charge of the meeting arrangements. Eichholz left the university to join a market research firm, and Heather Black began sitting in on meetings in the summer and officially joined the research team in August 2000, just in time for the meeting in New York.

Representing their countries at this two-day meeting were Chris Lawe-Davies, Australia; Guoliang Zhang and Jonathan Zhu, P.R. China; Maria Soledad Puente, Chile; Kavita Karan, India; and Tankred Golinpolski, Russia. Shoemaker, Skewes, and Black represented the United States; Cohen represented Israel, and Eichholz represented Germany. Mohammed Ali, representing Jordan, was unable to attend the meeting. A short time later Danie du Plessis became country director for South Africa, and Natalia Bolotina became country director for Russia. Cohen later met with Ali to go over the study materials. Shoemaker met with du Plessis and Bolotina. In addition, Carsten Reinemann joined the team to represent Germany and was trained by Cohen. Also of assistance in Russia were Yassen Zassoursky, dean, and Elena Vartanova, deputy dean for research, both at Moscow State University.

Data collection for the content analysis proceeded in November and December 2000, with coding and the focus groups in 2001. Tim Vos joined the research team in fall 2001, just in time to help Black plan an international symposium to present the early results in April 2002.

The *What's News? Symposium* included results from each of the ten countries, along with research papers competitively selected as a result of an international call for papers. Papers were given by scholars from more than twenty countries. In addition, scholars from thirteen other countries also attended.

It took much longer than expected to ensure that the ten-country data sets were error free and able to be merged into one large data set. Black and Vos bore this responsibility. The country directors wrote and submitted their chapters during 2002–2003, when Shoemaker was on research leave. In fall 2003, Pamela Morris and Jong-Hyuk Lee joined the research team. Morris worked with the country directors on final changes, while Lee worked with Shoemaker and Cohen on the data analyses for the deviance, social significance, and gatekeeping exercise chapter 5. In fall 2004, Gang (Kevin) Han and Donald Holeman became part of the team, reading the manuscript and suggesting many important changes. Josh Shear read page proofs in 2005.

In addition to those named already, we must thank Syracuse University, which agreed to contract with faculty at universities in the nine other countries to be studied. Joan Ruggaber, now the retired budget director in the Newhouse School, was sometimes surprised by Shoemaker's requests, but she always found a way to accomplish them. Sherry Coryell has ably followed Ruggaber as budget director. Christi MacClurg provided administrative and clerical services. Rubin continued to give his enthusiastic support.

The John Ben Snow Foundation funded the study, through its 1965 establishment of the John Ben Snow Professorship in the Newhouse School. We thank Jonathan Snow and Ann Scanlon for their support, and for Scanlon's welcoming remarks at the April 2002 *What's News? Symposium.*

We are pleased with the work everyone has done, yet we have to admit that it took a lot longer than we expected to complete each phase of the project, especially writing this book. But the passing of time is sometimes an advantage, and in this book we have made some important changes in our theory as we confronted the realities of news in ten countries. So in this case more time spent on the project has meant a deeper understanding of our subject. Thanks to editors Matthew Byrnie, Devon Sherman, and Robert Sims of Routledge for their patience and assistance.

Part 1
Theory and Method

1
Introduction

People are interested in news. Whether the news comes from other people or from the mass media, we like to know what's going on—in faraway places or in our neighborhood—we want to be in the know. Of course not everyone is equally interested in general news, and people may be more interested in one topic than another. But an underlying interest is there. How could people fail to attend to an earthquake in their region, or to a cure for cancer or AIDS?

In part 1, we propose that all humans monitor the world around them in order to find out what occurrences, either threatening or hopeful, are important, and that we share this with many animals. Lasswell (1960, 118) calls this the *surveillance* function of the mass media. "In some animal societies certain members perform specialized roles, and survey the environment. Individuals act as 'sentinels,' standing apart from the herd or flock and creating a disturbance whenever an alarming change occurs in their surroundings."

In human societies, these individuals may be town criers, gossips, or others who pass along information. In most societies, journalists also fill this role—acting as socially sanctioned, professional surveyors for the rest of us. While most people may survey the world for ideas, people, or events that have personal relevance to them, journalists tend to concentrate on the social realm. Journalists relate to the rest of us those things to which we should pay attention, but to which we do not otherwise have access.

The existence of the mass media is largely justified by this surveillance function, and while the mass media may be part of the economic and political systems, even this is a result of their role to give people the type of information in which they are more or less universally interested. Although news is a manufactured product and is subject to a wide-ranging set of influences (Shoemaker and Reese 1996), the basic form from which news comes is people's innate interest in two types of information: (1) people, ideas, or events that are deviant (either positive or negative); and (2) people, ideas, or events that have significance to the society.

Early news media research has set the stage for investigating the definition of *news* by focusing on the news production process, including influences on journalists and their organizations. White (1950) created the first gatekeeping study by reviewing how, as a wire editor, Mr. Gates's personal beliefs and knowledge of news routines influenced his selection of news items (see also Shoemaker 1991). Soon afterward, Breed (1955) wrote about how journalists are socialized in the newsroom, recognizing organizational policies and supervisors' influences.

Several scholars discuss *news* as a socially constructed product. Molotch and Lester (1974) analyzed news items in terms of being "routine," "scandal," or "accident." Tuchman (1978) explored the news-making process and found that the structure and labor division of news organizations influence the definition of *news*. Fishman (1980) focused on routines of local beat reports and found that journalists guide news by making decisions on their beat structure rather than on news values.

Phillips (1976, 88) fielded a participant-observation study of four news media organizations and reported that news practitioners do "not conceptualize their own experiences or place concrete particulars into a larger, theoretical framework." Gans (1979) championed an integrated approach to news in his combination of a content analysis of television and magazine news with participant observation at four media organizations. Cohen et al. (1996) looked at the production, news content, and news reception in eleven countries in connection with the European Broadcasting Union's News Exchange Service. Jensen (1998) and his colleagues in seven countries used content analysis, individual interviews, as well as household interviews to study themes that emerged in people's thinking about the news.

Research on international news is primarily centered on how news about foreign countries is distributed and structured around the globe (Hester 1973; Malek and Kavoori 2000; Pasadeos et al. 1998; Sreberny-Mohammadi 1984; Wallis and Baran 1990). For example, content-based studies include Gerbner and Marvanyi's (1977) analysis of foreign news coverage in selected newspapers in nine countries. Findings show that variation in the amount of foreign news coverage is correlated with political systems. Kim and Barnett (1996) reported that economic development is the most important determinant of a country's place in the network of international news flow. More recently, Wu (2000) reviewed foreign news in thirty-eight countries and suggested that coverage is primarily determined by economics and availability of news sources.

The definition of news within different cultural settings is not always considered; however, some studies try to understand the meaning of news within the culture and how meanings are similar and dissimilar across cultures. Galtung and Ruge (1965) reviewed coverage of three international crises in four Norwegian newspapers and proposed eight "culture-free" and four "culture-bound" news values that influence whether an event will become a news item. Harcup and O'Neill (2001) reviewed twelve factors to present a contemporary set of ten news values: the power elite, celebrity, entertainment, surprise, bad news, good news, magnitude, relevance, follow-up, and newspaper agenda.

Shoemaker and Reese (1996) proposed a hierarchical model suggesting that influences on news result from individual news workers; the routine practices with which news is collected, transformed, and disseminated; the characteristics of news organizations; social institutions outside of the media; and social system ideology and values. Weaver (1998) expanded upon these ideas in his research on journalists around the globe.

Any way we look at it, people seem to have an innate interest in information. Furthermore, people's interest in certain types of information tends to be the same, even while their interest in specific topics may differ. This book builds upon the primary idea that humans in all cultures and countries are basically interested in the two types of news information described already.

We wish to point out that this book—indeed, the whole project—is an example of the *deductive* model of social science research. This approach begins with theory, derives hypotheses from the theory, tests the hypotheses, and then revises the theory as necessary. In an ideal world, the deductive process continues until the theory is complete (or until scholars grow weary of it).

Although the deductive model is often taught, it is rarely observed in our literature. Journal articles, which usually allocate fewer than twenty pages, typically begin with some limited theoretical notions and test hypothesis, but the discussion is often too short for a comprehensive interpretation of the results and for proposals to change the theory in anything more than a "the next step ought to be" fashion. If scholars were able to immediately follow one journal article after another, showing how their theory has been revised, how new hypotheses are derived, tested, and so on, then we would probably have more and better theories in our field. Such a set of journal articles could eventually contain the theory in its entirety. But the world of scholarly publishing does not encourage the development of theory-driven research programs; manuscripts are judged as unique products and not as the second or third iteration in the logical and deductive process of theory building. The result is that theory construction is discouraged.

While we do not claim to have derived a complete theory of news and newsworthiness, this book does present the stages of the deductive process that we went through. Accordingly, chapter 2 presents the initial theory and hypotheses. The methodologies are presented in chapter 3 in which we discuss how in the first stage we selected ten countries for the study and how we completed quantitative content analyses of sixty newspapers and television and radio news programs for seven days across seven weeks. The second stage involved eighty focus groups in twenty cities, representing journalists, public relations practitioners, and high and low socioeconomic-status audience members. These groups were used to find out what type of information people want and, later, what sorts of news events they thought were most memorable. In stage three, we used quantitative data secured at the end of the focus groups, which asked people to rank-order newspaper headlines.

The hypotheses are tested in several ways across countries in chapters 4–7 and detailed within countries in chapters 8–17. In chapter 18 we interpret these results and conclude that not only are some of our original assumptions

and sensationalism. These indicators of news tell us about something different from our day-to-day lives. For example, when the former Princess Diana of Great Britain died in an automobile crash in Paris, the subsequent outpouring of sympathy and curiosity occupied the news for many days in many countries.

Later we present four dimensions of *social significance*, although it can generally be defined as that which has relevance for the social system—whether the social system is as large as the world or as small as a neighborhood. Journalistic criteria that involve elements of social significance include importance, impact, consequence, and interest. For example, when terrorists skyjacked airplanes to destroy the World Trade Center in New York City and to ram into the Pentagon near Washington, D.C., this was news of worldwide social significance. Although nearly every country in the world sent messages of condolence to the United States, another, perhaps more important, measure of social significance was a heightened feeling of exposure to terrorist attacks in many countries that do not have a history of much terrorism.

Comparing the death of Princess Diana with the attacks on the World Trade Center and the Pentagon shows us how the concepts of deviance and social significance may interact. Diana's death was of little social significance outside of Great Britain and possibly even of limited significance there, since she was divorced from Prince Charles and no longer in line to become Queen. Her death—and in fact her life—was an example of deviance to most people. She was novel, controversial, and sensational. Her divorce involved conflict. On the other hand, the terrorist attacks in New York and near Washington, D.C. had economic and political significance as well as an effect on the general level of well-being of the public. Yet these attacks were also clearly deviant: airplanes targeting skyscrapers and causing them to collapse; thousands of people dying. And there was also deviance in that a fourth airplane—possibly headed for the White House—was not successfully skyjacked by terrorists. Passengers apparently overwhelmed the terrorists, even though the airplane still crashed, killing everyone aboard. Heroism is rare and therefore deviant.

We propose that deviance and social significance are separate predictors of newsworthiness, and that the combination of these two dimensions—when both have intense values—results in an accentuated level of newsworthiness. As Figure 2.1 shows, newsworthiness is highest when an event has both intense deviance and intense social significance, as was the case with the U.S. World Trade Center story. However, when the intensity of deviance is high and the intensity of social significance is low, newsworthiness is still high, as is the case with the story about Princess Diana's death. On the other hand, when deviance is less intense, newsworthiness is lower, but the event may be moderately newsworthy if social significance is intense, as is the case when legislators pass a nation's budget. The event is important to the country, but passing a budget is a routine activity of government. Finally, when both social significance and

Social Significance	Deviance: Low intensity	Deviance: High intensity
Low Intensity	Low Newsworthiness *(Routine city council meeting)*	High Newsworthiness *(Death of Princess Diana)*
High Intensity	Moderate Newsworthiness *(Passage of national budget)*	Highest Newsworthiness *(Attack on U.S. World Trade Center)*

Fig. 2.1 Theoretical typology to predict how newsworthy events (also people and ideas) are perceived to be. Events are selected as news according to how intensely deviant and socially significant they are (adapted from Shoemaker, Danielian, and Brendlinger 1991, 783).

deviance are less intense, people, ideas, and events should be of the lowest newsworthiness, and most of them will never pass through the news "gate" to become a news story. In fact, most of the billions of events that occur each day fall into this category (Shoemaker 1991).

Biological Evolution

We propose that two forms of evolution—biological and cultural—profoundly influence the form news takes around the world. People have been biologically influenced to attend to deviance and culturally influenced to attend to social significance. The basis for deviance as a part of news is based in biological evolution.

Attention to certain ideas, people, and events is a cognitive act, and theories of cognition must begin with biology—with the ways in which the human brain evolved (Shoemaker 1996). As Newell (1990, 112) put it, "Evolution is the designer of the human cognitive architecture. It will pick and choose systems that will aid in the survival of the species." Evolutionary psychology suggests that the evolution of the mind occurred in a manner similar to that of the body (Malamuth, Heavey, and Linz 1993). Buss (1991) says that biological evolution may have affected the mind in a variety of ways. At one end of the continuum is the suggestion that general mechanisms, such as the ability to learn or reason, are the result of biological evolution. The opposite idea is that evolution has created several specific psychological mechanisms to solve various environmental problems. The latter interpretation is held by evolutionary psychologists, who see their research as the missing link between biological evolution and the explanation of behaviors (Cosmides and Tooby 1987). Cosmides and Tooby suggested that psychological processing is the mechanism through which biological evolution affects social behaviors. Cognitive programs that evolved to solve specific problems, such as avoiding predators, are called "Darwinian algorithms" (293). Darwinian algorithms can also result in outcomes such as building schemas or frames (299).

Electrical metaphors abound in theorizing about cognition and the brain. Lang (1985, 156) integrated biological, social learning, and condition-response theories. Some relationships between stimuli and responses are "firm wired" in the brain. Singer (1980, 37) said that human brains are "wired up" to automatically respond to unexpected events in the environment. Cosmides and Tooby (1987, 284) said that the psychological programs that direct behavior are "in some kind of neural 'hardware.'" Shoemaker (1996) proposed that biological evolution has resulted in humans being "hard-wired for news."

Darwinian Evolution

The first coherent theory of biological evolution lies in the work of Charles Darwin ([1860] 1936a, [1871] 1936b), whose greatest contribution is a plausible explanation for how evolution works. All organisms reproduce, individuals vary slightly within a species, and all organisms compete for survival. When the environment changes or a new organism confronts an established one, the organism most adapted to the new situation will gradually outbreed the one less well adapted. Variations between the organisms or changes in the environment will therefore give a reproductive advantage to some members of a species and a reproductive disadvantage to others.

The concept of *adaptation* has proved crucial to Darwin's theory. Adaptation is defined as any structure, process, or behavioral pattern that makes an organism more likely to survive and reproduce when compared with other organisms in its population. Simply put, adaptive traits give people a higher probability that their genes will be represented in subsequent generations and that "the favored genes will spread throughout the population, and the trait will become characteristic of the species" (Wilson 1978, 32–33). Although some specific genes have been identified to cause certain diseases, for example, it is more likely that combinations of genes, some "off" and some "on," cause social behaviors. As our understanding of the human genome increases, we will be better able to address genetic explanations of social behaviors. Presently there are nearly daily discoveries concerning our genetic structure, many of which change what has been known before. The exact way in which the characteristics of genes may interact is not yet known.

A surveillance function, the gathering and dissemination of news, would have been quite adaptive for early humans, with those who were on the lookout for predators being more likely to survive and to reproduce than others in the same population. Early humans must have displayed many variations, whether because of diet, geography, natural disasters, or mutations. Likewise, some were more interested in or more adept at surveying their environment than others. Over the years, the surveillance function would be present to some degree within virtually all members of the population, perhaps explaining the existence of what some anthropologists have called "cultural universals, for example, the emotions fear, anger, depression, and satisfaction" (Kemper 1987, 263). Fear and anger could have interacted within early humans to stimulate the surveillance mechanism because these emotions "energize the organism to undertake urgent activity for survival purposes when faced with danger or with threat from others" (268–269).

Bad News, Good News

Wilkins and Patterson (1987) suggested that, to a journalist, the best news story is someone else's disaster. This is reinforced by the prevailing audience opinion that the media prefer bad news to good news, and indeed the news is filled with scandals, conflict, crime, and disasters. Why? Because people are interested in such events. While occasionally there is a cry for good news, people actually pay less attention to good news than to bad news.

Psychologists who study how people pay attention have found that novel things seem to be better remembered than routine events (Nelson 1989). Attention to a novel or unusual stimulus lasts as long as it creates a difference between what is seen and the person's schema (Rovee-Collier 1989). Exposure to a large amount of information not currently in the schema can cause the person to be startled or afraid (Singer 1980). In their study of the effects of negative television news on memory, Newhagen and Reeves (1992) showed that information is most likely to be remembered if it follows rather than precedes a compelling negative image.

Nisbett and Ross (1980) suggested that humans are more likely to attend to and remember *vivid* rather than *pallid* information. Although all vivid information is not necessarily bad news and all bad news is not necessarily vivid, there does seem to be a connection between the two. Even when the media are presented with a relatively pallid bad news item, such as a rise in the unemployment rate, journalists tend to personalize such statistical events by investigating how they affect individual people, thus making the message more vivid. We are more likely to pay attention to vivid information than pallid information because vivid information fires our imagination in one of three ways: (1) it is interesting and involves our emotions; (2) it is concrete and evokes images; or (3) it is information about something that is proximate—if not in terms of geographical closeness, then in terms of time or our senses.

Using this definition, vivid (and bad) news includes such events as coups, earthquakes, murders, and scandals. Because bad news is also generally news about deviance and since deviant things require more cognitive processing than routine things (Newhagen and Reeves 1992, 26), one significant question is why the acquisition and dissemination of information, especially deviant information, has become so important for humans. People pay more attention to bad or deviant news items because it is in their best interest to do so. When people survey the environment and think about deviant stimuli, they increase the likelihood that they will adapt in order to neutralize or diminish threats to the status quo (Newell 1990; Newhagen and Reeves 1992). When confronted with images that are deviant and very negative, emotions may be triggered that help us perform well when confronted with dangerous situations (Lang 1985).

Cultural Evolution

If people's attention to deviant news is not innate and biologically derived, then a news-consuming habit must be acquired after birth. There is no question that as children we are taught that surveillance is an important behavior ("Look both ways before crossing the street"). We are taught to attend to the world around us and to cognitively process information about possible threats and potential benefits. Especially where a deviant event, or "bad news," is concerned, the identification of deviant people ("Stay away from strangers"), events ("Don't use drugs"), and ideas ("Who told you that?") is functional not only for maintaining the status quo, but also for the safety of the individual involved. We learn what can hurt us. We learn what is right and wrong by our culture's standards. We learn what our culture emphasizes as noteworthy.

Culture has been defined as "a pool of at least somewhat organized information of various kinds, socially transmitted within and across generations" (Barkow 1989, 112). As Van Dijk (1988) suggested, such social learning provides our general norms and values, and when deviance is of concern, our culture teaches us about outcasts and their punishments, about which deviant behaviors or ideas need attention.

Culture is something we learn from one another. When one generation teaches another, each generation may inherit the knowledge, attitudes, and behaviors of its ancestors through nongenetic means. The learned material constitutes a cultural inheritance. Such an approach emphasizes that social learning, not biological evolution, is the basis of culture (Rogers 1988). A social learning explanation (Bandura 1971) of why humans pay attention to news suggests that social behaviors such as surveillance are acquired after birth through the process of cultural socialization. The primary agents of socialization are the family, school, peers, and the mass media, with each teaching children about the culture in which they live. Learned material resides in long- and short-term memory, where it may be retrieved as needed. Information about specific threats to the self and environment are socially learned, as is (according to this theory) the need for general environmental surveillance.

Although the term *socialization* is most often used to explain how children learn the culture of their birth, the term *acculturation* describes a very similar process: the learning of a second or substitute culture. The process through which each of these occurs is assumed to be similar. Culture is learned as a set of interrelated cognitive representations after any genetic effect has been introduced. In most learning theories, the influence of biology on the acquisition of social behaviors is not considered (Kim 1977; Osenberg 1964; Shoemaker, Reese, and Danielson 1985).

Deviance

We stated earlier that people pay attention to deviant individuals, ideas, and events, which can be as common as a car darting in front of someone on a busy street to less frequent events such as invading armies. People may be alert to the possibility that their children may become drug users, that they themselves may be criticized in the workplace, or that deviant political ideas could change the nature of society.

This tendency is not new but has its roots in the emergence of human beings through thousands of generations. For early humans it was functional to pay attention to the environment, to warn others when a threat appeared, or to call attention to something that could be helpful. Warning others that a tiger was outside of the cave allowed our ancestors to avoid injury or death. Finding and describing a new food source to other people helped the group survive and prosper.

Deviance is defined in a variety of ways in the sociological, social psychological, and psychological literature. *Normative* deviance is what most people think of when they hear this term, because it refers to the breaking of norms and laws (Wells 1978). Criminals are deviant, but so are women who do not follow their culture's standards of dress. Normative deviance is generally thought of as negative, because its referents are the laws and norms of the existing social system. However, breaking a law in one country (negative to that country) might be considered a positive action in another.

Social change deviance was originally called *pathological.* A century ago when sociologists thought of society as an organism, some types of deviance were thought of as illnesses or threats to the health of the status quo as of the organism (Matza 1969). As an example, homosexuality was at that time thought to be a threat to the status quo—to the body public in the sense of a disease invading the body's cells. At present, however, there is quite a different view of homosexuality in countries where the older organic view of society has changed dramatically. Some argue that it is biologically caused; others claim it is caused by culture or by childhood occurrences. Others say it is just a personal preference.

Still, the idea that social systems have a status quo and that this status quo can be changed is valid, but as the term *pathological* has only negative connotations, we have renamed and reconceived this dimension as *social change* deviance, in which ideas, people, or events challenge the status quo of the social system, whether large or small. In the 1960s, when U.S. African Americans deliberately challenged laws of where they could go to university or where they could sit on buses, they challenged the status quo and ultimately changed the social system. They changed laws and norms, as well as some attitudes. That racism still exists in the United States is not in doubt, but much has changed. Many consider this a positive change in the status quo, but as with normative deviance, whether change is positive or negative depends substantially on each person's perspective. Revolution is negative to those in charge of the status quo being overthrown, but it is positive for the challengers if they win.

Statistical deviance is a third dimension, in which an idea, person, or event is very different from the average in being odd, unusual, or novel. Airplane crashes are statistically deviant because most airplanes do not crash. A person who scores extremely high or extremely low on a standardized test is statistically deviant, because the score is very far from the mean or average score. Car crashes, excessive rain in a desert region, a painting left on a subway but returned to the owner, and the oldest person in the world are all examples of statistical deviance.

In short, the dimensions of deviance used in this study include normative, social change, and statistical.[1] All three dimensions can be considered positive or negative depending on the context, and we propose that all three dimensions of deviance may be used to define newsworthiness.

Social Significance

Even as we see deviance determining much of the news (Shoemaker, Danielian, and Brendlinger 1991), it is clear that there is another major dimension. As people go through their days paying attention to some things and not to others, there is a tendency to watch out for things that will affect their lives, either on a personal or a public level. Carey (1987, 191) described significant news items as "signs that must be read, portents of something larger, events to be prized and remembered as markers, as peculiar evidence of the state of civilization, or the dangers we face or the glories we once possessed." Events such as elections have social significance, as do new laws, decisions of courts, cultural events, the closing of a museum, higher or lower inflation, devaluation of currency, war, terrorist attacks, a new AIDS medicine, the closing of newspapers, the beginning of cable television networks, the death of prominent people, and so on.

Previous studies have differentiated among economic, political, and cultural significance, arguing that these three dimensions explain an event's significance to a particular culture or society (Shoemaker, Danielian, and

Brendlinger 1991). We define a news item's *political* significance as the extent to which the content of a news item has potential or actual impact on the relationship between people and government or between governments, including the judicial, legislative, and executive subsystems. People, ideas, or events having political significance may range from local crimes to international agreements. So anything dealing with local laws (and the breaking of them), issues concerning prisons and their characteristics, and complaints to the city council have political significance. Thus, political significance is not limited to national or international concerns but can also result from local events.

The *economic* significance of a news item refers to the extent to which the content of the news item has potential or actual impact on the exchange of goods and services, including the monetary system, business, tariffs, labor, transportation, job markets, natural resources, and infrastructure. Local events having economic significance include a company going out of business and laying off workers, a labor union striking against a manufacturer for higher salaries, and flooding that keeps people from buying groceries. International events of economic significance could include a more favorable exchange rate for one country compared to another. The increase or decrease in manufacturing capabilities would have national economic significance, as would the discovery of a new natural resource.

Cultural significance is defined as the extent to which the content of a news item has potential or actual impact on a social system's traditions, institutions, and norms, such as religion, ethnicity, or the arts. For example, an increase in the number of people who regularly practice a religion has cultural significance, as does the looting of a national museum. On a local basis, the opening of a new library or music hall adds to the culture of the community.

The fourth dimension is *public* significance, which refers to the enhancements or threats a news item represents for the public's well-being. Generally, we assume that a high number of people affected by a news item indicates that it is more significant for the public's well-being, regardless of whether the influence is positive or negative. This dimension covers news items not addressed by the other three social significance dimensions, such as the discovery of a new medication, the conviction of a mass murderer, major traffic accidents, or long-term effects of an oil spill.

Based on these definitions, we assume that a news item rating as highly intense on all four dimensions of social significance should receive more news coverage in that specific culture than a news item of low intensity. It is important to note that the context in which social significance is judged will be the geographic unit in which ideas, people, and events occur. Thus, something that has much social significance in one country or culture may have no social significance in another.

The Joint Impact of Biological and Cultural Evolution on News

As noted already, there are at least two possible explanations for the development of a powerful surveillance behavior in humans. The first is that human beings' interest in news may have biological roots. A reproductive advantage inherent in news gathering and dissemination genetically predisposed increasing numbers of the population to survey and report on what was happening in the environment. The competing reason for our fascination with news is cultural: In many societies, people are taught from childhood that they ought to pay attention to the world around them. Teachers give current-events quizzes. Parents discuss politics and world events with their children. In the United States, an underlying assumption is that paying attention to news is good for democracy, with surveillance of deviant political ideas being more important than looking out for tigers.

The important distinction between the biological and cultural explanations is that the biological argument presumes humans are born with a predisposition to attend to news content. In contrast, the cultural explanation assumes that a predisposition to attend to news can be learned and is a function of the overall socialization of individuals to their community or country. The decision about specifically to what should be attended is also more likely to be culturally determined.

So, is the human propensity toward surveillance and the shaping of news to emphasize deviance and social significance more a function of genetic evolution or of cultural socialization? The best answer is almost surely that they interact, with each setting the bounds for the other (Lumsden and Wilson 1981). Our genetic heritage has given us the ability to see and hear only a small part of the physical world, but culture has further determined the phenomena to which we may choose to attend in that small window. Whereas genetic evolution operates over thousands—even millions—of years, cultural change can take place in the comparative wink of an eye. Although our genetic heritage may determine what types of social behaviors are possible and probable, culture determines the form and expression of those behaviors (Lumsden and Wilson 1981). As Malamuth, Heavey, and Linz (1993, 67) put it, "Current behavior is not a product of 'nature vs. nurture,' but of 'nature interacting with nurture.'"

In biological evolution, the concern is with how various genes and gene combinations, which are constructed from chemical bases, change to produce genetic evolution. When we talk about cultural evolution, on the other hand, we mean that mental constructs, such as ideas or symbols, take shape from human perspectives and memory traces to change or to integrate with other concepts, or both (Handwerker 1989). In cultural evolution, the brain generates and organizes "systems of understanding and meaning" (324), and therefore culture exists only ephemerally because each person creates and recreates her or his understanding of culture. Each person has a unique life experience, and "culture" becomes common only when people share their personal cultures.

This challenges the idea that culture is a characteristic of the social system, yet we know that individuals within a culture do share some common perspectives. As Wilkins and Patterson (1987) pointed out, news follows well-understood stylistic and cultural patterns. For example, news of disasters tends to be portrayed as melodrama, involving predictability and stereotypes. News is not something apart from culture; but it is "a social construct empowered by a cultural history and a traditional institutional practice based on material and cognitive realities" (Koch 1990, 190). To the extent that this includes deviance, therefore, we might expect that surveillance somehow has important cultural significance.

It is clear that we are biological beings—we are born, we grow, we reproduce, and we die. Lopreato (1984, 33) suggested that culture is human beings' way of expressing their human nature, which he defines as "a set of genetically based behavioral predispositions that have evolved by natural selection in part at least under sociocultural evolution." Evolutionary psychology suggests that an individual's manifest psychology and behaviors result from an "interaction between evolved psychological mechanisms and environmental factors that activate them, differently across individuals" (Buss 1991, 478). The newborn child is not a blank slate upon which to be written solely by culture but rather is programmed through its genetic heritage with the potential of behaving in certain ways. Through cultural socialization, these potentialities are either encouraged or discouraged. Culture can determine the tune to be played or whether any tune is played at all, but it cannot turn a piano into a trombone.

This interaction proposal suggests that cultural concepts cannot exist for a person unless "genes construct neural circuitry that allows information from one sense to be related to information from several or all senses" (Handwerker 1989, 317). However, genes do not determine the content of the construct. Each person is molded both by a genetic heritage and by a cultural environment.

What anthropologists have identified as "cultural universals" (Rindos 1986, 319) may well be the basic genetic template we all share. Such universals are made up of fairly consistent characteristics across time and space. Our genetic heritage may give us a set of behavioral predispositions that form the foundation for culture (Wozniak 1984). These may include incest taboo, bond formation, territoriality, hierarchy, and ethnocentrism (Lopreato 1984), or aggression dominance, status, sexual behavior, and sex roles (Gove 1987), in addition to environmental surveillance. The observed variations in beliefs and premises among cultures may derive from a common base of the same logical processes (Cole and Scribner 1974). Similar ideas have been suggested by Chomsky (1968), who proposed that human grammar may operate cross-culturally, and by Piaget (1972), who argued that the intellectual growth of a child is at its base the result of universal human biological processes, with environmental influences operating on what biology has produced.

Eckland (1982) offered a model to show the interaction between biological and cultural evolution. He argued that biology and culture mutually influence one another through continuous intergenerational feedback. New patterns invariably emerge through the interaction of heredity and environment. Genes restrict the possible range of human development, and within these limits humans alter their environment or cultural arrangements in order to change the distribution in the next generation, which then enables them to carry out even more change. Genes prescribe a set of possible biological processes whose exact form is shaped by culture and the physical environment.

If the contributions of culture to social behaviors are not considered, the resulting theory could justifiably be accused of "genetic determinism" (Geiger 1990). Likewise, if the contributions of biology to social behaviors are not considered, the resulting theory—representing the bulk of social science—could be described as "cultural determinism." If across cultures we tend to see the same social behavior taking place, this may be evidence of genetic transmission of the social behavior. In contrast, significant variation of social behaviors among cultures could be evidence of cultural transmission of the behaviors.

An interest in news is probably the result of an interaction between these two pure models, with humans both innately interested in deviant events and socialized to attend to events that have some significance to their particular culture and society. Can culture escape its biological basis? Wilson (1978, 167) said that "genes hold culture on a leash. The leash is very long, but inevitably values will be constrained in accordance with their effects on the human gene pool. The brain is a product of evolution."

Hypotheses and Research Questions

Interest in news as a concept is not new. What is news, and how do we explain its content? Although numerous scholars have addressed factors that impact the production of news content (Gans 1979; Tuchman 1978), our theory looks at how people may have been conditioned by biological and cultural evolution. Biological evolution provides a foundation on which all other news production factors build. If we accept that human brains are hard-wired to survey their environment and to prefer news about deviant and otherwise threatening events and ideas, we can then more fully understand how journalists' selection of deviant or bad news may reflect a basic biological disposition to such news and not a peculiarity of journalists.

This theoretical approach challenges the age-old charge that the mass media inappropriately cover too much bad news, suggesting instead that humans are innately interested in information about deviance (especially environmental threats), which is often synonymous with bad news. If people's interest in news is genetically programmed, then this explains why news content includes so much bad news. Although news content is shaped by a wide variety of forces (Shoemaker and Reese 1996), they cannot overcome the fact that news is produced by human beings who are the product of biological evolution.

An interest in news, particularly deviant news, has become widespread in the population across human history. Attending to environmental threats and other forms of deviance gave early humans a reproductive advantage, thus making surveillance adaptive in the Darwinian sense of the term and facilitating its use as a social behavior throughout the population. Although variation among individuals in the population—a key part of Darwin's theory—has occurred, there is still reason to expect that some fundamental goals, such as surveillance, would be shared by humans the world over (Buss 1991).

The project presented here is designed to address some of these ideas. For example, applying this theory to the study of news suggests that human beings in general pay more attention to deviant events than to nondeviant ones and to events that have the most intense deviance. This perspective suggests that people in different countries share an interest in deviance and social significance, and that their news is more alike than different. We base this on two assumptions: (1) journalists are people; and (2) people are interested in deviance. On the other hand, one would expect that, among deviant events, people from two different cultures might attend to different kind of events. This theory also suggests that the more intensely deviant or socially significant an event is, the more prominently it is covered by the news media regardless of time and space. That is, the effect should be observable in news media, both longitudinally and across cultures. A genetic predisposition for surveillance, compounded by cultural and time-bound variables, may determine the types of deviant events given the most news coverage.

Within this theoretical framework, we have developed the following research questions and hypotheses. The first set of research questions addresses the kinds of topics in the news and which are the most deviant and socially significant.

RQ1: Which news topics are most common?

RQ2: Which news topics include the most intense deviance and social significance?

RQ3: How similar are the countries' intensity of deviance and social significance in news topics?

RQ4: How much deviance and social significance are in the news?

RQ5: To what extent do the countries have news items that are similar in how intense their deviance and social significance are?

RQ6: How similar is the intensity of deviance and social significance in the news of major and peripheral cities?

RQ7: How similar is the intensity of deviance and social significance among the three media—newspapers, television, and radio news?

RQ8: How similar are the deviance and social significance intensity levels of visual and verbal content?

In addition, we address the following questions in the country-by-country analyses presented in chapters 8–17. The country directors analyze the focus groups with journalists, public relations practitioners and the public.

RQ9: What kind of information catches people's attention, and why?

RQ10: What types of news items are most salient in people's minds, and why?

RQ11: How do journalists, public relations practitioners, and news consumers differ in their definition of news?

We test these hypotheses using quantitative data from newsworthiness exercises held at the end of the focus groups.

H1: There are positive relationships between audience, journalist, and public relations practitioner assessments of how newsworthy events are.

H2: There are positive relationships between how the three groups of people rate event newsworthiness and how much prominence newspapers give the events.

On one hand, if news has been defined over the millennia by biological and cultural evolution, then people of all types (whether they are involved in mass communication or another career) should rank news items in a similar way. Similarly, if people's newsworthiness assessments of events are very similar to their newspapers' coverage of the events, then this could lend support for the idea that (1) newspapers give audiences what they want, or (2) audiences have been trained over the years by newspapers to want what newspapers give them. The absence of a relationship (or the presence of a negative relationship) would indicate that forces other than individuals' opinions about newsworthiness (whether journalists, public relations practitioners, or the audience) are more important in shaping the news (Shoemaker and Reese 1996).

Finally, Hypotheses 3 and 4 are tested in the quantitative content analysis of news items.

H3: The more intensely deviant a news item is, the more prominently it is covered in the media.

Shoemaker, Chang, and Brendlinger (1987, 334) suggest that "deviance creates a common focus for group emotion against threats to the status quo; it clarifies the rules for everyone else without actually testing the rules themselves." They argue that deviance underlies many of the indicators of newsworthiness and may be a better and more important dimension of the concept of newsworthiness (410).

H4: The more intense the social significance of a news item is, the more prominently it is covered in the media.

Due to the obligatory recording responsibilities of the media, some socially significant events will get media coverage no matter how ordinary the event, such as a community meeting. However, space and time limitations will naturally filter events based on how ordinary or routine they are for the more socially significant events to gain more prominence in terms of space and time allotments (Shoemaker, Chang, and Brendlinger 1987, 356).

3
Methodology

This study was conducted in ten countries: Australia, Chile, China, Germany, India, Israel, Jordan, Russia, South Africa, and the United States. The countries were selected to include large, medium-sized, and small countries; so-called western and eastern countries; countries in both the Northern and Southern hemispheres; developed and developing countries; and countries with different dominant religions, different political systems, different languages, different cultures, and varying gross national products.

The project directors solicited the help of a senior communication researcher in each country. These country directors assisted in making the decisions concerning the sampling of media and other issues in their respective countries. Several meetings were conducted with the country directors during the course of the study, and numerous phone calls, emails, and faxes were sent in developing the research tools and in overseeing its various stages.

Two primary methods were used in the ten countries of the study:

1. A quantitative content analysis was conducted of a sample of newspapers as well as television and radio news programs.
2. Qualitative focus group discussions were held with journalists, public relations practitioners, and audience members.

In addition, data from part of the content analyses were used in the focus groups to create an interrelationship between the two methods. Quantitative data were collected at the end of the focus groups in a gatekeeping exercise. Thus, the chronological ordering of the data collection was critical; it was necessary that the content analyses precede the focus groups.

Each country's research director chose two cities as the research sites: a major city and a peripheral city. Some sociologists, notably Shils (1988), draw a distinction between center and periphery societies, arguing that members are consciously aware of differences between the two. In this study, the rationale for choosing two such cities in each country was that the media environment in the major city would typically be relatively rich, whereas in the peripheral city it would be relatively poorer. Also, the nature of the events taking place in the two cities could be somewhat different. Finally, the centrality of national and international news in each of the cities' media might also be different.

The selection of each country's major city was based on the number of inhabitants in its metropolitan area, as well as its strategic, political, and cultural importance. In most cases the major city was in fact the capital city, but in Australia, Israel, South Africa, and the United States another city was selected. In contrast, the peripheral city in each of the ten countries varied in size from approximately half a million inhabitants to less than 100,000 residents. In addition to size, the peripheral city was chosen based on its relative distance from and relative lack of contact with the major city. The definition of the *major* and the *peripheral* cities is relative, of course, and given the variability in the size of the countries, even a peripheral city could be quite large.

The goal was for each city to have a locally published daily newspaper, as well as local radio and television stations, both of which should broadcast at least one daily news program. Some deviations from this general plan were made given the circumstances in some of the countries. In any event, the media outlets selected in each city were to be those the population read, viewed, and listened to the most. This was determined by the country directors based on the news media's reputations or circulations or reach.

In each city, the leading newspaper was selected. The paper had to be published in the city, meaning that the main editorial offices were located there. The same logic was followed for television and radio news programming. In each country, the director selected the television or radio station providing the leading local newscast. In cities where no local television or radio newscast was available, the television or radio station providing the leading regional television newscast was selected. In some cases, in which no television or radio station provided a local or a regional newscast, the station providing the leading national newscast was selected.

The Content Analysis

Sampling for the content analysis was based on a composite week, including seven days, with one day selected in each of seven consecutive weeks.

Newspapers

Newspaper content was sampled first. Starting with a Monday, seven newspaper issues based on a composite week of newspaper content were collected—that is, Monday of the first week, Tuesday of the second week, Wednesday of the third week, and so forth. If the newspaper was not published on a particular weekday, then the next available issue following that day was collected. For example, if a daily newspaper was not published on a Sunday (that had been selected as part of the composite week), then the following Monday was selected. The rationale for this decision was that Sunday's news would logically appear in the Monday paper. The sampling scheme assumed that most leading newspapers are published daily. However, in cases where a city's leading local newspaper was published weekly, seven issues of the weekly paper were collected over a seven-week period. See the country chapters for details on such deviations. The dates selected for the content analyses appear in Table 3.1.

Television and Radio

As in the case of newspapers, the specific newscasts sampled were considered to be the most important—usually the most widely watched—television and radio newscast for each day of the composite week.

Because newspapers and broadcast media have different news cycles, the collection of television and radio news content took place on the days previous to the selected newspaper sampling days. Thus, because the newspaper sample began on a Monday, the television and radio sample started on a Sunday, with every following broadcast sample day being the day previous to the respective newspaper sample day. In locations where there was no daily news broadcast, a weekly broadcast was selected and repeated for seven consecutive weeks on the same day of the week.

Unit of Analysis

All of the newspapers and newscasts were analyzed in their entirety. The unit of analysis was the news item, defined in general as a set of contiguous verbal and visual content elements (verbal only in the case of radio). Thus, a story or event could be represented in any given medium by more than one news item. For example, in the case of the terrorist events in the United States on September 11, 2001, one item might have been a report from the scene of the World Trade Center in New York City and another from the Pentagon near Washington, D.C.; a third event could have been an item on reactions from Saudi Arabia.

More specifically, a newspaper news item consists of a set of verbal (e.g., text or article) and/or visual (e.g., photo or map) content elements that relate to the same event. It was decided to analyze all the editorial material in the entire newspaper—that is, everything except advertisements, classified ads, and paid-for obituaries counted as news. In cases such as gossip columns, each reference to a different celebrity or event was considered as a separate item.

TABLE 3.1 Dates of the Content Analysis

Newspapers		Television and Radio	
Monday	November 13, 2000	Sunday	November 12, 2000
Tuesday	November 21, 2000	Monday	November 20, 2000
Wednesday	November 29, 2000	Tuesday	November 28, 2000
Thursday	December 7, 2000	Wednesday	December 6, 2000
Friday	December 15, 2000	Thursday	December 14, 2000
Saturday	December 23, 2000	Friday	December 22, 2000
Sunday	December 31, 2000	Saturday	December 30, 2000

The most difficult selection decisions usually had to be made on a newspaper's front page and section fronts, because in many cases the materials on these pages were "teasers" referring to items in inside pages. Likewise, a television news item could include any combination of verbal content (e.g., anchor's or reporter's spoken words) and/or visual content (e.g., video clips or still photos). For radio, of course, only verbal content existed. As with the case of the newspaper items, in order for different content elements to be considered part of the same news item they had to belong together based on some common event.

Each newspaper item was clearly marked and was given a unique identification number on the newspaper page with a colored marker so that it could be retrieved if necessary. News items that spread over more than one page were so marked.

Coding Newspaper Items

Several rules were developed for coding the newspaper items:

- Teasers (e.g., a headline, sentence, or photo) and brief news summaries (e.g., a paragraph) were separate items—even if they included a referral to other pages, and were therefore coded separately from the story that appeared elsewhere in the newspaper.
- All weather information that appeared on the same page was treated as one news item, regardless of whether it was a front-page teaser with only a brief weather note or icon or an entire weather page with maps, text, and tables. If an article or item dealt with a specific weather event or phenomenon, then it was treated as a separate news item.
- All standard stock market tables were treated as one news item without regard for the number of pages over which the information is distributed. However, a general article or analysis of the stock market (or a specific stock) was treated as a separate item.
- A horoscope with all its subsections was treated as one news item.
- All comic strips in a newspaper edition were treated as one news item. In contrast, editorial cartoons were considered as separate items, whether single or multiframe.
- Each part (e.g., paragraph) of a gossip column was treated as a separate item.
- Hiring and appointments listed in the same column were treated as one item.
- Standing material (such as weather tables, stock reports, horoscopes, comics, hiring and appointments) included in every issue and which rarely change were coded using standard values established by the researcher within his or her cultural context. Out-of-the-ordinary news stories relating to these subjects, however, were coded individually.

- A combination of a major headline followed by two minor headlines with separate stories was treated as three news items. Because the major headline is a separate design element with no accompanying text, it was treated as a separate item.
- If different reporters or reporter teams wrote stories on a given event, they were treated as separate news items.

Coding Television and Radio Items

Several coding rules were developed for the coding of the television news items:

- An anchor's introduction and the film or audio clips belonging to the same event represented one item.
- A major event such as the election of a new head of state was often treated in various substories—each of which could include video or audio clips from reporters or sources. Each of these substories was treated as a separate item.
- News briefs and short summaries (usually moderated or commented on by the anchor) were considered as separate items.
- An interview between anchor and source(s) or reporter and source(s) was treated as one item.
- All introductory or transitional film clips, background pictures, and similar elements at the start or break of a newscast were treated as one news item.

The Variables Coded

A list of variables to be coded in the three media was created. First a theoretical definition for each variable is provided, followed by the development of an operational definition. These are designed to be as compatible as possible for the news items of the three media.

The first few variables deal with the identification of the various items, followed by the determination of the main topic of each item. A coding scheme developed by Cohen, Adoni, and Bantz (1990) in their research on social conflicts was adapted for use here. Each of the twenty-six major topic categories has subcategories, thereby enabling detailed coding. (See Appendix A for a list of all topics and subtopics.)

The following variables measure the prominence of the news items: space (in newspapers); length (in television and radio); and the placement of the items, namely, on the front page or other pages (for newspapers) and in the top, middle, or last third of the newscasts (in the case of television and radio). A composite index of news item prominence was ultimately created for each item by multiplying its space or time by its placement.

The final variables measure deviance and social significance. As noted in chapters 2 and 3, three dimensions of deviance are defined: statistical deviance, social change deviance, and normative deviance. Four dimensions of social significance are defined: political, economic, cultural, and public. Each is rated at one of four intensity levels, ranging from extremely intense to no intensity. Because each news item's visual and verbal content were evaluated separately, this results in six deviance variables and eight social significance variables (see Appendix A).

The last four variables are the different kinds of social significance—political, economic, cultural, and public—as explicated in chapter 2. As in the case of the deviance measures, each of the four kinds of social significance is coded at one of four intensity levels, ranging from very intense to not at all intense.

Coder Training

Given the complex international and cross-cultural context of the study, the training was designed as two stages. First, the ten country directors met in New York City for an intensive workshop, in which the codebook was explained in detail and several sample items were analyzed as a group. Some revisions to the codebook were made as a result. Second, once it became evident that there was common understanding of the variables, each person returned to her or his country and trained university students who were paid to code news items from their country's newspaper and television or radio news. In a few cases, the country directors differed between the meeting in New York City and the beginning of work on the study, requiring the two study directors to train new country directors (and sometimes student coders as well). Shoemaker trained coders in Russia and South Africa, and Cohen trained coders in Germany and Jordan.

Each of the ten country directors oversaw all content coding at their own universities. The country directors trained the coders by using news items outside of the actual study sample but from the same general time frame. Practice coding sessions were held until the desired agreement was achieved on the study variables. The goal was to achieve a Scott's pi coefficient, which measures intercoder reliability near or above .80.[2]

Reliability and Validity

To give our theory the most stringent test, we selected ten countries that vary as much as possible on a number of dimensions: geographic, economic, political, cultural, language, size, and relative influence on other countries. Two cities were selected for study in each country: one a major city, and the other a peripheral one. In all of these selections, the logic was to create variability among the countries and between the cities in much the same way as one would manipulate the values of the independent variable in an experiment to enhance variance between the treatment groups. Had we studied countries in one region, they would probably have shared similar cultural, economic, and political systems and would therefore have similar media systems. We chose countries as different from one another as possible.

This methodological strategy makes it easier to find differences than similarities. If there are differences among countries, then selecting very different countries would reveal them. In addition, a statistical requirement forced us to test hypotheses of differences among the countries rather than hypotheses of similarity. Statistics can only support a hypothesis showing a difference; there is no statistical test that can support a hypothesis of no difference. Thus, although our ten countries show only small differences in how deviant or socially significant their news is, these differences tend to be statistically significant, at least partially because we studied more than 32,000 news items. This is an artifact of the way in which statistical significance is calculated: The more cases (in our study, *news items*), the easier it is to reach statistical significance.

So finding differences ought to be easy, while finding similarities should be more difficult. As we designed the separate segments in this project, we found ourselves establishing protocols not only to guarantee similarities in how the countries carried out the research but also to preserve their unique country contexts. For example, we decided that we (the authors) would not conduct the content analyses from the United States or Israel, even though we had access to students who are native speakers of the countries studied and who could have served as coders for our study. This might have created artificial similarities among the countries due to the coders being removed from their cultural milieu by being located in foreign countries. So instead, to guarantee that we would find differences (if they exist), we contracted with one scholar in each country to direct all of the content analyses and focus groups. Although we established project guidelines, coding instructions, and protocols that were standard across the countries (which would make similarities easier to find), the country directors made decisions about how these would be carried out in their own countries, thereby guaranteeing that their unique cultural contexts were maintained. In short, our aim was to obtain functionally equivalent data and analyses.

There has been debate about whether cross-cultural research can be performed with both high reliability and validity—that the same concepts are measured in the same way. Study designs such as ours stretch these ideals to the maximum: Although we created standardized protocols, we enhanced variability in both our selection of countries and cities. Country directors translated materials into their own languages and trained their own coders, such that intercoder reliability was achieved within each country.

There was no attempt to assess intercoder reliability across countries because of our parallel goal of achieving validity. We could have trained all coders in the ten countries to reliably code news items from the *New York Times*, for example, but what would that have shown about their ability to code stories from their own countries' newspapers? To what extent would the news in their countries be like the news in the *Times*? We could have achieved reliability but would have lost validity.

Because one of the theories underlying our work is that of cultural evolution, it was absolutely vital to maintain the countries' cultural contexts when measuring variables. In other words, to achieve validity it was necessary to measure the variables within each country's culture by members of that culture. Measures of intercoder reliability within each country demonstrate the reliability of our standardized protocols, whereas the use of country directors to make decisions relevant to their own cultures enhances our study's validity.

This gives the most stringent possible test to our notion that news is much the same around the world. Granted, individual countries (and individual news media) cover the day's events differently, but we believe that such differences are superficial. There are commonalities among these ten countries' news that can be accounted for by two theoretical constructs: how (1) deviant and (2) socially significant the events, ideas, or people are.

Focus Groups

In each of the 20 cities, four focus groups were conducted with the following categories of participants: journalists, public relations practitioners, low socioeconomic status (SES)[3] news consumers, and high SES news consumers. Deviations from this general model occurred in some countries and are spelled out in the country chapters. The focus groups were moderated by the country director or by a trained assistant.

Identifying and Recruiting Focus Group Participants

Given the lack of statistical representation in the composition of focus groups on the one hand and the richness of the potential data on the other hand, an attempt was made to select participants who might shed interesting perspectives on the issues under investigation. Prior to conducting the focus groups in the ten countries, a pilot study was done in Germany and Israel in order to determine how the group participants would treat the topics. Several modifications in the plans as well as in the strategy of recruiting the participants were derived from the pilot.

Journalists In each city, about ten journalists were approached and asked to participate in the group discussion. Generally, three to four journalists from each of the three media organizations sampled for a city's content analysis were invited from the newspaper, the television station, and the radio station. In some countries, other journalists were invited as a supplement when an insufficient number could be recruited from the studied media organizations. In order to achieve the greatest variance possible, reporters, editors, and producers were invited. However, care was taken to avoid recruiting from the same news organization both reporters and the editors to whom they reported; most of the participants were in fact reporters. An attempt was also made to vary the gender, age, and professional experience of the participants.

Public Relations Practitioners In each city, approximately ten public relations practitioners were invited, ideally up to three from each of the following four professional areas: public relations agencies, corporate public relations departments, governmental public affairs departments, and nonprofit public relations offices. Given the objective of the research, an attempt was made to recruit public relations practitioners who deal with media relations activities, such as the writing and distribution of news releases and organization of press conferences, mainly with the local media. Thus, for example, an attempt was made to recruit such persons as the local police spokesperson, the public relations manager of a local museum, or a hospital's media relations person. As with the case of the journalists, invitations were also extended by gender, position, and professional experience.

News Audience In each city, two separate focus groups of news consumers were conducted: one with high socioeconomic status (SES) people and one with low SES people. For details concerning the specific composition of the groups in each country, see the country chapters. In addition to the SES level, the ten or so people invited to participate in each group were balanced by gender, age, and occupation.

Focus Group Arrangements and Procedures

The members of the two audience groups were presented with a modest monetary compensation. No compensation was given to journalists and public relations practitioners. Since it was not desirable to tell the participants in advance exactly what the study was about, they were only given a general idea of what the discussion would center on without mentioning the term *news*. Prospective participants who asked detailed questions about the topics to be discussed in the focus group were told that, since the objective was to obtain spontaneous responses, no further information could be given prior to the convening of the group.

The focus groups were conducted at the country director's university, in a hotel, in the workplace of one of the journalists or public relations practitioners, and in the homes of some of the news consumers. Two weeks prior to the focus group, an official invitation was sent to each participant, including the location, date, and time as well as instructions for how to get to the location. Two or three days before the scheduled group, a telephone reminder was made to each participant.

For each of the groups, special measures were taken to ensure participation. In the case of journalists, who have flexible and ever-changing schedules, contact was sometimes made with the head of the organization to solicit approval. In fact, in some cases the editor-in-chief required some of the staff to attend. In addition, a surplus of journalists was invited in order to guarantee a minimal number of participants. The fact that the study had an international component was emphasized to increase the interest among the journalists. This point was also made to the public relations practitioners.

Based on the experiences of the pilot study, it was decided to recruit the high SES groups using the snowballing technique, by asking a person who agreed to attend to invite some of her or his relatives and friends. The focus was on creating curiosity among them, as the monetary compensation did not seem to be of importance. An important task of the recruiter was to help people overcome their fear of talking in front of a larger group. Focus group prospects were told that the groups would take place in a very informal setting, without prolonged speeches, and with other interesting people.

Lower SES people were generally more difficult to recruit despite the monetary compensation. In some locations, invitational flyers were put up at bus stations, vocational schools, unemployment offices, and hospitals. The flyer headline read, "Money For Your Opinion!" Here, too, the snowballing technique was used.

The objective of the focus group discussions was to examine various individual-level perceptions and explanations of news and related questions in depth, including:

1. What kind of information catches people's attention and is more salient in their minds? While news items span the full spectrum of human life, they vary in terms of their relevance to individuals' daily lives. What is of interest to one person may be totally irrelevant to another, whereas at times a particular item might be of central importance to an entire nation. The study sought to determine what kinds of news items people in different societies consider to be newsworthy, both regarding extremely unusual situations as well as on an ordinary random and uneventful day.

2. Are people more interested in positive or negative news? This controversial issue often comes up in discussions of news. People often express a desire to obtain more good or positive news in the media, while at the same time they admit that bad or negative news is more exciting and important. What is it that makes negative news attractive, and why do people wish to avoid it? Does this phenomenon occur universally, or is it culturally bound?

3. How do journalists, public relations practitioners, and news consumers differ in their definitions of news? Each of these three groups of people relate to news. Journalists produce news based on various sources. Public relations practitioners often provide journalists with information with the hope that it will be incorporated in the news. And consumers read, watch, and listen to the news. Given these different positions vis-à-vis news, it is important to understand whether the three groups understand the concept of news in the same way or whether they have different perspectives regarding it.

Listing the Most Significant News Events

During the next stage of the focus group discussion, the moderator gave each participant a blank page and requested that each participant write down the three most important news events that occurred during her or his lifetime. When all the participants were done, they were asked to indicate which events they had noted and to explain why they made their particular choice. The purpose of this task was to look at whether there is any commonality in the types of events people in the ten countries named.

Linking the Content Analyses with the Focus Group Data

Toward the end of each focus group session, the participants were asked to take part in a gatekeeping exercise to see the extent to which they agreed on how newsworthy the news stories were. This task was based directly on the content analysis of the local newspapers and that had to be completed before the focus groups could be conducted.

Upon completion of the content analysis of each newspaper, the prominence scores for each news item were calculated as previously noted based on the size of the newspaper items (including both verbal and visual content) and their position within the newspaper. Three of the seven days were randomly selected. For each day, ten items were identified based on the ranking of their prominence scores from the highest to the lowest score. The ten items were identified as follows: the item with the highest prominence score (ranked as #1), the item closest to the 10th percentile of the prominence distribution (ranked as #2), the item closest to the 20th percentile (ranked as #3), and so forth until the item closest to the bottom percentile (ranked as #10).

Once the three lists of ten items were created, the headline and subhead (if it existed) of each item were printed on the cards, with one news item's headline per card. The indexes cards were color-coded by day. No other information appeared on the cards. Each set of ten cards was shuffled prior to the focus group so that the order in which the cards appeared in each set was random for each person.

After the qualitative discussion was completed, the group moderator handed each participant a set of ten cards from one of the three days. The participants were told that all of the items appeared in the local newspaper on a certain day some time ago, and they were asked to arrange the cards in the order in which they, had they been the newspaper editor, would have ranked them for publication, based on the degree of importance they personally assigned to each item.[4] In other words, they were asked to indicate which item, in their view, was the most newsworthy and deserved to receive top priority, the next priority, and so forth, for all ten of the items. When respondents completed the task, they gave their sets of cards to the moderator. When all of the participants were done, they were given the second set of cards and were asked to repeat the procedure. This was done again for the third set of cards, thus resulting in three sets of newsworthiness rankings for each person in each focus group. Shortly after the focus group discussion ended, the moderator recorded the order in which the cards were arranged by each of the participants. Once the rankings were recorded, the indexes cards were reshuffled for the next focus group.

Conclusion

This study collected data from ten countries during a period of more than two years. The first phase of the study was a content analysis of a composite week of newspaper issues as well as television newscasts and radio newscasts in two cities in each of the countries. In the second phase, four focus groups were conducted in each of the cities: one with journalists, one with public relations practitioners, and two with audience members of low and high socioeconomic status, respectively. In the interim between the two phases, the prominence of news items from focus group members' local newspapers was calculated in order to create a gatekeeping exercise that was conducted after the end of each focus group.

Part 2
Data Analyses across the Countries

4
Topics in the News

In this chapter we look at the topics defined as news by the mass media in our ten countries. *Topic* is an interesting variable when looked at from the perspectives of both biological and cultural evolution. On the one hand, biological evolution might predict that the same news topics would be of innate interest to humans all around the globe. On the other hand, cultural evolution might predict differences among countries based on culturally defined ideas of what is important. As it turns out, we find both significant similarities and differences.

Our content analysis yields 25,886 newspaper news items, as well as 2,947 in television and 3,251 in radio news programs. We coded the news items into twenty-six main topic categories, with each having several subtopics. Appendix A includes all topic categories and subcategories.

News Topics Around the World

If we combine the news from television, radio, and newspapers in all ten countries, we see that sports news items are the most prevalent, with stories dealing with politics within each country ("internal politics") being nearly as frequent (see Table 4.1). Other frequent topics include cultural events; business, commerce, and industry; international politics; internal order; and human interest. The least common topics in these countries' news are stories about the environment, labor relations and trade unions, energy, fashion and beauty, and population.

If we look at the countries individually, we see that sports stories are most frequent in Australia, South Africa, and the United States but are least frequent in China. Internal politics is quite a bit less frequent in Australia but is the most frequent news topic in Chile, China, India, and Russia. In Germany, most news items deal with cultural events, whereas this topic is less common in news from China, India, South Africa, and the United States. The percentage of stories dealing with business topics is largest in the United States and China and smallest in Jordan, Russia, and Israel. In Jordan, nearly a quarter of all news items deal with international politics, and about one in ten stories in China is about international politics. Stories about internal order are most common in India, South Africa, Israel, and Russia.

TABLE 4.1 Distributions of News Item Topics in the Three Media

Topics	Australia	Chile	China	Germany	India	Israel	Jordan	Russia	S. Africa	U.S.	Total
Sports	21.3	12.0	4.5	12.8	11.5	15.5	10.0	11.4	19.0	18.7	14.1
Internal Politics	5.2	18.1	16.6	14.2	18.5	10.5	11.0	16.1	10.9	12.4	13.3
Cultural Events	13.2	13.5	6.5	16.8	8.7	12.7	11.8	9.9	4.9	7.7	11.1
Business/Commerce/ Industry	10.7	6.4	12.8	11.3	8.7	2.6	5.0	3.9	11.4	13.5	9.0
International Politics	5.1	5.4	11.4	2.6	5.9	2.9	24.3	8.7	5.2	5.3	7.4
Internal Order	4.9	4.6	4.9	6.0	11.0	9.9	5.2	7.8	10.8	6.3	6.8
Human Interest	10.2	4.2	6.5	4.8	8.2	3.8	2.0	5.8	7.7	7.1	6.1
Economy	2.6	5.4	4.9	2.4	2.4	2.6	6.1	5.5	4.7	2.0	3.7
Entertainment	2.6	2.1	1.2	1.9	2.1	16.5	3.7	2.5	2.8	1.9	3.2
Health/Welfare/Social Services	2.2	3.9	5.0	2.2	2.1	2.8	4.7	1.8	2.8	2.9	3.0
Education	.9	2.8	4.8	1.8	3.9	4.6	2.8	.8	3.4	2.3	2.7
Transportation	2.8	3.1	2.7	2.4	1.0	1.3	.9	1.4	2.7	4.0	2.4
Disasters/Accidents/ Epidemics	2.9	2.1	1.0	4.7	2.2	1.0	.6	2.9	4.8	1.4	2.3
Communication	3.7	2.2	1.7	2.4	1.7	2.9	1.0	4.7	.8	2.0	2.3
Ceremonies	2.0	2.8	2.0	4.6	1.6	.8	1.4	3.0	1.6	1.2	2.2
Weather	1.6	1.2	1.9	2.0	1.0	.0	1.1	3.3	.6	3.8	1.7
Social Relations	1.3	1.5	.7	1.3	1.6	1.6	3.3	1.5	1.3	1.2	1.5
Housing	1.4	1.2	1.6	2.0	.8	2.6	.1	1.5	.5	.8	1.2
Military and Defense	1.1	1.4	1.2	.5	1.0	1.9	2.1	2.3	.6	.7	1.2

Science/Technology	1.0	1.1	3.2	.9	1.2	.2	1.0	2.4	.5	1.0	1.2
Environment	1.5	.6	3.0	.6	1.1	1.0	.5	.7	.6	1.2	1.0
Labor Relations/Trade Unions	.5	2.3	.5	.4	1.2	1.3	.7	.7	1.3	.4	.9
Energy	.1	.5	.8	.4	1.5	.0	.1	.6	.4	1.1	.6
Fashion/Beauty	.7	.4	.4	.2	.6	.9	.1	.5	.1	.3	.4
Population	.3	.1	.4	.3	.4	.0	.4	.4	.0	.2	.3
Others	.1	10.0	.0	.6	.4	.0	.0	.0	.4	.6	.4
Total[a]	100.0	100.0	100.0	100.0	100.0	100.0	100.0	100.0	100.0	100.0	100.0
(N)	(4245)	(4630)	(2210)	(3933)	(3858)	(2033)	(3351)	(1545)	(2025)	(4073)	(31907)

Note: Distributions given in percent.

[a]Total percentage may not actually be 100.0 due to rounding error.

News Topics in Newspapers, Television, and Radio

If we look at news topics within each medium, we see that there are interesting differences. Newspapers include more stories about sports; cultural events; internal politics; business, commerce and industry; and human interest (see Table 4.2). Overall, when newspapers cover politics, it is more likely to be internal politics rather than international politics. The most common topics in Chinese newspapers are international and internal politics, business, and human interest (46%). In Jordan, they are international politics and internal politics, plus cultural events and sports (53%). Australian newspapers emphasize sports, cultural events, business, and human interest (58%), whereas in Chile the dominant topics are cultural events, sports, and internal politics (44%). Both German (51%) and U.S. (54%) newspapers emphasized business, internal politics, cultural events, and sports. South African newspaper articles are primarily about sports, business, and internal order (43%). Newspaper stories in India and Russia are about a more diverse set of topics, with the leading topics being internal politics and sports in India (28%) and internal politics and cultural events in Russia (27%). Finally, in Israel the leading topics are entertainment, sports, and cultural events (51%).[5]

Television differs from newspapers. Across the ten countries, television carries more stories about internal politics, sports, and internal order (see Table 4.3). Internal politics stories are most common in the television news of Chile, China, Germany, India, Russia, and South Africa, and international politics dominate television news in Jordan—four out of ten Jordanian television news stories are about international politics. In contrast, television news programs in Germany and the United States contain few news items about international politics, and this topic is also less common in Australia, Chile, and China. Television news about sports is so common in Australia and the United States that it comprises more than one-fourth of all stories. Throughout the ten countries' television news programs, just over 10% of stories are about maintaining internal order, except in China and Jordan, where these stories are quite rare. Television news about culture is more common in Germany and Israel than in the other countries. Although about 10% of U.S. television news is about the weather, there is virtually no weather news on television in Chile, China, Israel, and South Africa.

Thus, we can see that television news in Australia is mostly about sports, human interest, and disasters or accidents (58%). Chilean and Chinese television news covers diverse news topics, with the most common being stories about internal politics and internal order (38%) in Chile and internal politics and business (40%) in China. In Germany, the most common topics are internal politics, internal order, and cultural events (40%). Three topics—internal politics, international politics, and internal order—dominate Indian television news (64%). Israeli television news concentrates on cultural events, sports, and stories about internal order (48%), and in Jordan the major topics are international politics, internal politics, and sports (66%). Russian television

TABLE 4.2 Distributions of Newspaper Items by Topic

Topics	Australia	Chile	China	Germany	India	Israel	Jordan	Russia	S. Africa	U.S.	Total
Sports	20.7	12.5	6.7	13.5	11.6	15.9	11.1	9.3	20.5	17.2	14.4
Cultural Events	15.2	15.9	7.1	17.7	9.8	13.8	14.3	11.8	5.7	10.5	13.1
Internal Politics	4.8	15.2	12.3	14.2	16.4	8.1	10.1	15.0	8.5	11.6	11.8
Business/Commerce Industry	11.8	7.2	12.5	11.7	9.2	3.1	5.7	4.6	12.4	14.7	9.7
Human Interest Stories	10.5	4.9	9.0	4.9	9.5	3.8	2.0	7.1	8.8	8.6	6.9
International Politics	5.0	5.3	11.7	2.7	4.9	.7	17.7	7.7	3.2	6.3	6.3
Internal Order	3.7	3.5	4.1	4.9	10.5	8.3	5.8	7.4	10.3	5.6	6.1
Economy	2.9	5.8	3.4	2.5	2.5	2.2	6.9	6.2	5.3	2.1	3.9
Entertainment	2.8	2.5	1.9	2.0	2.4	21.5	4.5	3.2	3.1	2.7	3.9
Health/Welfare/Social Services	2.0	3.6	4.7	2.1	2.2	3.3	5.3	1.7	2.8	2.6	3.0
Education	.9	2.9	6.7	1.8	4.4	5.0	2.8	.9	3.5	2.3	2.9
Communication	4.0	2.6	1.9	2.5	1.8	3.6	1.1	6.2	1.0	1.9	2.5
Ceremonies	2.1	2.7	1.3	5.0	1.3	.2	1.3	2.9	1.7	1.1	2.2
Transportation	2.6	3.0	1.9	2.3	1.1	1.5	1.1	1.6	2.1	2.4	2.1
Disasters/Accidents/Epidemics	1.9	1.7	1.0	4.1	2.0	.6	.7	1.9	4.4	1.1	2.0
Social Relations	1.4	1.3	.9	1.1	1.7	1.5	3.7	1.7	1.4	1.4	1.6

(Continued)

TABLE 4.2 Distributions of Newspaper Items by Topic (*Continued*)

Topics	Australia	Chile	China	Germany	India	Israel	Jordan	Russia	S. Africa	U.S.	Total
Housing	1.6	1.2	1.6	1.8	.9	3.0	.2	1.7	.5	1.0	1.3
Science/Technology	1.0	1.2	3.4	1.0	1.4	.2	1.3	2.1	.6	1.1	1.2
Military and Defense	1.0	1.4	1.0	.5	1.1	.8	2.0	1.5	.8	.8	1.1
Environment	1.5	.5	3.2	.6	1.0	.8	.5	.7	.5	1.3	1.0
Weather	.8	1.1	1.4	1.2	.5	.0	.4	2.7	.6	1.1	.9
Labor Relations/ Trade Unions	.4	1.4	.1	.4	1.0	1.2	.8	.7	.9	.3	.7
Energy	.1	.5	.9	.3	1.2	.1	.1	.5	.5	1.0	.5
Fashion/Beauty	.7	.5	.7	.2	.6	.6	.1	.6	.2	.5	.5
Population	.3	.1	.2	.3	.3	.1	.5	.3	.0	.1	.2
Others	.1	1.3	.0	.7	.5	.0	.0	.0	.5	.8	.5
Total[a]	100.0	100.0	100.0	100.0	100.0	100.0	100.0	100.0	100.0	100.0	100.0
(N)	(3638)	(3808)	(1339)	(3569)	(3297)	(1563)	(2760)	(1170)	(1663)	(2906)	(25713)

Note: Distributions given in percent.

[a]Total percentage may not actually be 100.0 due to rounding error.

TABLE 4.3 Distributions of Television Items by Topic

Topics	Australia	Chile	China	Germany	India	Israel	Jordan	Russia	S. Africa	U.S.	Total
Internal Politics	5.1	27.5	29.8	14.5	33.0	6.6	17.3	20.1	19.0	10.8	17.9
Sports	26.0	7.5	.6	5.6	8.4	18.0	9.2	3.9	14.6	29.1	13.8
Internal Order	11.1	10.2	3.8	14.1	17.6	10.4	2.7	13.0	13.1	10.8	10.4
International Politics	4.5	5.4	9.5	1.2	13.2	4.7	39.1	17.5	14.6	.7	10.2
Business/Commerce/Industry	4.3	3.6	12.1	8.8	6.6	1.4	3.1	2.6	7.3	5.2	5.6
Cultural Events	1.8	4.2	7.6	11.6	.7	19.9	.0	3.9	1.5	1.8	4.6
Human Interest Stories	11.1	2.1	1.9	4.4	.4	8.1	3.4	1.3	2.9	6.3	4.5
Weather	7.1	.3	.6	5.6	5.1	.0	6.1	3.2	.7	9.6	4.3
Disasters/Accidents/Epidemics	9.8	2.7	.6	7.6	2.9	2.8	.3	3.2	6.6	2.2	4.0
Health/Welfare/Social Services	3.3	5.1	7.0	3.2	1.5	1.4	3.7	2.6	2.2	4.5	3.7
Transportation	5.8	3.6	6.0	3.2	.0	.9	.3	.6	3.6	4.7	3.3
Ceremonies	1.8	5.7	2.5	2.0	1.5	2.8	3.4	5.8	1.5	1.6	2.7
Economy	.5	3.3	6.0	2.0	.4	2.4	2.7	3.9	2.9	2.2	2.5
Education	.8	3.9	1.0	2.0	.0	3.8	4.1	.6	2.9	2.9	2.2
Labor Relations/Trade Unions	1.0	6.0	.3	.4	.7	.9	.0	.6	3.6	.9	1.5

(Continued)

TABLE 4.3 Distributions of Television Items by Topic (*Continued*)

Topics	Australia	Chile	China	Germany	India	Israel	Jordan	Russia	S. Africa	U.S.	Total
Environment	1.3	1.2	2.5	.8	1.1	4.3	.0	.0	1.5	1.3	1.4
Social Relations	.3	2.4	.3	3.2	1.1	2.8	1.7	.6	.0	.4	1.2
Housing	.0	1.5	1.3	5.2	.4	2.4	.0	1.3	.0	.4	1.1
Military and Defense	1.0	.6	.6	.4	.7	.5	.7	9.1	.0	.4	1.0
Communication	1.0	1.5	1.0	1.6	.7	1.4	.7	.6	.0	.9	1.0
Science/Technology	.5	.9	3.5	.4	.0	.5	.0	3.2	.0	1.6	1.0
Energy	.0	.3	1.3	1.6	2.9	.0	.7	.6	.0	1.1	.8
Entertainment	1.8	.6	.0	.0	.0	.0	.7	.0	1.5	.0	.5
Fashion/Beauty	.3	.0	.0	.0	.4	3.8	.0	.0	.0	.0	.3
Population	.0	.0	.0	.4	.7	.0	.0	1.3	.0	.4	.2
Total[a]	100.0	100.0	100.0	100.0	100.0	100.0	100.0	100.0	100.0	100.0	100.0
(N)	(396)	(334)	(315)	(249)	(273)	(211)	(294)	(154)	(274)	(446)	(2946)

Note: Distributions given in percent.

[a]Total percentage may not actually be 100.0 due to rounding error.

news carries primarily stories about internal politics, international politics, and internal order (51%). South Africa also emphasizes these topics, plus sports (61%). American television news shows a world full of sports, internal politics, issues concerning internal order, and the weather (60%).

Finally, fully one-third of all radio news items in the ten countries are about politics, either internal or international (see Table 4.4). The emphasis on politics in radio news is particularly evident in Chile, China, India, Israel, Jordan, and South Africa. In Jordan, seven of ten radio news items are about international politics, whereas news about internal politics is most common on Chilean, Indian, Israeli, and South African radio. Sports stories are most common in Australian, Russian, and U.S. radio. The maintenance of internal order shows up most frequently in German and Israeli radio, and business news is most frequent in China and the United States.

Conclusion

When journalists look at the many events that occur in the world each day and decide which will become news, one of the first criteria they consider is the topic of the event. Is it political? Does it involve business? What happened? Who is involved? Are people hurt or killed? Knowing only that an event involves issues relating to transportation or the environment or science or education helps us predict whether the event becomes news—and if it is about one of these topics, it probably will not be covered by the news media. These are among the least covered topics in each of our ten countries.

The ten countries we studied differ in many ways: geography, language, economy, political systems, religion, culture, population size, and so on. Yet there is remarkable agreement across the countries on what kinds of events, ideas, and people should constitute *news*. Around the world, an event, person, or idea is most likely to become news if it deals with sports, international or internal politics, cultural events, business, internal order, or human interest. It is least likely to become news if it deals with science and technology, the environment, labor relations and trade unions, energy, fashion and beauty, and population.

Figure 4.1 shows that around the world, the range of topics available to people limits their news diets. We studied twenty newspapers, twenty television news programs, and twenty radio news programs for seven days in the first. Addressing research question one, we see that, although there are some between-country differences, there is also strong agreement among the countries about which topics constitute *news*. Only seven topics account for two-thirds of our 32,000+ news items. The other nineteen topics account for the remaining third.

Our tables show some variation in the relative importance of these seven topics across the ten countries, but the amount of agreement far outweighs the minor differences observed. If biological evolution has hard-wired people to want certain kinds of information, then we would expect just what we see in

TABLE 4.4 Distributions of Radio Items by Topic

Topics	Australia	Chile	China	Germany	India	Israel	Jordan	Russia	S. Africa	U.S.	Total
Internal Politics	12.3	34.3	19.4	12.2	29.2	27.8	13.5	18.6	29.5	16.5	21.5
International Politics	6.6	6.3	11.5	2.6	10.1	14.3	70.4	8.1	14.8	4.3	13.8
Sports	23.7	11.2	1.4	7.8	12.8	11.6	.0	27.6	4.5	18.3	11.9
Internal Order	14.2	8.7	7.4	20.9	10.8	19.3	1.7	6.3	13.6	6.2	9.1
Business/Commerce/Industry	3.8	1.8	13.7	7.0	5.2	.8	.7	.9	4.5	14.0	7.0
Weather	6.2	2.8	4.0	18.3	2.4	.0	3.0	6.3	.0	11.0	5.5
Transportation	.9	3.3	2.5	2.6	.3	.8	.0	.5	10.2	10.0	3.7
Disasters/Accidents/Epidemic	7.1	4.3	1.3	15.7	3.8	1.5	.0	8.1	5.7	2.2	3.5
Economy	1.9	3.7	7.9	.9	2.1	5.0	2.0	3.2	.0	1.5	3.4
Health/Welfare/Social Services	3.3	5.7	4.7	2.6	1.4	.4	.0	1.8	3.4	2.9	3.0
Military and Defense	3.3	1.4	2.0	2.6	.3	9.3	4.4	1.8	.0	.3	2.2
Labor Relations/Trade Unions	2.4	6.7	1.6	.0	3.8	1.9	.3	.9	1.1	.3	2.1
Education	.9	1.4	2.2	1.7	2.1	3.1	2.0	.5	3.4	1.8	1.8
Ceremonies	.0	1.2	3.2	.0	4.5	2.7	.3	1.8	.0	1.4	1.8
Human Interest Stories	4.3	.6	3.1	3.5	.3	.4	.7	2.3	2.3	1.8	1.8
Cultural Events	.5	.8	4.3	.0	2.8	.0	.0	4.1	.0	.3	1.5
Communication	3.3	.2	1.4	.0	1.4	.0	.0	.0	.0	3.3	1.4

Environment	1.9	1.0	2.7	.0	1.4	.0	.3	1.4	1.1	.7	1.2
Social Relations	1.4	2.4	.4	.9	1.4	.8	.7	.9	3.4	.7	1.1
Science/Technology	.5	.4	2.5	.0	.0	.0	.0	3.2	1.1	.6	.9
Energy	.0	.2	.4	.0	3.5	.0	.0	.9	.0	1.4	.8
Housing	.5	1.0	1.6	.9	.0	.4	.0	.5	1.1	.1	.6
Population	.5	.4	.9	.0	.3	.0	.0	.5	.0	.4	.4
Fashion/Beauty	.5	.0	.0	.0	.0	.0	.0	.0	.0	.0	.0
Total[a]	100.0	100.0	100.0	100.0	100.0	100.0	100.0	100.0	100.0	100.0	100.0
(N)	(211)	(492)	(556)	(115)	(288)	(259)	(297)	(221)	(88)	(721)	(3248)

Note: Distributions given in percent.

[a]Total percentage may not actually be 100.0 due to rounding error.

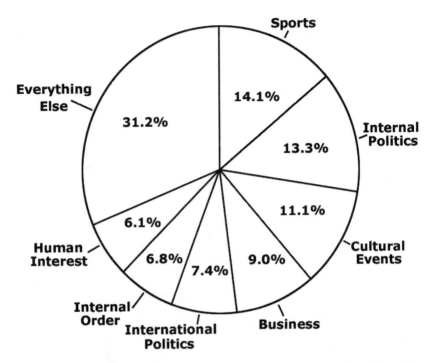

Fig. 4.1 Two-thirds of the world's news menu is comprised of only seven topics. These define what is important and the type of information diet that the audience can consume.

this chapter—substantial agreement between the countries in the topics that make up the news. Still, variations due to cultural evolution are also expected and this explains why the Jordanian media put such an emphasis on news about international politics while Australians get a big serving of sports news. Yet both value news about international politics and sports news.

5
Deviance in the News

In this chapter, we look at the intensity of deviance in the news of the ten countries. As mentioned in chapters 2 and 3, we use three measures of deviance: statistical deviance, social change deviance, and normative deviance.

Statistical deviance concerns things that are unusual, odd, or novel, as well as anything that is either above or below average. For example, either higher- or lower-than-expected results in the world's stock exchanges makes the news. Other examples include a person surviving weeks after a tsunami, a record-setting home run, and the birth of septuplets.

Social change deviance in the news refers to threats to the status quo, for example, civil demonstrations that bring about a new presidential election or an election in a country divided by ethnic and religious constituencies in which terrorists threaten and sometimes kill the candidates or their supporters.

Normative deviance deals with violations of laws and norms, such as crime or changes in the way in which things are or ought to be done. Examples can range from acts of terror to the violation of cultural traditions such as the way people dress, work, or eat.

In addition to these three kinds of deviance, we are also interested in *verbal* and *visual* depictions of news. The verbal content of newspapers includes headlines (or captions) and printed columns of text. On television and radio, verbal content includes spoken reports, printed text that appears as part of a news item, text captions for a visual element, as well as text scrolling across the screen. The visual content in newspapers includes photographs, charts, graphs, tables, figures, cartoons, drawings, maps, logos, or any other graphic device. In television news, all items have a visual component, even if it is only the face of the speaker (e.g., anchor, reporter, interviewee). Radio news, of course, has only verbal content.

In many recent content analyses of the news (e.g., Carter, Fico, and McCabe 2002; Nataranjan and Xiaoming 2003), visual and verbal content elements are combined, but we believe that deviance may be carried differently by visual elements (images) and verbal elements (words).[6] A very deviant photograph, such as one of dead bodies after an earthquake or tsunami, may give the reader a much stronger impression than words alone do, perhaps causing the reader to interpret the item as being more deviant than if it had appeared without a photograph. In contrast, an item about a company president embezzling money might be interpreted as being less deviant if accompanied by a head-and-shoulders photo of the president—or no photo at all—rather than seeing him in handcuffs.

Given the three measures of deviance and the distinction we draw between verbal and visual elements, our analyses deal with six measures: (1) statistical deviance of verbal content; (2) statistical deviance of visual content; (3) social change deviance, verbal; (4) social change deviance, visual; (5) normative deviance, verbal; and (6) normative deviance, visual.

As indicated in the previous chapter, we include in our definition of news both traditional stories as well as items that are not normally considered as news but that do appear in the media, especially in newspapers. We therefore include items such as weather maps, stock market reports, and gossip columns, along with (traditional) hard news. Normally, items such as stock reports and weather maps are coded as having no deviance. (There could be deviant news about the stock market or weather, but this information would be presented as a separate news item.) Comics, horoscopes, and gossip columns may provide information that is somewhat deviant, but as a whole they are coded as having no deviance.

How Many News Items Contain Deviance?

To answer this question, we contrast the number of news items that have no deviance with those that have some deviance on one or more of the three deviance dimensions—statistical, social change, and normative. The left-hand column of Table 5.1 indicates the number of dimensions on which a news item is scored as being deviant. A score of 0 indicates that a news item is coded as having no deviance on any of the three dimensions. In contrast, a score of 3 indicates that the news item is coded as deviant on all three dimensions. Thus, Table 5.1 gives us some sense of the distribution of deviance dimensions in the news.

In the 32,000+ news items analyzed, we see that about two-thirds of the verbal content and just under one-half of the visual content contains some type of deviance. Considering that few of the 32,000+ news items are made up of solely visual content, we can conclude that at least two-thirds of the news in

TABLE 5.1 Distributions of News Items Coded as Deviant on the Three Dimensions of Deviance

Number of Deviance Dimensions in News Items	Verbal News Items	Visual News Items
0	34.8	55.6
1	32.9	28.4
2	20.7	9.7
3	11.7	6.3
Total[a]	100.0	100.0

Note: A score of 3 indicates that the news items are deviant on all three dimensions—statistical, social change, and normative. Distributions given in percent.

[a]Total percentage may not actually be 100.0 due to rounding error.

the ten countries is composed of some type and amount of deviance. Although the theory does not predict that news items are deviant on more than one dimension, Table 5.1 demonstrates that one-third of verbal news items are deviant on only one dimension, one-fifth on two dimensions, and just over a tenth on all three dimensions. For visual content, just over one-fourth of news items contain at least one deviance dimension, about a tenth are coded as having two dimensions of deviance, and 6% of visuals show all three types of deviance.

That there is less deviance in visual news content than in verbal content (text) is not surprising, given the frequently bland visual diet of head-and-shoulder shots in newspapers, and anchor shots and reporter stand-ups in television news. Still, just under half of visual content is coded as having some amount of deviance, and the same is true for two-thirds of the verbal content. Further, it is interesting to note that as many as one-third of verbal news items are coded as deviant on two or three dimensions, and 17% of visuals are rated the same. This supports our idea that deviance is a multi-dimensional construct: Some news items are deviant on one dimension and not others, whereas other news items show multiple types of deviance. News items that contain multiple deviance dimensions may differ in not only the amount of deviance but also in the quality or nature of the news and its topics.

The following analyses look at statistical, social change, and normative deviance separately to find out which dimension of deviance is most common and most intense. To measure intensity, we calculate the mean level of deviance on each of the three dimensions; however, because one-third of verbal content and one-half of visual content contain no deviance (Table 5.1), the mean levels of deviance are on the low side—lower than one might expect given the fact that two-thirds of news items include some sort of deviance.

Deviance and News Topics

Table 5.2 shows that, although sports is a very common news topic, sports stories tend to have less intense deviance. In contrast, stories about internal order—including crime stories—tend to be more intensely deviant than stories about other topics. Finally, stories about international and internal politics are also more intensely deviant than most other topics, with international items being consistently more deviant than domestic ones (research question 2).

In looking at the three media separately (Tables 5.3–5.5), television and radio news items generally tend to be more intensely deviant than newspaper items, mainly regarding statistical deviance. More specifically, deviance is less intense in newspaper items than in radio or television news, with the exception of more intense deviance in stories about internal politics, international politics, and internal order.

TABLE 5.2 Mean Intensity of Deviance Scores of the Most Common News Topics in the Three Media

Most Common Topics	Statistical Deviance		Social Change Deviance		Normative Deviance		Index	
	Visual	Verbal	Visual	Verbal	Visual	Verbal	Visual	Verbal
Sports	1.4	1.6	1.1	1.1	1.1	1.1	1.2	1.3
Internal Politics	1.7	2.1	1.3	1.6	1.2	1.5	1.4	1.8
Cultural Events	1.5	1.6	1.0	1.1	1.1	1.1	1.2	1.3
Business/Commerce/ Industry	1.5	1.8	1.1	1.3	1.1	1.2	1.3	1.4
International Politics	1.8	2.4	1.6	1.7	1.5	1.7	1.6	2.0
Internal Order	2.0	2.3	1.5	1.7	1.7	2.5	1.7	2.1
Human Interest	1.6	1.8	1.1	1.1	1.1	1.2	1.3	1.4
Economy	1.5	1.8	1.2	1.4	1.1	1.2	1.2	1.5
Entertainment	1.3	1.2	1.1	1.0	1.1	1.1	1.1	1.1
Health/Welfare/ Social Services	1.6	1.9	1.3	1.4	1.2	1.4	1.3	1.6

TABLE 5.3 Mean Intensity of Deviance Scores of the Most Common News Topics in Newspapers

Most Common Topics	Statistical Deviance		Social Change Deviance		Normative Deviance		Index	
	Visual	Verbal	Visual	Verbal	Visual	Verbal	Visual	Verbal
Sports	1.4	1.6	1.0	1.1	1.1	1.1	1.1	1.3
Internal Politics	1.5	2.0	1.2	1.6	1.1	1.5	1.2	1.7
Cultural Events	1.5	1.5	1.0	1.1	1.1	1.1	1.2	1.3
Business/Commerce/ Industry	1.5	1.7	1.1	1.3	1.1	1.1	1.2	1.4
International Politics	1.6	2.2	1.3	1.6	1.3	1.7	1.4	1.8
Internal Order	1.9	2.2	1.4	1.6	1.6	2.4	1.6	2.1
Human Interest	1.6	1.7	1.1	1.1	1.2	1.1	1.3	1.3
Economy	1.4	1.8	1.2	1.4	1.1	1.2	1.3	1.5
Entertainment	1.2	1.2	1.1	1.0	1.1	1.1	1.1	1.1
Health/Welfare/ Social Services	1.5	1.9	1.2	1.3	1.1	1.3	1.2	1.5

Aside from stories about sports and cultural events, stories in television generally have more intense statistical deviance (see Table 5.4). Television news about internal order and international politics also has intense normative deviance, as do international politics stories that challenge the status quo (social change deviance).

TABLE 5.4 Mean Intensity of Deviance Scores of the Most Common News Topics in Television

Most Common Topics	Statistical Deviance		Social Change Deviance		Normative Deviance		Index	
	Visual	Verbal	Visual	Verbal	Visual	Verbal	Visual	Verbal
Sports	1.3	1.5	1.1	1.1	1.1	1.1	1.2	1.2
Internal Politics	2.0	2.4	1.4	1.7	1.4	1.7	1.6	1.9
Cultural Events	1.6	1.9	1.1	1.1	1.1	1.2	1.2	1.4
Business/Commerce/ Industry	1.7	2.1	1.2	1.4	1.3	1.4	1.4	1.6
International Politics	2.2	2.6	2.0	2.2	1.8	2.1	2.0	2.3
Internal Order	2.0	2.4	1.6	1.9	1.9	2.6	1.8	2.3
Human Interest	1.7	2.0	1.1	1.1	1.2	1.4	1.4	1.5
Economy	1.7	2.1	1.1	1.4	1.1	1.2	1.3	1.6
Entertainment	2.1	1.8	1.3	1.1	1.1	1.0	1.5	1.3
Health/Welfare/ Social Services	1.8	2.2	1.4	1.5	1.3	1.5	1.5	1.7

TABLE 5.5 Mean Intensity of Deviance Scores of the Most Common News Topics in Radio

Most Common Topics	Statistical Deviance	Social Change Deviance	Normative Deviance	Index
Sports	1.6	1.1	1.1	1.2
Internal Politics	2.4	1.7	1.4	1.8
Cultural Events	2.2	1.2	1.1	1.5
Business/Commerce/ Industry	2.0	1.4	1.2	1.5
International Politics	2.7	1.8	1.8	2.1
Internal Order	2.5	1.6	2.6	2.2
Human Interest	2.5	1.3	1.2	1.6
Economy	2.2	1.6	1.5	1.8
Entertainment	–	–	–	–
Health/Welfare/ Social Services	2.4	1.8	1.4	1.9

In radio news, all topics except sports have relatively intense statistical deviance scores (Table 5.5); thus, radio news stories are more likely to be about odd or unusual events, people, or ideas. In addition, stories about internal order—the most deviant topic category—have especially intense normative deviance, but have less social change deviance.

Verbal Versus Visual Deviance

As for verbal and visual elements, across the board deviance in verbal content is more intense than in visual content; thus, verbal text tends to carry the deviance. Visuals in both television and newspapers are on average less intensely deviant than verbal text.

Accordingly, within each type of deviance—statistical, social change, and normative—deviance in visual content (e.g., photographs or video, figures, graphs) is less intense than in verbal text. This may reflect editors' and producers' hesitance—mainly on television but also in newspapers—to show horrifying video or photographs, even if they discuss the event in the verbal content of the news item. For example, verbal descriptions of airplane crashes or terrorist bombings often include descriptions of body parts at the scene, yet body parts are rarely shown; on the other hand, an overall prohibition does not seem to exist against showing pictures of dead bodies in their entirety. Gatekeepers of images may be concerned about upsetting the audience and about causing audience members to stop viewing or reading; therefore, they put information about deviance more in verbal content than in visuals. Television news departments may also present more graphically intense images relating to events outside their country or of the "enemy" than to domestic events or to depictions of "our own" casualties.[7]

Similarities and Differences among the Countries

In addition to the overall picture across the ten countries, we are also interested in similarities and differences among them. We therefore look at the topics in each country that are coded as having the most intense deviance scores—that is, those coded as either 3 or 4 on any of the six, 4-point deviance scales. Table 5.6 addresses research questions two and three, that, across all ten countries, the most intensely deviant stories are about internal politics, internal order, and international politics and that this is especially true of Jordan, Israel, Russia, and the United States, with India and South Africa close behind. Stories about internal politics have relatively intense deviance in all the countries except for Australia. Deviant Australian news is most likely to be about internal order, international politics, or sports, whereas in Germany deviance is more intense in stories about internal politics, cultural events, internal order, sports, and business. In Chile, nearly a quarter of the most intensely deviant stories are about internal politics. In China, deviant stories are most likely to be about internal or international politics and business. Finally, deviant stories in Indian and South African news are about internal politics and internal order.

Tables 5.7–5.9 show the same basic pattern when each medium is examined separately. The most intensely deviant stories are about internal or international politics and internal order, and this is true in most countries. There are notable points about some of the countries: Business stories in newspapers are among the most intensely deviant in Australia, China, and

TABLE 5.6 Distributions of Very Intense Deviant News Items in the Three Media

Most Common Topics	Australia	Chile	China	Germany	India	Israel	Jordan	Russia	S. Africa	U.S.	Total
Sports	10.0	7.7	3.5	10.6	9.1	4.2	1.1	4.7	7.4	5.9	7.2
Internal Politics	7.5	24.0	17.0	19.4	22.6	19.6	12.3	19.6	14.1	16.9	18.4
Cultural Events	5.1	9.5	3.6	13.3	5.4	.7	1.2	6.0	1.3	1.9	6.4
Business/Commerce/Industry	7.9	6.1	11.7	10.3	6.4	2.4	2.4	3.6	6.4	5.7	6.7
International Politics	12.7	8.8	17.5	4.2	9.0	10.3	44.0	14.0	9.2	12.2	13.0
Internal Order	14.3	8.9	6.8	11.5	16.3	27.9	10.4	15.8	23.8	24.6	14.2
Human Interest	6.2	5.5	3.6	3.7	6.0	1.2	1.3	3.0	7.4	4.3	4.5
Economy	3.5	2.5	7.4	2.4	2.1	4.6	4.4	3.9	2.7	.8	3.1
Entertainment	.6	.4	.0	.5	.9	.2	1.9	.6	.4	.2	.6
Health/Welfare Social Services	4.5	2.8	5.1	2.3	1.9	6.1	5.1	1.4	3.7	3.7	3.3
Others	27.7	23.8	23.8	21.8	20.3	22.8	15.9	27.4	23.6	23.8	22.6
Total[a]	100.0	100.0	100.0	100.0	100.0	100.0	100.0	100.0	100.0	100.0	100.0
(N)	(859)	(1911)	(693)	(1770)	(1492)	(409)	(970)	(698)	(703)	(629)	(10134)

Note: Very intense deviant news items include those that received a rating of 3 or 4 on any of the six 4-point deviance scales. Distributions given in percent.
[a]Total percentage may not actually be 100.0 due to rounding error.

TABLE 5.7 Distributions of Very Intense Deviant News Items in Newspapers

Most Common Topics	Australia	Chile	China	Germany	India	Israel	Jordan	Russia	S. Africa	U.S.	Total
Sports	10.6	8.1	5.5	11.9	9.9	7.0	1.7	5.4	8.5	6.8	8.5
Internal Politics	7.3	21.2	12.8	19.5	18.9	16.7	10.4	19.6	12.0	17.3	16.8
Cultural Events	6.6	12.2	3.8	14.0	6.9	1.3	2.0	6.6	1.3	2.7	8.1
Business/Commerce/Industry	10.0	7.0	10.8	10.5	6.9	3.9	3.1	4.4	6.2	5.1	7.5
International Politics	13.7	9.4	19.1	4.4	7.8	1.8	34.2	13.2	5.3	19.7	11.1
Internal Order	11.1	7.2	8.3	9.8	16.0	25.0	15.1	15.6	23.6	27.8	13.3
Human Interest	5.4	6.9	4.5	3.6	8.0	1.8	1.4	3.4	8.7	3.1	5.2
Economy	4.6	2.7	6.8	2.5	2.3	6.1	6.0	4.0	3.1	.3	3.4
Entertainment	.2	.6	.0	.6	1.3	.4	2.9	.8	.5	.3	.8
Health/Welfare/Social Services	4.5	2.5	3.5	2.2	2.2	10.5	6.8	1.4	4.5	2.7	3.3
Others	26.0	22.2	24.9	21.0	19.8	25.5	16.4	25.6	26.3	14.2	22.0
Total[a]	100.0	100.0	100.0	100.0	100.0	100.0	100.0	100.0	100.0	100.0	100.0
(N)	(648)	(1422)	(398)	(1577)	(1109)	(228)	(588)	(499)	(551.)	(295.)	(7315)

Note: *Very intense deviant news items* include those that received a rating of 3 or 4 on any of the six 4-point deviance scales. Distributions given in percent.
[a]Total percentage may not actually be 100.0 due to rounding error.

TABLE 5.8 Distributions of Very Intense Deviant News Items in Television

Most Common Topics	Australia	Chile	China	Germany	India	Israel	Jordan	Russia	S. Africa	U.S.	Total
Sports	9.5	4.6	1.1	.8	6.6	.0	.5	.9	3.7	5.8	3.9
Internal Politics	8.8	31.5	24.1	19.2	36.7	3.1	18.9	18.3	20.4	13.0	21.9
Cultural Events	.7	3.2	1.1	12.5	.5	.0	.0	5.5	1.9	2.2	2.7
Business/Commerce/ Industry	2.0	4.1	9.2	9.2	4.1	3.1	1.6	2.8	7.4	2.2	4.2
International Politics	6.8	5.5	24.1	2.5	14.8	21.9	45.8	20.2	24.1	1.4	16.3
Internal Order	20.3	14.2	.0	22.5	21.4	53.1	3.7	16.5	25.9	29.7	17.9
Human Interest	10.8	1.8	1.1	4.2	.0	3.1	2.1	1.8	1.9	8.0	3.4
Economy	.0	.5	8.0	2.5	.5	3.1	2.6	2.8	1.9	.7	1.8
Entertainment	2.7	.0	.0	.0	.0	.0	.5	.0	.0	.0	.4
Health/Welfare/Social Services	5.4	4.1	3.4	3.3	1.0	3.1	4.7	1.8	.0	5.1	3.3
Other	33.0	30.5	27.9	23.3	14.4	9.5	19.6	29.4	12.8	31.9	24.2
Total[a]	100.0	100.0	100.0	100.0	100.0	100.0	100.0	100.0	100.0	100.0	100.0
(N)	(148)	(219)	(87)	(120)	(196)	(32)	(190)	(109)	(108)	(138)	(1347)

Note: Very intense deviant news items include those that received a rating of 3 or 4 on any of the six 4-point deviance scales. Distributions given in percent.
[a]Total percentage may not actually be 100.0 due to rounding error.

TABLE 5.9 Distributions of Very Intense Deviant News Items in Radio

Most Common Topics	Australia	Chile	China	Germany	India	Israel	Jordan	Russia	S. Africa	U.S.	Total
Sports	4.8	8.1	.5	.0	7.0	.7	.0	5.6	2.3	4.6	3.7
Internal Politics	6.3	32.2	22.1	19.2	29.4	27.5	11.5	21.1	25.0	18.9	22.8
Cultural Events	.0	.4	4.3	.0	1.6	.0	.0	3.3	.0	.5	1.2
Business/Commerce/ Industry	.0	2.6	14.4	8.2	6.4	.0	1.0	.0	6.8	9.2	5.3
International Politics	15.9	8.1	11.5	2.7	10.7	20.8	72.4	11.1	22.7	8.7	19.4
Internal Order	33.3	13.3	6.7	30.1	12.8	26.8	2.6	15.6	20.5	16.3	14.7
Human Interest	3.2	1.1	2.9	4.1	.5	.0	.5	2.2	4.5	3.6	1.8
Economy	.0	3.0	8.2	.0	2.7	2.7	1.6	4.4	.0	1.5	3.0
Entertainment	.0	.0	.0	.0	.0	.0	.0	.0	.0	.0	.0
Health/Welfare/ Social Services	3.2	3.7	8.7	4.1	1.1	.0	.0	1.1	2.3	4.1	3.1
Other	33.3	27.5	20.7	31.6	27.8	21.5	10.4	35.6	15.9	32.6	25.0
Total[a]	100.0	100.0	100.0	100.0	100.0	100.0	100.0	100.0	100.0	100.0	100.0
(N)	(63)	(270)	(208)	(73)	(187)	(149)	(192)	(90)	(44)	(196)	(1472)

Note: *Very intense deviant news items* include those that received a rating of 3 or 4 on any of the six 4-point deviance scales. Distributions given in percent.
[a]Total percentage may not actually be 100.0 due to rounding error.

Germany; in Israel, deviant newspaper stories are often about health and welfare (Table 5.7). As for television, in Australia, human interest stories are more intensely deviant, and the same is true for cultural events in Germany (Table 5.8). Finally, regarding radio, the percentage of intensely deviant business, commerce, and industry stories on the radio stations in the two Chinese cities is relatively high (Table 5.9).

Evaluating the Intensity of Deviance in the News

In the ten country chapters that follow, deviance across all 32,000+ news items is on average of low intensity—with mostly common or slightly common information, minimal or no threat to the status quo, and minimal violation of norms—even though two-thirds of the news items are coded as having some amount of one or more dimensions of deviance.

Research questions 2–4 ask about similarities in the amount of deviance across the ten countries, the three types of media, and between large and small cities. The tables in this chapter show only minor and often trivial differences; the intensity of deviance in the news is fairly uniform throughout the ten countries, the three news media, and between the major and peripheral cities. Still, the calculation of statistical significance is based partly on the sample size; as noted, when dealing with thousands of cases, even small differences between means are statistically significant. Therefore, instead of presenting a myriad of statistical comparisons between means, we show the results of Scheffé post hoc tests, which group the countries according to their mean scores on each variable. These results should not be interpreted as high-intensity or low-intensity deviance scores, since as noted all are on the less intense side. Rather, the results indicate whether there is relatively more or less deviance among the countries, cities, and media.

Differences among the Ten Countries

Table 5.10 shows the groups of countries that presented more or less intense deviance, based on the Scheffé post hoc tests.

Statistical Deviance Russian verbal text presents the most statistical deviance, while Israeli and American verbal text presents the least. The findings for the statistical deviance of visuals are somewhat different, with the most intense statistical deviance appearing in the visual news of China, India, Russia, and Australia. The least statistically deviant visuals are found in Israeli news.

Social Change Deviance Russian verbal text also includes the most intense social change deviance, whereas it is least intense in the United States, Israel, and Jordan. For visuals, the most intense social change deviance is in Indian and Australian news, and the least is in visuals from the United States, Israel, and Germany.

TABLE 5.10 Grouping of Countries according to Intensity of Deviance in Visual and Verbal Content

Intensity	Statistical Deviance		Social Change Deviance		Normative Deviance		Index	
	Visual	Verbal	Visual	Verbal	Visual	Verbal	Visual	Verbal
Most	China India Russia Australia	Russia	India Australia	Russia	China India	China Russia Germany	China India Australia	Russia
Least	Israel	United States Israel	United States Israel Germany	United States Israel Jordan	United States Israel Chile Germany	United States	Israel	United States

Note: Grouping of countries based on Scheffé post hoc tests.

Normative Deviance The most intense normative deviance is in verbal news text from China, Russia, and Germany, and the least is in U.S. news. For news visuals, the most intense normative deviance is found in China and India, and the least intense normatively deviant visuals are in the United States, Israel, Chile, and Germany.

Finally, two additive indexes—one for intensity of the verbal contents deviance and the other for the intensity of the visual contents deviance—provide an overall assessment of the differences among the countries on the deviance variables.[8] The verbal deviance index reveals that the most intensely deviant news is in Russia, whereas the least intensely deviant news is in the United States. The visual deviance index shows that the most intensely deviant visuals are found in China, India, and Australia, while the least intensely deviant visuals are in Israel.

Differences between the Major and Peripheral Cities

There are very small differences between the intensity of deviance in the news media of the major and peripheral cities. Table 5.11 demonstrates that there is no particular tendency of peripheral or major cities to have more intense statistical deviance. Major city news items have slightly more intense social change deviance in both the verbal and visual texts and more intense normative deviance in the verbal text, whereas peripheral cities have more statistically deviant visuals that are also more normatively deviant. However, if we look at the indices of verbal and visual deviance, we see that major city news is more intensely deviant in the verbal text, whereas no difference is found for the visual elements.

TABLE 5.11 Grouping of Major and Peripheral Cities according to Intensity of Deviance in Visual and Verbal Content

Intensity	Statistical Deviance		Social Change Deviance		Normative Deviance		Index	
	Visual	Verbal	Visual	Verbal	Visual	Verbal	Visual	Verbal
Most	Peripheral city	No difference	Major city	Major city	Peripheral city	Major city	No difference	Major city
Less	Major city		Peripheral city	Peripheral city	Major city	Peripheral city		Peripheral city

Note: Difference between cities based on Scheffé post hoc tests.

Differences among the Three Media

As Table 5.12 shows, across all the verbal and visual news texts, newspaper content has less intense deviance than news content from television and radio. However, this may be an artifact of the study design, since, as noted already, all newspaper content that is not advertising is treated in this study as *news*. In newspapers especially this includes content not traditionally considered to be news. Accordingly, when we calculate the overall deviance intensity score for newspapers, we must consider that the means of newspaper deviance may be diluted by items not traditionally considered as being news.

Variability of Deviance

In many cases, the average intensity of deviance found in the ten countries is low, but the mean value of a variable can hide the range of values. Are all news items of low intensity? Do some news items have more intense deviance? What sorts of news items are more intensely deviant, and what sorts are of lower deviance?

Appendix B presents bar charts of the distributions of deviance intensity scores for statistical, social change, and normative deviance for both verbal and visual news items—six variables per country. The top bar in each figure provides an overall view of the results from the ten countries combined. Here we see that some news items, whether verbal or visual, extend to even the highest deviance intensity. The most common form of deviance is statistical, with a considerable number of news items being described as somewhat to extremely unusual. Although there are also news items with intense social change deviance and intense normative deviance, their frequency is much lower than that of statistically deviant news.

The other bars, one per country, indicate the variability among them. Looking at the intensity of statistical deviance of the verbal elements, we see on the one hand that the modal (or most common) value in Germany, Israel, Jordan, and the United States is "common information." On the other hand, in Australia, Chile, China, India, Russia, and South Africa the mode for verbal statistical deviance is actually "somewhat unusual information," with a substantial number of news items scoring "unusual" or "extremely unusual" intensity. A different pattern emerges for social deviance change and normative deviance, where a smaller number of cases are scored as being beyond the lowest deviance intensity.

Conclusion

These findings show that there is deviance in the news of the ten countries. Two-thirds of the verbal content contains deviance, whereas just under half of visual content is deviant. Television and radio news items are, in general, more deviant than newspaper items. We suspect that this is due to time constraints on television and radio news that limit the number of words or images able to

TABLE 5.12 Grouping of Types of Media according to Intensity of Deviance in Visual and Verbal Content

Intensity	Statistical Deviance		Social Change Deviance		Normative Deviance		Index	
	Visual	Verbal	Visual	Verbal	Visual	Verbal	Visual	Verbal
Most	TV	Radio	TV	TV Radio	TV	TV	TV	Radio TV
– – – – – TV – – – – – – – – – – – – – – – – – Radio – – – – – – –								
Least	NP		NP	NP	NP		NP	NP
		NP				NP		

Note: Grouping of types of media based on Scheffé post hoc tests. NP = newspaper. TV = television.

be included; at the same time, television seeks "good" or "powerful" images to bolster its newscasts.

Finally, we must point out that the aggregation of data from ten countries should not be interpreted as representing the intensity of deviance in "world" news, because as noted in chapter 3 the countries were purposively selected. No ten countries can be said to represent the world as a whole. The sampling plan of this study, however, is designed to include as much variability as possible in these dimensions: geographic location, economic system, political system, language, culture, and religion. Thus, these ten countries do provide an estimate—albeit without information about sampling error—of what the news is like in many such countries, if not in the whole world. This provides more information about the nature of news around the world than if we had studied only one country or region, such as the United States or the Middle East.

6
Social Significance in the News

Having shown that deviance is present in the news from ten countries, in this chapter we look at the extent to which social significance is present. As mentioned in chapter 3, we use four measures of social significance:

Political significance consists of anything involving the political system, including elections, government activities, the passing of laws, or the breaking of laws. *Economic significance* covers business and commerce, including unemployment, exports and imports, currency rates, or budgets. *Cultural significance* includes social institutions such as religion, ethnicity, or language, and more ordinary events such as the opening of a museum. Finally, *public significance* relates to events that affect the well-being of the citizenry, including issues of health, the environment, and natural disasters.

As with the measures of deviance (see chapter 5), we examine visual images separately from verbal text because information is carried differently by these two modes. We thus have a total of eight measures of social significance: (1) political significance of verbal content; (2) political significance of visual content; (3) economic significance verbal; (4) economic significance, visual; (5) cultural significance, verbal; (6) cultural significance, visual; (7) public significance, verbal; and (8) public significance, visual.

As with deviance, we include in our content analysis both traditional stories and items not normally considered news, such as weather maps and gossip.

How Many News Items Contain Social Significance?

As in the previous chapter, we contrast the number of news items that have no deviance with those having low, medium, or high deviance on one, two, three, or all four social significance dimensions. Table 6.1 shows that about 80% of verbal news content is socially significant, as is just over half of the visual content. In contrast, 19% of verbal content has no elements of social significance, and the same is true for 47% of visual content. Since there are few solely visual news items, we can conclude that at least 80% of our 32,000+ news items contain some type of social significance.

As with the prior analysis of deviance in the news, we lack a theoretical explanation as to why news items should be socially significant on more than one dimension, yet about 43% of verbal news content is coded as being socially significant on two, three, or four dimensions. About 38% of verbal content is

TABLE 6.1 Distributions of Verbal and Visual News Items Coded as Socially Significant

Number of Social Significance Dimensions in News Items	Verbal News Items	Visual News Items
0	19 .1	47.4
1	37.7	28.4
2	26.3	15.0
3	13.2	7.1
4	3.7	1.9
Total[a]	100.0	100.0

Note: A score of 4 indicates that the news items are socially significant on all four dimensions—political, economic, cultural, and public. Distributions given in percent.

[a]Total percentage may not actually be 100.0 due to rounding error.

coded as having only one dimension. Just under a quarter of visuals are coded as having two or more dimensions of social significance, whereas 28% are noted as being significant on only one dimension.

Also, as with deviance, we see fewer social significance dimensions present in visual news content. Presenting three dimensions of social significance are 7% of news items, and fewer than 2% are coded as having all four dimensions. The ubiquitous head-and-shoulder visuals convey little about the social significance of the content. Even a photo of a president may have little social significance by itself, whereas the text could convey, for instance, a scandal, a decision to run or not run for re-election, or high approval ratings in public polls. Still, Table 6.1 shows that more than 50% of visual content includes some sort of social significance, as does over 80% of verbal content.

The fact that many news items are coded as having two or more social significance dimensions supports our claim that the social significance construct is comprised of multiple dimensions. Some news items are socially significant on only one dimension, whereas a substantial number show multidimensional social significance. Multidimensional news items may differ from unidimensional items not only in the number of dimensions involved but also in the qualitative nature of the news. News items characterized by only one dimension of social significance are probably quite different than those that can claim two to four types of social significance.

The remainder of this chapter is comprised of analyses using the mean values on the four social significance dimensions—from having no significance to having low, medium low, medium high, and high significance. These analyses are intended to assess the intensity of social significance on each of its four dimensions. We know that news items showing no social significance (19% for verbal content and 47% for visual content) cause the means to appear lower than we might expect, given the large number of news items (especially verbal content) that have some aspect of social significance. Analyses given in this chapter assess the nature of social significance by including only news items that show the most intense amount of the construct.

The Social Significance of News Topics

Across the ten topics (Table 6.2), there is a significant number of cases in which the mean values are halfway between 2.0 and 3.0 (on the 4-point scale), thus indicating social significance of moderate to high intensity. In fact, these cases provide good indications of construct validity:

- stories about politics rate high on political significance;
- business and economic stories rate high on economic significance;
- cultural events rate relatively high on cultural significance;
- several topics (including human interest) carry elements of public significance.

Across the three media and four types of social significance, stories about internal politics, international politics, and the economy have the most intense scores. Conversely, while sports is the most common news topic, the analyses show that these stories have low social significance, as do stories dealing with entertainment and human interest.

Furthermore, as is the case with deviance, television and radio news items generally tend to have more intense social significance than newspaper items (Tables 6.3–6.5). Compared to radio, television items have a tendency to have higher scores for economic significance, whereas the overall trend indicates higher scores for radio regarding cultural significance. There is no clear differential pattern between television and radio and the intensity of political significance or public significance.

Looking at each of the media separately, newspaper items clearly have the least intense social significance, both in verbal and visual content. Other than topics pointing to construct validity (such as those with relatively high political significance in items dealing with internal and international politics), five topics receive relatively intense public significance scores: internal order; health, welfare, and social services; international politics; internal politics; and items on the economy (Table 6.3).

A similar pattern appears for television, but six topics receive high intensity scores on public significance (the same five topics as in newspapers plus business, commerce, and industry [Table 6.4]). Also notable are the relatively intense scores coded for the visual measures on the inherently relevant topics:

- Political significance: internal and international politics
- Economic significance: business, commerce, and industry
- Cultural significance: cultural events
- Public significance: international politics

Finally, the same basic pattern appears for the radio items (Table 6.5).

Across the board, social significance is more intense in verbal content than in visual content. In general, the verbal text in a news items carries the

TABLE 6.2 Mean Intensity of Social Significance Scores for Most Common News Topics in the Three Media

Most Common Topics	Political Significance		Economic Significance		Cultural Significance		Public Significance		Index	
	Visual	Verbal	Visual	Verbal	Visual	Visual	Visual	Verbal	Visual	Verbal
Sports	1.0	1.0	1.1	1.1	1.4	1.5	1.2	1.3	1.2	1.2
Internal Politics	1.9	2.6	1.1	1.4	1.1	1.2	1.4	1.9	1.4	1.8
Cultural Events	1.1	1.1	1.1	1.1	1.7	2.2	1.2	1.4	1.3	1.5
Business/Commerce/Industry	1.1	1.2	1.8	2.3	1.1	1.2	1.3	1.6	1.3	1.6
International Politics	2.0	2.5	1.3	1.5	1.2	1.2	1.7	2.1	1.5	1.8
Internal Order	1.4	1.6	1.1	1.3	1.2	1.3	1.7	2.1	1.4	1.6
Human Interest	1.1	1.1	1.1	1.2	1.5	1.6	1.4	1.5	1.3	1.3
Economy	1.2	1.5	1.7	2.6	1.1	1.1	1.3	2.0	1.3	1.8
Entertainment	1.1	1.1	1.1	1.1	1.3	1.2	1.3	1.3	1.2	1.2
Health/Welfare/Social Services	1.1	1.3	1.2	1.4	1.1	1.3	1.6	2.2	1.3	1.6

TABLE 6.3 Mean Intensity of Social Significance Scores for Most Common News Topics in Newspapers

Most Common Topics	Political Significance		Economic Significance		Cultural Significance		Public Significance		Index	
	Visual	Verbal	Visual	Verbal	Visual	Visual	Visual	Verbal	Visual	Verbal
Sports	1.0	1.0	1.1	1.1	1.3	1.4	1.2	1.3	1.2	1.2
Internal Politics	1.7	2.5	1.1	1.4	1.1	1.2	1.3	1.9	1.3	1.8
Cultural Events	1.1	1.1	1.1	1.1	1.7	2.2	1.2	1.4	1.3	1.5
Business/Commerce/Industry	1.1	1.2	1.7	2.3	1.1	1.2	1.3	1.6	1.3	1.6
International Politics	1.8	2.3	1.2	1.4	1.1	1.2	1.4	2.0	1.4	1.8
Internal Order	1.3	1.6	1.1	1.3	1.2	1.3	1.7	2.1	1.3	1.5
Human Interest	1.1	1.1	1.1	1.2	1.5	1.6	1.4	1.5	1.3	1.3
Economy	1.2	1.5	1.7	2.6	1.1	1.1	1.3	1.9	1.3	1.8
Entertainment	1.1	1.1	1.1	1.1	1.3	1.2	1.3	1.3	1.2	1.2
Health/Welfare/ Social Services	1.1	1.3	1.1	1.4	1.1	1.3	1.4	2.1	1.2	1.5

TABLE 6.4 Mean Intensity of Social Significance Scores for Most Common News Topics in Television

Most Common Topics	Political Significance		Economic Significance		Cultural Significance		Public Significance		Index	
	Visual	Verbal	Visual	Verbal	Visual	Visual	Visual	Verbal	Visual	Verbal
Sports	1.0	1.0	1.0	1.1	1.5	1.5	1.3	1.4	1.2	1.2
Internal Politics	2.2	2.8	1.2	1.4	1.2	1.2	1.7	2.1	1.6	1.9
Cultural Events	1.1	1.2	1.1	1.2	2.0	2.3	1.3	1.5	1.4	1.6
Business/Commerce/Industry	1.3	1.5	2.0	2.4	1.1	1.2	1.5	1.9	1.5	1.8
International Politics	2.4	2.8	1.5	1.6	1.3	1.2	2.2	2.4	1.8	2.0
Internal Order	1.5	1.7	1.2	1.3	1.2	1.3	1.8	2.2	1.4	1.6
Human Interest	1.1	1.1	1.1	1.1	1.3	1.4	1.3	1.6	1.2	1.3
Economy	1.3	1.7	1.7	2.7	1.0	1.2	1.4	2.1	1.3	1.9
Entertainment	1.0	1.0	1.0	1.3	1.0	1.1	1.8	1.7	1.2	1.3
Health/Welfare/Social Services	1.2	1.3	1.3	1.5	1.2	1.3	1.9	2.5	1.4	1.7

TABLE 6.5 Mean Intensity of Social Significance Scores for Most Common News Topics in Radio

Most Common Topics	Political Significance	Economic Significance	Cultural Significance	Public Significance	Index
Sports	1.0	1.0	1.8	1.4	1.3
Internal Politics	2.8	1.4	1.3	2.1	1.9
Cultural Events	1.1	1.1	2.4	1.7	1.6
Business/Commerce/ Industry	1.2	2.2	1.1	1.6	1.5
International Politics	2.8	1.4	1.2	2.2	1.9
Internal Order	1.7	1.2	1.4	2.0	1.6
Human Interest	1.2	1.2	1.7	1.7	1.5
Economy	1.6	1.6	1.1	2.2	1.9
Entertainment	–	–	–	–	–
Health/Welfare/ Social Services	1.4	1.5	1.3	2.4	1.7

social significance, although as noted, there are some instances of relatively intense scores for visual social significance as well.

We also look at the topics that have the most social significance intensity (coded as 3 or 4). Table 6.6 presents the combined findings of the three media. It shows that across all ten countries the stories with the most intense social significance are (1) internal politics; (2) international politics; (3) business, commerce, and industry; (4) cultural events; and to a lesser degree, (5) internal order. Internal politics is the leading news topic in eight of the ten countries, ranging from 32% in the United States to 19% in China. Only in two countries are other topics ranked first in terms of social significance intensity: In Australia, the most socially significant news topic is business, commerce, and industry; in Jordan, it is international politics.

We see similar patterns when we look separately at each of the three news media (Tables 6.7–6.9). Looking across the ten countries, internal politics has the most intense social significance. Radio news shows this the most, with 31.4%, while television has 26.2% and newspapers 18.7%. International politics is second highest in terms of social significance intensity—mostly radio news (20.3%) and television (16.1%). In newspapers, however, international politics occupies fourth place, with 9.2% of all items having intense social significance. In other words, it seems that the broadcast media present items of intense social significance on fewer topics compared with newspapers, which spread social significance more evenly across the different topics.

When examining the individual countries, we see in Table 6.7 that the most intense social significance is in newspaper stories about internal politics in six of the countries (Australia, Germany, India, Israel, Russia, and South Africa). In television (Table 6.8) and radio (Table 6.9), nine of the countries—all but Jordan—show the most social significance in that same topic. The second most intense topic in terms of social significance is spread among five different

TABLE 6.6 Distributions of Very Intense Socially Significant News Items in the Three Media

Most Common Topics	Australia	Chile	China	Germany	India	Israel	Jordan	Russia	S. Africa	U.S.	Total
Sports	9.0	2.0	1.4	2.4	6.7	.9	3.9	3.1	6.8	2.9	4.0
Internal Politics	10.3	27.8	19.1	20.3	24.1	27.9	10.7	21.6	17.8	32.2	21.2
Cultural Events	11.6	9.9	3.9	18.7	7.8	4.6	17.5	9.8	4.6	4.5	10.7
Business/Commerce/Industry	16.9	7.8	13.5	17.1	9.3	3.3	5.5	5.2	12.4	17.9	11.1
International Politics	12.7	8.4	18.0	4.3	7.8	8.8	31.9	13.1	9.5	11.2	11.4
Internal Order	5.3	5.7	4.2	4.2	11.3	9.9	3.5	9.9	11.7	3.2	6.9
Human Interest	2.7	2.8	2.3	3.4	5.8	.7	1.9	2.4	4.2	3.2	3.3
Economy	7.0	7.6	7.7	3.6	3.1	5.9	8.7	8.3	7.3	4.8	6.0
Entertainment	.6	.4	.1	.3	1.6	.2	2.8	.5	.8	.2	.8
Health/Welfare/Social Services	2.8	3.4	6.2	2.3	2.0	6.8	1.7	2.2	4.4	3.5	3.1
Others	21.1	24.2	23.6	23.4	20.5	31.0	11.9	23.9	20.5	16.4	21.5
Total[a]	100.0	100.0	100.0	100.0	100.0	100.0	100.0	100.0	100.0	100.0	100.0
(N)	(1129)	(2156)	(854)	(2210)	(2195)	(545)	(1189)	(976)	(888)	(625)	(12767)

Note: *Very intense socially significant* news items include those that received a rating of 3 or 4 on any of the eight 4-point social significance scales. Distributions given in percent.

[a]Total percentage may not actually be 100.0 due to rounding error.

TABLE 6.7 Distributions of Very Intense Socially Significant News Items in Newspapers

Most Common Topics	Australia	Chile	China	Germany	India	Israel	Jordan	Russia	S. Africa	U.S.	Total
Sports	9.7	1.6	2.2	2.5	5.5	1.6	5.7	1.7	8.8	4.1	4.3
Internal Politics	8.5	25.2	13.1	20.2	21.6	23.8	9.2	21.1	14.7	21.9	18.7
Cultural Events	13.6	12.2	4.8	19.4	9.3	7.4	25.6	11.6	5.8	7.0	13.2
Business/Commerce/Industry	18.9	9.3	15.3	17.6	10.0	5.5	6.8	6.3	14.6	24.6	12.8
International Politics	12.2	8.5	19.7	4.5	6.9	1.6	19.4	12.2	5.2	12.7	9.2
Internal Order	3.7	4.8	5.4	3.9	11.0	8.4	4.2	10.1	10.4	2.7	6.4
Human Interest	3.2	3.5	2.8	3.3	7.4	1.3	1.8	2.8	4.9	4.3	3.9
Economy	8.0	8.7	6.6	3.6	3.5	5.8	11.6	9.5	8.5	4.6	6.6
Entertainment	.7	.4	.2	.3	2.0	.3	3.9	.7	1.0	.3	1.0
Health/Welfare/ Social Services	2.0	3.2	4.8	2.0	2.0	10.9	1.5	2.1	5.1	3.8	2.9
Others	19.5	22.6	25.1	22.7	20.8	33.4	10.3	21.9	21.0	14.0	21.0
Total[a]	100.0	100.0	100.0	100.0	100.0	100.0	100.0	100.0	100.0	100.0	100.0
(N)	(958)	(1662)	(498)	(2005)	(1723)	(311)	(813)	(715)	(673)	(370)	(9728)

Note: *Very intense socially significant* news items include those that received a rating of 3 or 4 on any of the eight 4-point social significance scales. Distributions given in percent.

[a]Total percentage may not actually be 100.0 due to rounding error.

TABLE 6.8 Distributions of Very Intense Socially Significant News Items in Television

Most Common Topics	Australia	Chile	China	Germany	India	Israel	Jordan	Russia	S. Africa	U.S.	Total
Sports	7.1	5.3	.8	.7	7.6	.0	.0	.0	.0	3.6	3.0
Internal Politics	16.1	29.1	32.0	19.6	35.6	20.5	17.7	21.6	25.3	39.8	26.2
Cultural Events	.0	4.4	3.3	17.5	.9	4.5	.0	4.0	1.3	2.4	3.7
Business/Commerce/ Industry	8.0	3.5	9.8	11.9	8.0	.0	4.6	3.2	5.1	6.0	6.3
International Politics	12.5	7.5	16.4	2.1	12.9	15.9	41.7	19.2	24.1	2.4	16.1
Internal Order	13.4	7.5	.0	4.2	16.4	4.5	3.4	13.6	19.0	10.8	9.8
Human Interest	.0	.9	.8	4.2	.0	.0	3.4	.0	2.5	1.2	1.4
Economy	.9	4.0	7.4	3.5	.4	4.5	4.0	4.8	5.1	9.6	4.0
Entertainment	.0	.9	.0	.0	.0	.0	.6	.0	.0	.0	.2
Health/Welfare/ Social Services	8.9	5.7	7.4	4.9	1.8	4.5	4.6	1.6	1.3	3.6	4.2
Other	33.1	31.2	22.1	31.4	16.4	45.6	20.0	32.0	16.3	20.6	25.1
Total[a]	100.0	100.0	100.0	100.0	100.0	100.0	100.0	100.0	100.0	100.0	100.0
(N)	(112)	(227)	(122)	(143)	(225)	(44)	(175)	(125)	(158)	(83)	(1414)

Note: *Very intense socially significant* news items include those that received a rating of 3 or 4 on any of the eight 4-point social significance scales. Distributions given in percent.

[a]Total percentage may not actually be 100.0 due to rounding error.

TABLE 6.9 Distributions of Very Intense Socially Significant News Items in Radio

Most Common Topics	Australia	Chile	China	Germany	India	Israel	Jordan	Russia	S. Africa	U.S.	Total
Sports	1.7	1.5	.0	.0	14.2	.0	.0	13.2	1.8	.0	3.6
Internal Politics	28.8	42.7	25.2	22.6	31.2	36.3	10.4	24.3	33.3	50.6	31.4
Cultural Events	1.7	.4	2.1	.0	3.2	.0	.0	5.9	.0	.0	1.4
Business/Commerce/ Industry	1.7	2.6	11.5	12.9	5.7	.5	1.0	1.5	7.0	9.3	5.0
International Politics	20.3	8.2	15.4	4.8	9.7	18.9	73.6	12.5	19.3	12.2	20.3
Internal Order	16.9	9.7	3.8	12.9	8.5	13.7	1.0	5.9	7.0	.6	7.1
Human Interest	.0	.4	2.1	4.8	.4	.0	1.0	2.2	.0	1.7	1.1
Economy	1.7	4.1	10.3	1.6	2.4	6.3	1.0	5.1	.0	2.9	4.2
Entertainment	.0	.0	.0	.0	.0	.0	.0	.0	.0	.0	.0
Health/Welfare/ Social Services	5.1	3.0	8.5	4.8	1.6	.5	.0	2.9	5.3	2.9	3.1
Other	22.1	27.4	21.1	35.6	23.1	23.8	12.0	26.5	26.3	19.8	22.8
Total[a]	100.0	100.0	100.0	100.0	100.0	100.0	100.0	100.0	100.0	100.0	100.0
(N)	(59)	(267)	(234)	(62)	(247)	(190)	(201)	(136)	(57)	(172)	(1625)

Note: Very intense socially significant news items include those that received a rating of 3 or 4 on any of the eight 4-point social significance scales. Distributions given in percent.

[a]Total percentage may not actually be 100.0 due to rounding error.

topics in the newspapers—once again indicating relatively more variability—whereas in nine of the ten countries (all except Germany), the second most intense topic is international politics and internal order. Finally, regarding television, in six of the countries the second highest social significance is in international politics; in two countries it is in internal order; and in two countries it is in another topic (sports in India and internal politics in Jordan). In summary, there is more communality among the countries than variability.

Evaluating the Intensity of Social Significance in the News

As with deviance in chapter 5, we evaluate the intensity of social significance in the news in three ways: (1) by looking at the percentage of news items coded as showing social significance on at least one dimension; (2) by looking at each of the ten countries separately; and (3) by combining the data sets to arrive at an overall measure of social significance.

For the verbal component of content, 80% of news items are composed of at least one social significance dimension. More than half of visuals contain some element of social significance. Although the means for the four dimensions are on the low side, because nearly half of visual items include no social significance, the amount of social significance in the news is fairly uniform throughout the ten countries, the three media, and between the major and peripheral cities.

Due again to the large sample size, small differences tend to be statistically significant. So, once again we present the results of Scheffé post hoc tests that group the countries according to their mean intensity scores on each variable. As we warned previously, these results should not be interpreted as being "high" or "low" social significance but rather as signifying whether a given country has relatively more or less intense social significance scores compared with the other countries, cities, or media.

In newspapers, within each type of social significance there is somewhat less social significance in visual content—photographs or video, figures, graphs, and so on—than in verbal content. This may reflect the routine use of what are basically uninteresting photographs and visuals (e.g., head-and-shoulder shots, people shaking hands, people walking or talking). While these provide the visual interest audiences may want, these sorts of visuals convey very little information on their own but rather are window-dressing for the text.

What we can see, however, is that the visual content of news items in television generally carries more information about social significance than the visual content in newspapers. Perhaps this reflects the tendency in television news to require "good visuals" as a prerequisite to covering a story, whereas most newspapers routinely publish pictures (often from archives) of people walking, talking, shaking hands, and posing in uninteresting ways.

There appears to be no overall difference between the amount of social significance portrayed in small- and large-city news.

Table 6.10 reveals slight between-country differences among the four dimensions of social significance based on the Scheffé post hoc tests.

Political Significance

India's news has the most intense political significance, in both its verbal and visual news content. Russian news media also include relatively more politically significant news when compared with other countries. The least intense political significance is found in visual content from Israel and verbal content from Australia and the United States.

Economic Significance

Half of the countries group in the highly intense economic category: Australia, China, and South Africa for visuals and Germany and Russia for words. Israel and the United States have the least intense economic news in both visual and verbal content.

Cultural Significance

News of cultural significance is most intense in visuals from Australian and Indian media, and is least intense in verbal content in German media. The least intense cultural significance is found in news from Israel (visuals only) and Jordan.

Public Significance

Again, the Indian news media carry the most intense public significance, both in verbal and visual content. Israel and the United States present news with the least public significance.

Two indexes—social significance intensity for verbal content and social significance intensity for visuals—give an overall assessment of the differences among the countries on the social significance variables.[9] The verbal content with the most intense social significance is found in India and Russia, whereas the least socially significant verbal content is found in the United States. The visual index shows that visual content with the most intense social significance is found in India, with the least socially significant visual content found in Israel.

Social Significance between Cities and Media

There are only small differences in the intensity of social significance between major and peripheral city news media. However, the Scheffé post hoc tests (see Table 6.11) show that there is a slight tendency for peripheral city news media to carry more intense socially significant news. As Table 6.12 shows, for the political, economic, and public social significance variables, newspaper news

TABLE 6.10 Grouping of Countries according to Intensity of Social Significance

Intensity	Political Significance		Economic Significance		Cultural Significance		Public Significance		Index	
	Visual	Verbal	Visual	Verbal	Visual	Verbal	Visual	Verbal	Visual	Verbal
Most	India	India Russia	China S. Africa Australia	Russia Germany	India Australia	Germany	India	India	India	Russia India
Least	Israel	USA Australia	USA Israel	USA Israel	Israel Jordan	Jordan	USA Israel	USA	Israel	USA

Note: Grouping of Countries based on Scheffé post hoc tests.

TABLE 6.11 Grouping of Major and Peripheral Cities according to Intensity of Social Significance

Intensity	Political Significance		Economic Significance		Cultural Significance		Public Significance		Index	
	Visual	Verbal	Visual	Verbal	Visual	Verbal	Visual	Verbal	Visual	Verbal
More	No difference	Major	No difference	Peripheral	Peripheral	Peripheral	Peripheral	Peripheral	Peripheral	Peripheral
Less		Peripheral		Major	Major	Major	Major	Major	Major	Major

Note: Difference between cities based on Scheffé post hoc tests.

TABLE 6.12 Grouping of the Media according to Intensity of Social Significance

Intensity	Political Significance		Economic Significance		Cultural Significance		Public Significance		Index	
	Visual	Verbal	Visual	Verbal	Visual	Verbal	Visual	Verbal	Visual	Verbal
Most	TV	Radio	TV	TV NP	NP	NP	TV	TV Radio	TV	Radio TV
	— — — — — TV — — — — —									
Least	NP	NP	NP	Radio	TV	TV Radio	NP	NP	NP	NP

Note: Grouping of media based on Scheffé post hoc tests. NP = newspaper. TV = television.

content has less intense social significance than television or radio, but newspapers have more intense cultural significance than television and radio.

The same pattern emerges when we analyze just those news items categorized as hard news, soft news, or editorials. That is, with the exception of cultural significance, television and radio tend to carry news items with more intense political, economic, and public significance than do newspapers. Newspapers carry news items of more intense cultural significance. This is consistent with the analysis of all news items, not just those that are "traditional" news.

Variability of Social Significance

When a mean value is low, we may wonder whether the mean value of a variable is accompanied by a small or broad range of values. As with deviance, we present in Appendix C bar charts of the social significance intensity scores (political, economic, cultural, and public), for both text and images—eight variables per country. The first figure provides an overall view of the results from the ten countries combined. Here we see that some news items, whether text or images, extend to even the highest intensity of political, economic, cultural, or public significance. The most intense form of social significance is that involving the public welfare in news text.

However, by combining the ten countries, we cannot see differences among them, so bar charts for each country are also included in Appendix C. Looking at the political significance of text information, we see that the modal (most common) value among the ten countries is "no political significance." In economic significance and cultural significance, the ten countries show that "no social significance" is the most common value. Visual news items show the same pattern across all ten countries in political, economic and cultural significance. Yet in the public significance of text news items, China, India, and Jordan show that the value of "minimal public significance" is most common. In public significance of visual items, Australia and India show "minimal public significance" most commonly.

Social Significance in Text and Images

As we discussed previously, newspaper and television news differ in the amount of visual content presented. There is less visual content in newspapers, and whereas some newspaper articles include visuals (e.g., photographs, maps, figures), most do not. Television news items virtually always include both verbal and visual content. Radio is obviously omitted. Overall, the intensity of social significance in visual content is less than in verbal content. Thus social significance of a news event is carried more in verbal than visual content.

Conclusion

The intensity of deviance and social significance is similar across news of the ten countries. The differences discussed in the Schéffe tests are minor and distinguishable only because of the large number of news items studied thus answering research questions 2–8.

Nearly 80% of verbal news items are characterized by at least one dimension of social significance, whereas this is true of more than half of the visuals. The most intense social significance is carried in television and radio, probably because newspapers can publish diverse stories including many that could be classified as entertainment. Newspapers, television, and radio may carry the same socially significant stories, but newspapers have the freedom to publish more news items, including those that have both low social significance and low deviance.

7

The Perception of Newsworthiness

In our explication of the basic concepts underlying this study, we discuss the two major concepts of deviance and social significance as predictors of newsworthiness. We also identify three dimensions of deviance—statistical, normative, and social—and four dimensions of social significance—political, economic, cultural, and public. We suggest that the more intense the deviance and social significance of events, individuals, or ideas—each of the two dimensions alone and in combination—the greater the prominence. That is, stories concerning these phenomena are more likely to be given more and better placement in newspapers, and more time and earlier position in the line-up of broadcast news.

While we do refer to people in general, the main thrust of our argument so far is concerned with how journalists think and practice their profession. In this chapter we bring public relations practitioners as well as news consumers into the picture to focus on research questions 9–11. The rationale for this is that, although journalists (mainly editors or producers and to some extent reporters) determine which stories will be in the news, public relations practitioners and news consumers have something to say about this, too.

The major role of public relations practitioners involved in media relations is to get information concerning their organization or client—private or public, for profit or not-for-profit—into the media, mainly into the news. This is the case both when they wish to promote their organizations or clients as well as when they find themselves defending their clients or organizations during a crisis. Thus, public relations practitioners offer their journalistic colleagues information they hope will be included in the news in the most prominent ways. The evolution of paper press releases to video news releases often makes it easier for a television news program to include information on the events.

News consumers also have a vested interest in what gets into the news. People read, watch, and listen to the news in order to obtain information that is important and meaningful to them. When consuming the news, people are selective; they do not pay the same amount of attention to every item that appears in the news. Media organizations often seek to learn, by conducting research, what interests their consumers and how they can provide the audience with what it wants. Thus, news consumers can influence, albeit indirectly, the way newspapers, as well as radio and television stations, make some editorial and programming decisions.

This chapter looks at the degree of correspondence between the perceptions of newsworthiness among journalists, public relations practitioners, and the news-consuming public. To what extent are the implicit definitions of news and newsworthiness shared by these different groups of people? What is the degree of correspondence between how the various groups perceive newsworthiness and the news that actually appears in the newspaper and on television and radio news programs? How satisfied are journalists, public relations practitioners, and news consumers with what appears in the news?

Returning to the cross-national component of this study, we are also interested in the extent to which these degrees of correspondence are similar or different across the countries in the study. In other words, we want to know whether—despite the variability among the political and economic systems and the cultural traditions of the ten countries—journalists, public relations practitioners, and people of high and low socioeconomic status (SES) perceive the newsworthiness of events similarly.

This chapter presents findings of a gatekeeping exercise conducted with focus groups in the study's twenty cities. As noted in chapter 3, four focus groups were conducted in each city with the following persons:

1. Journalists (mostly reporters and in some cases editors or producers, but in no case supervisors who oversaw the work of the participating reporters)
2. Public relations practitioners from the private and public sectors
3. High SES news consumers
4. Low SES news consumers

The Procedure

As noted in chapter 3, we sampled a composite week of one newspaper in each of two cities in each country: a major and a peripheral city. Table 7.1 presents the cities and newspapers used in the study. The size of each item in square centimeters was calculated for the entire article, even if it was printed on more than one page. Also, the placement of the item was determined, weighting news items as follows: (1) a score of 3 for an item that appeared (or had a teaser) on the front page; (2) a score of 2 if the item appeared (or began) on the first page of an interior section of the newspaper; and (3) a score of 1 if the item appeared elsewhere in the newspaper. Once the coding of the newspapers was completed, a prominence score was calculated for each news items by multiplying its physical size by its placement weight.

As noted, the gatekeeping exercise was based directly on the content analysis of the newspapers completed prior to conducting the focus groups. Using the prominence scores for each newspaper item, all items from three randomly selected days were ranked from the most prominent to the least prominent.

TABLE 7.1 Cities and Newspaper Titles

	Major Cities	Peripheral Cities
Australia	Sydney	Brisbane
	Sydney Morning Herald	*Courier Mail*
Chile	Santiago	Concepción
	El Mercurio	*El Sur*
China	Beijing	Jinhua
	People's Daily	*Jinhua Evening Post*
Germany	Berlin	Mainz
	Berliner Zeitung	*Allgemeine Zeitung Mainz*
India	New Delhi	Hyderabad
	Hindustan Times	*Eenadu*
Israel	Tel Aviv	Beer Sheba
	Ha'ir (weekly)	*Kol Nanegev* (weekly)
Jordan	Amman	Irbid
	Al-Rai	*Shihan (weekly)*
Russia	Moscow	Tula
	Izvestia	*Tula Izvestia*
S. Africa	Johannesburg	Bloemfontein
	The Sowetan	*Die Volksblad*
U.S.	New York City	Athens, Ohio
	New York Times	*Athens Messenger*

For each of these days, ten items from the participants' local newspapers were then identified based on their prominence percentile score, from least prominent to most prominent. Once the three lists of ten items were determined, each headline and subhead (if such existed) of each item was printed on an indexes card. The indexes cards were color-coded so that each day's set of ten headlines was printed on cards of different colors.

The goal was to have between eight and ten people in each group; however, some of the groups were smaller, while others turned out to be larger (see Table 7.2). Toward the end of the focus group, the group moderator handed each individual a pre-shuffled set of ten cards, representing one of the three newspaper days. The participants were asked to assume the role of the newspaper editor, and with this in mind to arrange the cards in the order in which they would rank them for publication, based on the degree of newsworthiness they personally considered each item to have. In other words, they were asked to indicate which item, in their view, was most newsworthy and deserved to receive top priority, followed by the next most newsworthy, and so forth, for all ten headlines. When participants were finished, they gave their cards to the moderator. The task was performed three times in each focus group, once for each of the three newspaper days sampled. After the focus group ended, the moderator recorded the order in which the cards were arranged by each of

TABLE 7.2 Composition of Focus Groups in the Ten Countries

	Major Cities				Peripheral Cities			
	Journalists	Public Relations	High SES[a] Audience Members	Low SES Audience Members	Journalists	Public Relations	High SES Audience Members	Low SES Audience Members
Australia	9	9	10	10	10	8	8	9
Chile	11	12	13	11	11	11	9	10
China	11	8	9	10	9	8	10	10
Germany	6	11	12	10	11	10	12	10
India	13	11	12	14	13	17	14	12
Israel	7	7	9	8	10	8	10	10
Jordan	9	10	10	10	7	10	10	10
Russia	8	8	10	12	8	7	12	11
S. Africa	12	10	8	10	9	10	10	11
U.S.	7	9	11	9	6	5	6	5

[a]SES = socioeconomic status.

the participants. To look for relationships between the rankings, we used the rank-order correlation coefficient, Spearman's rho, with the three days' cards combined for each person, resulting in ninety rank-ordered news stories for each person in each of the eighty focus groups.

Findings

Table 7.3 presents the mean Spearman's rho correlations between the rankings of the participants of the four focus groups in each of the cities in the ten countries. This allows us to compare, within each city and country, the similarity of journalists', public relations practitioners', and audience members' news rankings. Of the sixty correlations for the major cities, all but three in India are positive and statistically significant, indicating that the people in the four types of focus groups agreed with one another. The mean correlation per country ranges from a high of .83 in Germany to a low of .30 in India; the larger the coefficient, the more closely people agreed with one another. In the peripheral cities, of the fifty-seven correlations (excluding missing data for the Israeli high SES group), all but three each in India and Russia, two in Jordan, and one in South Africa are statistically significant. The mean correlation per country ranges from .86 in Chile to .27 in India.

So, by and large, a significant amount of agreement was found between the four types of participants in terms of how the items were ranked. It should be noted, however, that the average correlation coefficients in the major cities across the ten countries are higher than those in the peripheral cities.

TABLE 7.3 Average Spearman Rank Correlation Coefficients Among Focus Group Participants' Rankings of News Items

	Major Cities						Peripheral Cities					
	J/PR	J/LSA	J/HSA	PR/LSA	PR/HSA	LSA/HSA	J/PR	J/LSA	J/HSA	PR/LSA	PR/HSA	LS/HSA
Australia	.71^c	.64^c	.64^c	.82^c	.80^c	.75^c	.86^c	.63^c	.61^c	.65^c	.65^c	.78^c
Chile	.67^c	.62^c	.63^c	.65^c	.76^c	.38^a	.89^c	.89^c	.93^c	.81^c	.80^c	.82^c
China	.60^c	.58^c	.70^c	.71^c	.76^c	.70^c	.56^b	.71^c	.64^c	.75^c	.73^c	.63^c
Germany	.74^c	.82^c	.78^c	.87^c	.91^c	.86^c	.85^c	.52^b	.74^c	.59^b	.83^c	.67^c
India	.63^c	.11	.51^b	.07	.47^b	−.01	.18	−.13	.45^a	.31	.44^a	.37^a
Israel	.78^c	.90^c	.81^c	.84^c	.78^c	.80^c	.44^a	.65^c	NA	.75^c	NA	NA
Jordan	.75^c	.72^c	.76^c	.63^c	.48^b	.77^c	.72^c	.77^c	.22	.83^c	.40^a	.35
Russia	.50^b	.56^b	.40^a	.63^c	.66^c	.58^b	.50^b	.21	.05	.49^b	.35	.53^b
S. Africa	.87^c	.61^c	.77^c	.79^c	.88^c	.83^c	.88^c	.47^b	.60^c	.58^b	.70^c	.31
U.S.	.76^c	.55^b	.54^b	.65^c	.55^b	.56^b	.82^c	.79^c	.70^c	.88^c	.74^c	.73^c

Note: J = journalists; PR = public relations practitioners; LSA = low socioeconomic status audience (SES); HSA = high SES audience; NA = not available.
[a] $p < .05$; [b] $p < .01$; [c] $p < .001$.

In contrast, Table 7.4 presents the Spearman rho coefficients between the rankings of stories by the focus group participants and the actual ranking (prominence) of the news items used in the gatekeeping exercise. The correlation coefficients are based on showing agreement. Whereas the Spearman rank-order coefficients between the groups of participants are relatively high (as indicated in Table 7.3), the correlations with the actual coverage in the newspapers are much lower. Of the forty correlation coefficients in the major cities, only eighteen were statistically significant. A similar picture was obtained for the peripheral cities: of the thirty-nine correlations (excluding missing data from the Israeli high SES audience group), only six were statistically significant.

Furthermore, the size of the coefficients was relatively low, with the highest obtained in both Jordanian cities for the high SES groups and their newspapers: .69 in the major city and .73 in the peripheral city. In addition, eleven of the correlation coefficients for the major cities were negative (with three being significant) and five were negative (but not significant) in the peripheral cities. Finally, in the major cities in Jordan, Germany, and Israel, all four correlations were significant, thereby producing the relatively highest mean correlations, ranging from .57 in Israel to .42 in Jordan. In the peripheral cities, however, only Germany with a mean correlation of .42 was relatively higher than the other cities. Thus, we conclude that hypothesis one is supported: There is considerable agreement among the various focus groups. Hypothesis two is supported for some countries and not for others; however, even when statistically significant, when we compare how newsworthy people rank stories

TABLE 7.4 Spearman Rank-Order Correlation Coefficients between Focus Group Participants' Item Newsworthiness Rankings and Actual Newspaper Item Prominence

	Major Cities				Peripheral Cities			
	Journalists	Public Relations	High SES Audience Members	Low SES Audience Members	Journalists	Public Relations	High SES Audience Members	Low SES Audience Members
Australia	.29	.50[b]	.35	.39[a]	.18	.17	.11	.19
Chile	−.20	−.34	−.37[a]	−.14	.34 ·	.33	.24	.26
China	−.20	−.31	−.46[a]	−.37[a]	.01	.13	.04	.12
Germany	.48[b]	.49[b]	.47[b]	.48[b]	.58[b]	.38[a]	.35	.38[a]
India	.17	.38[a]	.19	−.04	.23	.38[a]	.33	.07
Israel	.57[b]	.56[b]	.49[b]	.67[c]	.25	−.16	NA	−.24
Jordan	.48[b]	.38[a]	.69[c]	.51[b]	−.13	.09	.73[c]	.05
Russia	.22	.05	.08	.34	.24	.03	−.08	.12
S. Africa	.06	.06	−.03	−.03	.13	.01	.12	−.16
U.S.	.28	.21	.04	.10	.42[a]	.35	.14	.25

Note: SES = socioeconomic status of the audience focus group members; NA = not available.
[a]$p < .05$; [b]$p < .01$; [c]$p < .001$.

with how prominently their newspapers presented them, the size of the Spearman's rho coefficients is low.

These results indicate that across the board people agree more with each other about the newsworthiness of the stories than with the newsworthiness decisions made by their cities' newspaper editors. This was particularly poignant in the major cities in China and Chile (where the mean correlations are negative) and in India, Russia, South Africa, and the United States. In the peripheral cities, only the German focus groups tended to agree slightly (.42) with the actual rankings of the newspaper items, while in all the other cities the mean correlations were between −.05 (Israel) and .29 (Chile).

On average, across the ten countries, the similarities between the prominence of the stories in the newspapers were highest for the journalists (.22 in major cities; .23 in peripheral cities), although not very high even here. There is no systematic pattern of agreement between newspapers and public relations practitioners, low SES audience members, or high SES audience members.

Discussion

The design of this exercise may be viewed as consisting of ten replications, one per country. Or, if we consider the fact that in each of the countries we dealt with two cities, we may thus have as many as twenty replications. Given the fact that each focus group participant performed the task three times—once for each of three newspaper days—we have quite a robust exercise and data set.

Since the overall mean Spearman correlation coefficient among the focus group participants in all ten countries was .66 for the major cities and .61 for the peripheral cities, it seems justifiable to consider these findings as quite similar and to combine them. Moreover, the mean overall Spearman rank coefficients between the different focus groups and the actual ranking of the newspaper items for the major and peripheral cities were both .17. This surely allows us to consider these data together.

Although seemingly different, the findings presented in tables 7.3 and 7.4 are definitely consistent with one another: While the various types of focus group participants in each country tend to agree with each other, they agreed less with the prominence stories were given in their respective newspapers. In other words, there seems to be a universal dissatisfaction with the prominence given to the various news items in newspapers. This is interesting, particularly with regard to journalists. They are definitely involved in the news production process, even though they do not typically set organizational priorities, determine the placement of specific items in the newspapers, or create headlines. In addition, whereas public relations practitioners try to influence the news from the outside, they are more in agreement with journalists than with the actual news content.

The empirical similarity in the cross-national findings raises the question as to why this phenomenon is so pronounced. To answer this question, we could consider a general deductive theoretical rationale, or look inductively at the idiosyncrasies of the individual countries (and cities) and attempt to arrive at an all-purpose explanation, despite the obvious variability among the countries.

Theoretically, we can speculate about a general sense of malaise or disappointment with the media as expressed by citizens—both media professionals and laypeople—around the world. Newspaper editors and publishers the world over may not be sensitive to the interests and needs of their readers. For example, a question raised and discussed in virtually all of the focus groups was the distinction drawn between "good" and "bad" news, and the extent to which each of these categories is presented in an exaggerated manner. Without going into details (which can be found in the country chapters that follow), suffice it to say that there was much consensus among the participants in the various countries about this subject and the general expression of a desire among people to be presented with more "good" news, even though many—if not most—people seem to understand the reasoning given by journalists for the abundance of "bad" news.

In contrast, looking at the countries individually may lead to interesting conclusions regarding each country, although the overall conclusion is the same. Thus, for example, in the major cities of Israel, Jordan, Germany, and Australia, the mean correlation coefficients are relatively high (ranging from .57 to .38) compared with the relatively lower mean correlations—negative or positive—in the major cities of Chile, China, India, Russia, South Africa, and the United States, ranging from −.34 in China to .18 in India. Could it be that in these countries for a variety of reasons there is greater detachment between the media and the people? We see further support for this idea in the peripheral cities, where the highest mean correlation is in Germany (only .42) followed by .29 in both Chile and the United States, .25 in India and much lower correlations, positive or negative, in the zero range in the remaining six countries. Could it be that, for varying sociopolitical reasons, people in these other countries are less trustworthy of their press?

In conclusion, within all of the countries the various groups of people seem to be consistent with each other in the ways in which they assessed the newsworthiness of their local news stories. Whereas some may have expected that journalists would have ranked the newsworthiness of stories differently than public relations practitioners, no such difference was found. Perhaps this is due to conventional wisdom that burned-out journalists often switch to public relations as a way to make more money doing a less stressful job. So, are public relations practitioners merely recycled journalists? And what about audiences? Although some have speculated that journalists are out of touch with their audiences, this study shows just the opposite. Perhaps journalists are people too!

The fact that these kinds of outcomes are replicated in major and peripheral cities, in big and small countries, in countries in the Northern and Southern hemispheres, in collective and individualistic cultures—in short, in countries around the world—lends credence to our theoretical argument that ideas about newsworthiness may be pervasive throughout humankind.

Part 3
Country-by-Country Analyses

8

What's News in Australia?

CHRIS LAWE-DAVIES AND ROBYNE M. LE BROCQUE

The Media System in Australia

Australia has operated as an English-speaking industrial society for only a little over 200 years, when English sailors first landed in Botany Bay in Sydney in 1788. Indigenous people have occupied the continental landmass for at least 40,000 years. Today the Australian population numbers almost 20 million people and is a modern democratic and multicultural society.

Newspapers have dominated the media scene since mechanical printing became popular in the mid-nineteenth century; radio broadcasting began in 1924 and television in 1956. However, it was with the establishment of the Australian Broadcasting Corporation (ABC) in 1932 as a government broadcaster, and commercial radio as "the other sector," that the modern "mixed economy" broadcasting sector began to develop: some of it government, some of it commercial. A third sector, community broadcasting, where government provides some of the infrastructure for otherwise voluntary and self-funding radio and television stations, was introduced in 1975, and currently there are more than 304 community radio stations and six television stations across the nation (ABC 2002, 7). These cater to general programming as well as specialist services in non-English, fine music, print disability, educational, access/experimental, and indigenous broadcasting.

While most mainstream metropolitan radio stations are commercial, ABC radio is highly influential. Outside its five-channel metropolitan (including an off-shore) services, which include a highly innovative youth channel, classical music, specialist news, and documentary, it has comprehensive coverage of regional and remote areas. ABC radio tends to be the main agenda setter with longer news bulletins and more current affairs than other broadcasters. Bulletins are local–national hybrids composed through an integrated digital distribution system where all newsrooms have all stories for the nation. Local news teams gather news and contribute to the national grid. News producers thus compile from that grid. The commercial radio stations are similarly networked, but not as comprehensively. Except for a few specialist talk AM stations, commercial stations have little news and no current affairs.

While there is a multiplicity of broadcast radio stations and narrowcast radio stations in each city, with up to twenty-five stations in each, television is a more sedate story. Each city has five free-to-air networks and fifty subscription (cable and satellite) channels, but only 23% of Australian homes take pay services, while penetration of free-to-air is almost 100%.

As with radio, the five television networks are traditionally split between commercial and government sectors (with a "sixth network" community license in each of the larger cities). The government-owned ABC has one channel, which again extends from metropolitan to remote markets; there are three commercial networks and one "hybrid," the Special Broadcasting Service (SBS), partly government funded and partly commercial but with a special brief to run programming catering to Australia's cultural diversity. SBS only began broadcasting in 1980—some quarter of a century after the other networks—but is a highly successful social and public policy experiment, transmitting in up to seventy languages other than English (subtitled) and with a commissioning structure for innovative productions (like Britain's Channel 4). ABC-TV achieves 15% audience share and SBS multicultural television less than 5%. However, both networks work on strict charters that limit their ability to build audiences. Instead, the aim is to extend reach of viewing, rather than frequency of viewing, as this enables both networks to achieve broad, diverse, and comprehensive social and broadcasting objectives (ABA 2002; SBS 2002).

Australian newspapers are pervasive but highly monopolistic in ownership. Indeed, it has been claimed that Australia has one of the highest levels of media ownership concentration in the world, after Ireland (McQueen 1978, 35). Rupert Murdoch's News Corporation titles predominate with nearly 70% circulation of the daily metropolitan and suburban markets. The News Corporation titles are generally in the high-circulation and tabloid end of the market, such as Brisbane's *Courier Mail*. The other major competitor is John Fairfax Publications, publisher of the *Sydney Morning Herald*, which operates more in the quality broadsheet end of the market.

Legislation enabling networks and individual market reach of 70% in television, providing no limits on press or radio, and imposing cross-media limits (ownership of one medium in each market) was introduced in 1987 by the Keating Labor government. Foreign-ownership limits on radio were lifted and eased slightly from 15% to 20% on television and newspapers. The resulting mixed economy means that Murdoch dominates 70% of newspapers but has no interests in radio or television. He has pay-television interests, and because he is not in broadcast, he is pressuring the government to either allow newspapers digital broadcasting through broadband or to lift the cross-media and foreign-ownership restrictions. The conservative Howard (Liberal) government has tried and failed for a second time to appease

Murdoch. They introduced a bill to the House of Representatives in 2002 and again in 2005 seeking to abolish cross-media rules, with special provisions preventing commonly owned newsrooms in press, radio, and television from working together. The Senate rejected the bill a second time. With Howard holding a Senate majority from July 2005, Australia is likely to see marked changes in media legislation, easing restrictions on cross-media ownership and foreign ownership.

The Study Sample

Sydney and Brisbane were the cities chosen for the study. Like most of Australia's larger cities, they are located on the nation's coastline, where the rainfall and river systems are sufficient to support larger populations. Sydney is the country's biggest metropolis, with a population of over 4 million. It has a strong manufacturing base and is a bustling port city. Brisbane is an important regional city 1,000 kilometers north of Sydney with a population of 1.7 million. The city is set in subtropical mangrove wetlands, and with sandy beaches to the north and south and a major coastal tourist hinterland it hosts, among other attractions, the Great Barrier Reef. Despite tourism, Brisbane is still capital of a large state that acts as a primary producer of agriculture and mining.

The newspapers selected for the study included were the *Sydney Morning Herald* and Brisbane's *Courier Mail*. The *Herald* is a state-based broadsheet that also tends to have a national profile. It is the oldest continually published newspaper in Australia, first appearing in 1831. The Brisbane newspaper, while broadsheet in format, tends more toward tabloid content, having more parochial state coverage. The two papers have separate ownership. The *Herald* belongs to a stable of quality broadsheets owned by the John Fairfax Publications, whose press interests are limited to about 20% of national circulation, although its influence extends farther than this. The Fairfax newspapers are regarded as quality agenda-setters in the news and current affairs market. The *Courier* is owned by Murdoch's News Corporation. Generally, the Fairfax papers have more advertising and consequently a greater number of pages. We analyzed a total of 1,976 items in the *Herald* and 1,663 items in the *Courier*.

The television news services chosen were the *Nine Network* services in both cities. A national network reaching 70% of the Australian market, Nine Network is owned largely by Kerry Packer's Consolidated Press Holdings, which does not have any newspapers at the moment but its magazine interests dominate the Australian market. With average ratings of over 30%, Nine Network at the time of the study was the market leader in general television coverage, particularly news and current affairs. Among its successes are local versions of U.S. television current-affairs shows such as *60 Minutes*. We analyzed 221 items from the *TCN-9* (Sydney) news service and 175 items from *QTQ-9* (Brisbane).

There is a marked difference between commercial and public radio in Australia. In both cities the ABC radio channels have the most comprehensive and influential news services. Commercial radio stations have short news bulletins—generally three minutes compared with the fifteen-minute main ABC bulletins. The ABC stations chosen were *2BL Sydney* and *4QR Brisbane.* Their morning news services average between 8% and 10% of listeners and were the bulletins chosen for analysis. We analyzed 113 items from Sydney and 98 items from Brisbane.

Topics in the News

Australia lives up to its reputation as a sporting nation, with over 20% of news items focusing on sports (see Table 8.1). News coverage of sports does not appear to vary much across cities or the different types of media. Coverage of sports—both national and international—was the most frequent topic in newspapers, television, and radio. Betting on horses and other races is also a large pastime in the Australian community, and details about race meetings and results are covered in the newspapers. Television and radio coverage of sports is also given a high priority, with a considerable visual component in television reports and description in radio newscasts.

Apart from sports, other topics varied in coverage across cities and news media. Cultural events were given high priority in the newspaper, comprising over 15% of news items in both Sydney and Brisbane. According to the results, newspapers are the main source of information about the many cultural events being held in these two cities. In contrast, cultural events received very low scores in television and radio news broadcasts. Topics covering business, commerce, and industry were also given high priority in newspaper coverage (Sydney, 14.5%, and Brisbane, 8.7%) but were not as frequently covered in television or radio news broadcasts.

Television news coverage concentrated on human interest stories; disasters, accidents, and epidemics; as well as internal order topics with small differences between cities being observed. Radio news coverage also concentrated on internal politics and internal order and was similar across both cities. Weather information also featured in a significant number of news items in both television and radio news. These results highlight structural differences between the different types of media in that television and radio news programs devote a large proportion of their content to these topics compared to newspapers.

The smaller regional city of Brisbane also had a higher number of stories featuring human interest in newspaper and television news broadcasts compared with Sydney. One possible explanation for this is the nature of the Brisbane market. For newspapers it means that the *Courier Mail* is more parochial in focus than the more nationally focused *Sydney Morning Herald.* Also, *Brisbane Channel 9*, although part of a national network, generates its own state-based news service, which also tends to have a more parochial focus.

TABLE 8.1 Distribution of General Topics of News Items by City and Medium

Topics	Newspaper		Television		Radio	
	Sydney	Brisbane	Sydney	Brisbane	Sydney	Brisbane
Sports	20.9	20.5	27.6	24.0	24.8	22.4
Cultural Events	15.4	15.0	2.7	.6	.9	.0
Business/ Commerce/ Industry	14.5	8.7	4.1	4.6	2.7	5.1
Human Interest Stories	7.9	13.5	9.5	13.1	4.4	4.1
International Politics	5.6	4.4	2.7	6.9	8.0	5.1
Communication	4.8	3.1	1.4	.6	4.4	2.0
Internal Politics	4.2	5.5	1.8	9.1	13.3	11.2
Economy	3.4	2.2	.0	1.1	.0	4.1
Transportation	2.8	2.4	8.6	2.3	1.8	.0
Internal Order	2.6	4.9	13.1	8.6	14.2	14.3
Ceremonies	2.3	1.9	3.2	.0	.0	.0
Health/Welfare/ Social Services	2.2	1.9	2.7	4.0	2.7	4.1
Entertainment	2.1	3.6	.9	2.9	.0	.0
Housing	1.9	1.2	.0	.0	.0	1.0
Environment	1.6	1.4	.9	1.7	2.7	1.0
Disasters/ Accidents/ Epidemics	1.4	2.5	10.9	8.6	7.1	7.1
Social Relations	1.4	1.3	.5	.0	.0	3.1
Military and Defense	1.0	1.1	.9	1.1	2.7	4.1
Education	.9	.9	.5	1.1	.9	1.0
Science/Technology	.8	1.4	.0	1.1	.0	1.0
Weather	.7	1.0	6.8	7.4	6.2	6.1
Fashion/Beauty	.6	.9	.5	.0	.0	1.0
Labor Relations and Trade Unions	.4	.4	.9	1.1	2.7	2.0
Population	.4	.2	.0	.0	.9	.0
Energy	.2	.1	.0	.0	.0	.0
Other	.2	.0	.0	.0	.0	.0
Total[a]	100.0	100.0	100.0	100.0	100.0	100.0
	(n = 1976)	(n = 1663)	(n = 221)	(n = 175)	(n = 113)	(n = 98)

Note: Distribution given in percent.

[a]Total percentage may not actually be 100.0 due to rounding error.

In broad terms, Brisbane television and newspapers tend toward the tabloid genre as opposed to the broadsheet genre.

Australia also has a consistent and fairly high interest in international news, which has ranked as one of the top-ten items for all media types across both cities. This reflects Australia's multicultural social ecology, with a strong focus on Asia and the Pacific region as well as Europe and America.

Statistically significant Spearman rank-order correlation coefficients were observed for news topics between the large city of Sydney and the smaller city of Brisbane for each of the media types (see Table 8.2). The correlation between the rankings of news topics in the two newspapers was high (.94). Correlation between the rank orders of items for Sydney and Brisbane radio was also high (.80). A moderate but significant correlation was also found between television news items in the two cities (.67).

Comparing Spearman rank correlations between the different media for Sydney showed that correlations of the ranking of news items were highest between television and radio (.76) and newspaper and television (.61). A significant but moderate correlation was found for news items presented in Sydney in the newspaper and on the radio (.46). This pattern was also seen in news items presented in Brisbane, with stronger correlations between news items presented on television and radio (.77) and newspaper and television (.62) and moderate correlations between Brisbane newspaper and radio (.47).

Deviance in the News

As Table 8.3 shows, of the thirty analyses of variance (ANOVAs) calculated, sixteen were statistically significant. Significant differences in prominence scores for news items were recorded for statistical deviance and social change

TABLE 8.2 Spearman Rank-Order Correlation Coefficients between Rankings of News Topics in Various Media

	Sydney Newspaper	Brisbane Newspaper	Sydney Television	Brisbane Television	Sydney Radio	Brisbane Radio
Sydney Newspaper		.94[c]	.61[b]	.49[a]	.46[a]	.39
Brisbane Newspaper			.70[c]	.62[b]	.53[b]	.47[a]
Sydney Television				.67[c]	.76[c]	.55[b]
Brisbane Television					.69[c]	.77[c]
Sydney Radio						.80[c]
Brisbane Radio						

[a]$p < .05$; [b]$p < .01$; [c]$p < .001$.

TABLE 8.3 Mean Verbal and Visual Prominence Scores for Intensity of Deviance by City and Medium

Intensity of Deviance	Sydney					Brisbane				
	Newspaper		Television		Radio	Newspaper		Television		Radio
	Verbal only (n = 1948)	Verbal plus Visual (n = 475)	Verbal (n = 221)	Visual (n = 221)	Verbal (n = 113)	Verbal only (n = 1614)	Verbal plus Visual (n = 625)	Verbal (n = 175)	Visual (n = 175)	Verbal (n = 98)
Statistical Deviance										
(1) common	264.4	604.2	52.5[a]	52.7[b]	21.4[c]	366.2	492.9[c]	60.2[b]	68.2[b]	21.9[c]
(2) somewhat unusual	288.3	542.3	92.9	99.9	94.3	336.2	590.3	122.3	116.7	103.0
(3) quite unusual	323.7	634.2	122.6	125.4	133.8	405.4	842.2	145.6	141.5	102.2
(4) extremely unusual	532.6	656.3	112.2	99.0	96.3	365.5	536.0	182.0	175.1	110.5
Social Change Deviance										
(1) not threatening to status quo	270.3	595.6	67.0[c]	55.6[c]	31.3[c]	355.4	556.9	89.9[c]	73.9[c]	51.2[c]
(2) minimal threat	295.6	540.9	136.8	126.2	107.5	344.5	664.5	128.9	126.1	113.3
(3) moderate threat	371.9	663.3	167.2	150.9	147.7	499.6	778.1	235.3	230.6	142.9
(4) major threat	313.0	–	58.0	–	–	585.0	585.0	–	–	–

[a] $p < .05$; [b] $p < .01$; [c] $p < .001$.

(Continued)

TABLE 8.3 Mean Verbal and Visual Prominence Scores for Intensity of Deviance by City and Medium (*Continued*)

Intensity of Deviance	Sydney					Brisbane				
	Newspaper		Television		Radio	Newspaper		Television		Radio
	Verbal only ($n = 1948$)	Verbal plus Visual ($n = 475$)	Verbal ($n = 221$)	Visual ($n = 221$)	Verbal ($n = 113$)	Verbal only ($n = 1614$)	Verbal plus Visual ($n = 625$)	Verbal ($n = 175$)	Visual ($n = 175$)	Verbal ($n = 98$)
Normative Deviance										
(1) does not violate any norms	278.0	573.3	83.4	85.9	80.4	375.1	586.1	80.4[c]	72.1[c]	63.5[c]
(2) minimal violation	319.6	645.0	109.0	92.4	83.7	296.8	594.1	146.2	141.5	122.5
(3) moderate violation	228.0	504.8	107.5	136.3	118.0	326.0	551.8	243.0	219.7	83.0
(4) major violation	264.6	759.0	161.8	126.7	153.3	269.5	469.7	163.5	258.0	186.3

[a]$p<.05$; [b]$p<.01$; [c]$p<.001$.

deviance in both television and radio media (and in Brisbane newspaper verbal plus visual ratings) but were not recorded for either form of deviance in the Sydney newspaper items or verbal ratings for the Brisbane newspaper items. A significant difference in normative deviance was only recorded for Brisbane television and radio news items.

In terms of newspaper prominence, no significant difference in prominence scores was recorded for more deviant items, with one exception—for the statistical deviance scores for verbal plus visual newspaper content in Brisbane. This suggests that, for stories in the Australian print news media, unusual or more deviant news items are not more likely to gain prominence contrary to the hypothesized relationship.

Other interesting results showed that significant differences were recorded for normative deviance scores in Brisbane television and radio. ABC radio news is a full fifteen-minute bulletin with no commercials. It is an agenda-setting medium, without pictures or sensation to distract from the text. Commercial television, on the other hand, barely gets fifteen minutes of stories in its half-hour bulletin, and the package must have elements of the quirky and bizarre to keep the audience engaged. Again, Brisbane is more parochial than Sydney, and the tabloid effect is therefore more pronounced.

Social Significance in the News

Of the forty ANOVA analyses undertaken, twenty-five produced statistically significant results (see Table 8.4). In the majority of these analyses, items with the lowest political, economic, cultural, and public significance also recorded the lowest prominence (twenty-three analyses out of twenty-five significant analyses). In sixteen out of the twenty-five significant analyses, the highest significance score was also the highest prominence score.

A dichotomy between news items with no political, economic, cultural, or public significance and items with any significance in these areas also appeared to exist in the television and radio news media. In ratings where there were significant differences in the mean scores, these differences tended to be driven by larger differences between items that had no political, economic, or public significance and items with minimal to major significance. This trend was evident for television and radio news items in both cities where differences between mean prominence scores for nonsignificant news items contrasted to the prominence scores of items with any significance. This suggests that for television and radio news, once an item has any political, economic, or public significance the item is allocated a certain amount of space and time (prominence) in the media.

Cultural significance did not affect prominence in television or radio news items in either city. This may, indeed, reflect the structural differences between broadsheet and tabloid press whereby broadcast news bulletins have little opportunity to present news items that cover cultural issues. The irony of this is that, outside the confines of the news bulletins and taking into consideration

TABLE 8.4 Mean Verbal and Visual Prominence Scores for Intensity of Social Significance by City and Medium.

Intensity of Social Significance	Sydney					Brisbane				
	Newspaper		Television		Radio	Newspaper		Television		Radio
	Verbal only (n = 1948)	Verbal plus Visual (n = 475)	Verbal (n = 221)	Visual (n = 221)	Verbal (n = 113)	Verbal only (n = 1614)	Verbal plus Visual (n = 625)	Verbal (n = 175)	Visual (n = 175)	Verbal (n = 98)
Political Significance										
(1) not significant	263.7[a]	575.1	72.5[c]	75.7[c]	62.8[c]	358.6	548.8[a]	81.8[c]	78.7[c]	61.0[c]
(2) minimal	332.6	599.1	148.0	146.8	104.7	326.7	729.7	164.4	157.4	102.0
(3) moderate	354.1	652.7	165.8	139.9	119.1	400.9	717.2	200.0	195.0	123.8
(4) major	401.3	586.3	188.3	209.7	133.3	421.7	690.8	219.3	251.4	160.6
Economic Significance										
(1) not significant	251.3[b]	575.6	72.5[c]	77.0[c]	78.3[a]	297.4[c]	546.6[c]	105.7	97.5	72.5
(2) minimal	330.6	644.4	134.4	167.4	119.3	386.6	566.8	119.4	127.7	99.1
(3) moderate	384.0	493.5	182.3	132.9	46.0	553.0	868.7	151.4	161.9	119.8
(4) major	347.4	585.0	–	–	–	234.0	585.4	–	–	–
Cultural Significance										
(1) not significant	267.1[b]	603.5	90.7	89.7	83.4	293.1[c]	553.4[c]	109.5	110.8	75.8
(2) minimal	297.3	595.4	92.7	118.8	–	427.9	582.9	110.6	102.8	83.5

(3) moderate	314.1	417.2	101.3	–	97.5	426.5	624.5	48.0	184.0	117.0
(4) major	782.8	643.5	–	–	177.0	832.0	2683.0	285.0	285.0	114.0
Public Significance										
(1) not significant	247.2[b]	542.2	49.8[c]	47.5[c]	24.6[c]	308.4[c]	495.3[b]	62.2[c]	73.8[c]	55.0[c]
(2) minimal	343.7	647.9	95.5	97.0	102.1	366.7	622.6	126.2	112.2	106.7
(3) moderate	344.6	573.3	179.3	157.5	146.1	540.5	746.4	226.6	213.4	119.7
(4) major	517.0	680.5	–	185.0	–	87.7	–	172.0	172.0	114.0

[a]$p<.05$; [b]$p<.01$; [c]$p<.001$.

the whole schedule of television and radio programming, these two media would have higher levels of cultural content than newspaper media.

A different pattern emerged for newspaper items. In Sydney, no significant difference was recorded for prominence scores for verbal plus visual items of differing political, economic, cultural, and public significance. Prominence of verbal plus visual items in the Sydney papers was therefore not affected by the political, economic, cultural, or public significance.

This again reflects the structural differences between the broadsheet and tabloid format for news presentation. The broadsheet provides a more informative coverage, while tabloids use more arresting visuals. Rating of economic significance did not appear to affect prominence scores for television and radio news items in Brisbane. However, at the time of the study there were no major economic crises affecting Australia.

In contrast, the verbal plus visual items in the Brisbane newspaper did show significant mean differences in the prominence scores for different political, economic, cultural, and public significance levels. The higher the cultural and public significance, the higher the mean prominence score. In terms of political significance, mean differences were again driven by low prominence scores for items that had no political significance, compared to higher scores on items that had minimal to major political significance. Both Sydney and Brisbane newspaper text showed significant differences in the mean prominence scores depending on the political, economic, cultural, and public significance scores. In these analyses, generally the higher the political, economic, cultural, and public significance scores, the higher the prominence allocated to the text. Prominence scores for the visual items in the Sydney newspaper were not affected by the political, economic, cultural, or public significance ratings.[10]

Deviance and Social Significance as Predictors of News Prominence

Although all the stepwise regression analyses produced significant results (see Table 8.5), in some of the cases only a very small amount of the variance in prominence was explained by these measures for newspaper (Sydney, 1% for both verbal and verbal/visual; Brisbane, 3% verbal and 4% verbal/visual). This suggests that something other than deviance and significance determines the prominence given to news items in newspapers.

On the other hand, deviance and social significance accounted for much of the variation in the size and placement of both television (Sydney, 17%; Brisbane, 30%) and radio (Sydney, 56%; Brisbane, 45%) news items. Yet results from these analyses show no real pattern in the way deviance and social significance affect the prominence allocated to individual news items. Different variables have significant beta weights in the various regression equations.

In terms of radio item prominence in Brisbane, statistical significance (.26), social significance (.32), and political significance (.28) were significant predictors of variation in radio news item prominence in this smaller city, with a

TABLE 8.5 Stepwise Regression Analyses of Intensity of Deviance and Social Significance on News Prominence

	Sydney								Brisbane							
	Newspaper Prominence Verbal only Total $R^2 = .01^b$ ($n = 1948$)		Newspaper Prominence Visual and Verbal Total $R^2 = .01^a$ ($n = 447$)		Television Prominence Total $R^2 = .17^b$ ($n = 221$)		Radio Prominence Total $R^2 = .56^b$ ($n = 113$)		Newspaper Prominence Verbal only Total $R^2 = .03^b$ ($n = 1614$)		Newspaper Prominence Visual and Verbal Total $R^2 = .04^b$ ($n = 575$)		Television Prominence Total $R^2 = .30^c$ ($n = 175$)		Radio Prominence Total $R^2 = .45^b$ ($n = 98$)	
Independent variables	r	Std. Beta	r	Std. Beta	r	Std. Beta	r	Std. Beta	r	Std. Beta	r	Std. Beta	r	Std. Beta	r	Std. Beta
Deviance																
– Statistical, verbal content	.05^a	ns	.08	ns	.18^b	ns	.62^c	.25^b	.01	ns	.15^c	ns	.28^c	ns	.49^c	.26^b
– Statistical, visual content	–	–	.02	ns	.21^b	ns	–	–	–	–	.15^c	ns	.25^b	ns	–	–
– Social change, verbal content	.05	ns	.02	ns	.30^c	ns	.70^c	ns	.02	ns	.11^a	ns	.29^c	ns	.60^c	.32^b
– Social change, visual content	–	–	.01	ns	.35^c	ns	–	–	–	–	.11^a	.11^b	.42^c	ns	–	–
– Normative, verbal content	.01	ns	.07	ns	.14^a	ns	.20^a	ns	–.05^a	ns	–.01	ns	.32^c	ns	.44^c	ns
– Normative, visual content	–	–	.04	ns	.11	ns	–	–	–	–	–.02	ns	.41^c	.28^c	–	–

$^a p < .05$; $^b p < .01$; $^c p < .001$; ns = not part of final stepwise regression equation.

(Continued)

TABLE 8.5 Stepwise Regression Analyses of Intensity of Deviance and Social Significance on News Prominence (*Continued*)

Independent variables	Sydney								Brisbane							
	Newspaper Prominence Verbal only $R^2 = .01^{[b]}$ (n = 1948)		Newspaper Prominence Visual and Verbal $R^2 = .01^{[a]}$ (n = 447)		Television Prominence $R^2 = .17^{[b]}$ (n = 221)		Radio Prominence $R^2 = .56^{[b]}$ (n = 113)		Newspaper Prominence Verbal only $R^2 = .03^{[b]}$ (n = 1614)		Newspaper Prominence Visual and Verbal $R^2 = .04^{[b]}$ (n = 575)		Television Prominence $R^2 = .30^{[c]}$ (n = 175)		Radio Prominence $R^2 = .45^{[b]}$ (n = 98)	
	r	Std. Beta	r	Std. Beta	r	Std. Beta	r	Std. Beta	r	Std. Beta	r	Std. Beta	r	Std. Beta	r	Std. Beta
Social Significance																
– Political, verbal content	.07[b]	ns	.10[a]	.10[a]	.31[c]	ns	.42[c]	ns	.01	ns	.13[b]	ns	.39[c]	ns	.48[c]	.28[b]
– Political, visual content	–	–	.05	ns	.29[c]	ns	–	–	–	–	.10[a]	ns	.42[c]	.26[c]	–	–
– Economic, verbal content	.09[c]	.07[b]	.05	ns	.30[c]	.19[b]	.20[a]	ns	.15[c]	.14[c]	.16[c]	ns	.08	ns	.23[a]	ns
– Economic, visual content	–	–	–.01	ns	.25[c]	ns	–	–	–	–	.15[c]	.16[c]	.17[a]	ns	–	–
– Cultural, verbal content	.05[a]	ns	–.02	ns	.01	ns	.13	ns	.12[c]	.10[c]	.06	ns	.04	ns	.11	ns
– Cultural, visual content	–	–	–.07	ns	.06	ns	–	–	–	–	.05	ns	.06	ns	–	–
– Public, verbal content	.09[c]	.06[b]	.09	ns	.36[c]	.30[c]	.73[c]	.56[c]	.10[c]	–.07[b]	.15[c]	ns	.40[c]	.23[b]	.43[c]	ns
– Public, visual content	–	–	.06	ns	.36[c]	ns	–	–	–	–	.13[c]	ns	.36[c]	ns	–	–

[a]p < .05; [b]p < .01; [c]p < .001, ns = not part of final stepwise regression equation.

model predicting 45% of the variance in item size and placement. In contrast, statistical significance (.25) and public significance (.56) were more likely to affect change in prominence for Sydney radio news items.

The influence of deviance and social significance in the television news media told another story again with public significance affecting the size and placement of news items in both cities (Sydney, .30; Brisbane, .23). In addition to public significance, economic significance (.19) also affected item size and placement in the Sydney television news media. Prominence of television news items in Brisbane was also affected by normative deviance (.28) and political significance (.26).

People Defining News

Eight focus groups were conducted, with two changes to the standard approach. First, the two audience groups in each city were not composed along socioeconomic status (SES) lines; rather each group was recruited as a mix of high SES and low SES. In retrospect, it turned out that the groups did coalesce around socioeconomic strata that resulted in a high and low SES audience group in each city. Second, the groups were conducted during the month following the attacks in the U.S. on September 11, 2001 (commonly known as 9/11), which necessitated eliminating this event from discussion, as it would have dominated and skewed the discussion based on media analysis of November–December 2001.

The focus groups tended to produce fairly homogeneous discussions within each group and therefore potentially distinctive and different behaviors across the eight groups. Important early definitions of news arose out of the first question asked of the groups: What is the most interesting thing you have heard today? In this sense, information important to the individual was not always news—it could have been personal. With this definition in mind we are able to "typify" each of the groups in terms of their attitude toward news.

Group 1: Audiences, Sydney

This turned out to be the "suspicious/conspiracy" group. They were somewhat worldly wise about the ways of the news industry, yet at the same time acknowledged the social role of news in helping society to cohere. They tended to be low SES. But social coherence did not necessarily organize itself around "important events," although the importance of political elections, for example, was acknowledged. Rather, finding out about how the microsystems of a society worked were important. For example, there was an interesting discussion about nonroutine events such as how to avoid a parking ticket if the parking meter got jammed. Similarly, for students in the group, the outside world mattered little to them outside of their exams, which were taking place at the time. These points focused on personal utility rather than on typical media news of a worldly nature. In this sense, the news media did not necessarily hold a privileged position in the information hierarchy; rather, they had to compete with other sources and forms of social knowledge.

On the other hand, the news media were viewed with some suspicion. At the time of the focus group, "boat people" and political refugees were in the news, including a tragic event of an overloaded refugee boat bound for Australia but sinking off the coast of Indonesia. One low SES female participant suggested that it was almost a daily occurrence that ships as "big as the Titanic" sink at least once a week somewhere in the world, "but we never hear about that on the news." This person also referred to television as her preferred news medium but disliked the negative emphasis of news, suggesting the solution was "you just turn it off and walk away from it." The same person also suggested that personalizing stories increases their impact and makes them more interesting. At one point it was suggested that during the 9/11 incident the second plane hitting the World Trade Center had been delayed in order to maximize news impact by stringing out the story. The group members seemed to be suggesting, "It's all about ratings ... for selling papers." This extreme cynicism characterizing the low SES people reflected anger toward the establishment from a position of low social power and esteem. Besides the solution of turning off the set, there was the Internet, or "reading the paper backward" to make sure the reader did not get side-tracked. With some cynicism and humor, the general tenor of the conversation was for consumers to beware. Intelligent sifting of the news media plays its social role of joining individuals to each other and to the broader social group.

Group 2: Audiences, Sydney

This group of higher SES people—the "personalizers"—can be summarized by suggesting that news or information is of significance when it speaks of one's personal world rather than of the world. Their response to "What is the most interesting piece of information you have heard today?" was generally personal and nonmedia: hearing of their boss's or colleague's personal crisis, anti-aging drugs, discovery of a bat colony—some people did not hear anything interesting because they were so busy or self-absorbed. One person who worked at a television station said, "Actually didn't hear any news at all!"

They tended not only to personalize the news initially but also to make elitist judgments about how (most) other people were more easily duped by the news. They tended to judge the media harshly and to switch it off if they thought the media were being sensationalist; at the same time they commented that many people did not switch it off. They tended to focus on relationships and to distinguish between reports and understanding of events by having time to think about them in terms of their own lives. Worldly events, dealing with death and destruction on a large scale, were mostly incomprehensible, and only became significant when empathy enabled group members to understand how they would feel in the same situation: "One person dying is a tragedy … many people dying is a statistic." Finally, stories dealing with children were more relevant to parents.

Group 3: Audiences, Brisbane

Members of this low SES group of various ages were the "validators." They provided various experiences in order to validate the news based on their own experiences and were generally impatient with news they saw as distracting, abstract, or overly gloomy.

One working-class student, on the way up, suggested that reporting war and unemployment first thing in the morning was "not a good way to start the day." This person's reaction was to switch channels to the *Teletubbies*. But this skeptical person was interested enough to "test" ideas he had encountered on the newscast either through his own life experience or by discussing them with his friends, whom he sometimes considered more credible. He told of a story about the relationship between fatty foods and brain function that had motivated him to have an interesting discussion with some of his fellow biology students. Even though he found the story lacking credibility, it made him think about his own diet. He also said news often acted as a trigger to social action. Another woman in the group who worked with prison inmates also said she used news as a trigger for social action. As her "clients" were often "in the news," she had difficulty reconciling reality and news reports. She said distortions and untruths in the news sometimes motivated her to seek justice for her clients. Another woman whose life was sometimes on the "wrong side" of the law relished the rare stories about "bent cops" because she felt poetic justice occurring. She was often angered when the media moved beyond her own experience into the general world, because she "doesn't believe a word of any of it." This feeling of disenfranchisement was also characteristic of low SES group members in Sydney.

The older people in the group seemed to prefer ABC radio, from which they gained much information. They tended to focus on positive information that appeared by implication to be in contrast with news, which was more negative. Toward the end of the discussion, there was some recognition of the need for news to attract attention, thus leading to negative and sensational coverage.

Group 4: Audiences, Brisbane

This was a higher SES group that could be characterized as the "free choice group": they were not saying that free choice was always available but that it was always an issue, particularly when it was not available. The choice paradigm was not only the province of audiences but of journalists as well. There was some concern that journalists often lacked choice and were working to a "media agenda." While there was some tendency to personalize the news, there was also greater balance between personal and public narratives. Most importantly, group members seemed interested in incorporating news within their own experiences. The most interesting things they had heard that day were about haircuts, computer software, new aspects of fellow workers' lives, and old friends catching up, but there was also interest in the anthrax threats and international trade issues following 9/11, sports stories, local political power plays, and various human interest issues.

The group tended to have a freewheeling conversational style, which meant they stayed with particular news topics rather than switching from one to the other. They also moved rather easily into making judgments about the news; again, this seemed to be a characteristic of high SES groups. Credibility of news services seemed to be a constant issue, with extremes of "I believe everything" to more cynical members using trustworthy friends to validate news stories. One young rural student expressed her appreciation for television news as having changed her life. Before going to college she had never really experienced television, but since purchasing a television set she became obsessed with watching the news. "I sort of watch it over and over, and hearing the way it's presented in different manners, even if it's just said with different words and a different face, you know, it's just a way of reinforcing it." Another female participant felt news connected her to the rest of the world. Yet another woman said the more information she had, the more she felt in control of her own life.

One male commented at the end of a discussion about how journalists work and how they were part of the entertainment industry, which often required them to be negative and sensational. "You've got a balance between … the negatives and the positives. It's no good believing in fairy tales … you've got to also know the negative side so you can arrive at a balanced decision in life." A woman responded that some images were simply so powerful she could not get them out of her mind. She spoke of a very positive story about a little disabled girl winning a trip to Disneyland, a story that deeply impressed her but that was quickly displaced as she flipped through the newspaper past an image of a man jumping from one of the Twin Towers on 9/11. She was suddenly awed by the image of the man jumping and imagined his biography—he was somebody's son or brother. That sense of loss overwhelmed the positive story of the little girl in Disneyland.

Group 5: Public Relations, Brisbane

Few of these participants had been journalists, and most were of the new breed of public relations people with training in communications. This group was surprisingly frank but at the same time seemed to have critical detachment from professional journalists as well as from their own industry. At times they referred to "spin" as if it was not something they were involved in but rather happened "out there" in the media.

Unlike the audience groups, the Brisbane public relations participants never talked about their own lives. The entire conversation operated at the "professional" level: the nature of news and its relation to public relations. The group spent much time discussing a particular case in the news at that time: The governor general (GG), the British Queen's representative, in his previous occupation as an Anglican Archbishop, mishandled a sexual abuse case. What interested the group was that the GG's current career was in jeopardy because he was not handling the media well. The GG was doing his own spin, and appeared not to be taking professional advice. In the view of one public relations person, the ongoing story was so out of control that it had become deviant news rather than instructive news of social significance, exploring proper adult–child relationships and the duty of public officials to preserve that relationship. One participant commented, "Because I'm more aware of the media than my next door neighbor … Like with any story such as this that's been beat up so much you just have to sit down and go, What the hell's going on here? Whose motivations are these and why is this so fascinating? And I quite frankly don't believe what I'm reading."

On other issues their attitudes toward the media, though still cynical, were more informed about why the media made their decisions. The main issue was the negativity of news values, and how public relations practitioners must craft their product along similar negative lines if they want to get stories published: "Good news doesn't sell papers." There was some brief discussion about the Internet and how at the moment it offered a less controlled environment. This view of the web, generally endorsed, was interesting because it suggested how public relations people as private citizens acknowledged the value of the Internet as being "beyond control," yet professionally this could only have frustrated them. Within this focus group they were happily straddling the two worlds—professional and private—even though they were in contradiction.

Again, non-news programming was singled out as being more effective in the realm of social significance. "I love listening to *Radio National,* you get new ideas and debates and … actually hear people talking about things that I'm interested in…." This comment opened up a negative set of observations about the news media: "Our organization inducted for the first time ever the first female president. She's a chemical engineer, and the press release I put out was really, really positive, but … every time we turned up to interview our president [journalists would] put a negative spin on it: 'What's it like for a female to be operating in a male-dominated industry?'" Clearly, public relations professionals cannot put a negative spin on their work simply to get a run. The other strategy is to choose their journalists carefully! And sometimes just being realistic is the best course of action.

Group 6: Public Relations, Sydney

This was a younger group with several members studying part time and some with media training and experience. Unlike the previous public relations group, the most important information these participants had heard that day was invariably personal. When this was pointed out to them, they were somewhat defensive and stressed the importance of staying abreast of news for work. "Keeps you in the loop ... it's a conversation starter." Similar attitudes seemed to come through in relation to politics. "I'm not into politics; however, I think as an Australian citizen I need to know what's going on." Sometimes popular television—soap opera and reality television—played the same role of keeping them in the loop and providing conversation starters. But while this kind of television was socially effective, it was also amusing. Indeed, what seemed to emerge from the group conversation was a different slant on the social utility of information: It was not only what you learned from information that made it important and desirable; it was also to some extent the social act of sharing it. But for different situations, the sharing can be manifested differently. Regarding 9/11, for example,

> the kinds of conversations I have at the school [while dropping off children] is very different from the sorts of conversations I have at work. I get to work with a whole bunch of professionals who are glued to television, but just previously I've just dropped my daughter at school and had a number of parents tell me that they turned off the TV. They didn't want to know ... It was just too horrifying, too difficult, they didn't know how to explain it to the kids ... All I needed [was] to get it off my chest, and it was almost like there was this sort of denial ... So I think it depends on whether, you know, the best way to look at that kind of information, how it socializes you ... Those [are times] when you get together, not just to share information, but for support, and their child defines them, and what they need to know defines them as well, and that's where all the stories are swapped and all that kind of stuff. You very rarely would talk about politics in a group like that.

That is at the very personal level. There was another sense of the personal that came out in the discussion—how they responded as media workers individually to media. "It can make you a lot more kind of cynical about things ... You forget the average person who doesn't have that level of understanding of the fact that you know ... [A] story generated by PR links or whatever, they just don't read it for that...." Some called for greater transparency in the media so that audiences could better understand the subjective nature of how stories are generated. They thought that English newspapers probably did this well.

Group 7: Journalists, Sydney

Being "on call" makes it difficult to achieve a full complement of journalists in a focus group session. Several sessions and some individual interviews were required. The first group consisted of three newspaper journalists and a television journalist. The second consisted of several commercial television journalists interviewed individually, and the third involved three ABC (public) radio journalists. Their most important news of the day was a mix of public and private: a refugee ferry sinking off the coast of Indonesia, the death of a friend, the quickening of a first baby in utero, and dealing with a real estate agent. For the three ABC radio journalists, it was the same public story—the collapse of Ansett Airlines. Even those who led with private topics were aware of the public, but they distinguished work issues from private ones. "There are almost two parts to that for a journalist. It's a commodity for us … and the first thing you do is compartmentalize that as something you would use." Another, who also gave priority to the personal, said that when he heard about the ferry sinking, as a police reporter he thought "aha, a quiet day for me." A big enough competing story would push his beat off the page.

The two issues—the kind of news that is of interest and the positive versus negative news question—yielded a broad range of responses. The radio journalists were very focused on work issues, saying that "people lying" was what attracted their attention. As journalists, being able to expose liars would have social significance. A negative story about corruption can have a positive outcome in terms of social reform. Predictably, definitions of news were weak, but there was a feeling that negativity was a necessary ingredient, since journalists must provide alternative explanations for the status quo. However, positive versus negative was too categorical, as stories often contain elements of each. Just as the world is not black and white, so journalists should focus more on the gray area, particularly in newspapers. One junior television reporter stressed the importance of good pictures, suggesting news was largely about emotional responses. He thought the governor general story was not about the facts of the case. "I couldn't tell you myself what were the specifics of the case … but people are just outraged he's covered up all this sexual abuse.… I often get my scripts cut because I've got too much information." The other commercial television journalist interviewed was more senior and experienced, and although she worked for the third-ranking network (*Ten*) her favorite information service was *ABC Radio National*. As a news producer she commented, "It's taken me a few years in the industry to realize how much rubbish goes to air, [but] if your product is working well why change it.… It's a business."

Group 8: Journalists, Brisbane

Again, the journalists had to be surveyed in several sessions: four senior and investigative journalists mainly from newspapers with some television, three radio journalists of mixed experience, and two television journalists (one junior and one senior). The gender mix and age spread was roughly even.

In terms of what is news, there was a wide difference of opinion. The senior freelance and investigative journalists were somewhat cynical about news agendas, as they were privy to many "inside" stories that either could not be published because of timing issues or more importantly could not find publications willing to give them a run. While "also-ran" freelancers often express this view, this group had a sound track record. One of them did an investigative series in the 1980s that brought down a long-standing state government on the grounds of police corruption. His view was that the media culture had become far more conservative. Another freelance reporter, who worked "on assignment" in various newsrooms covering various projects, was also dismayed at the changed media culture, especially the various deals by Murdoch's News Corporation and their impact on the Australian media scene in general, and on the situation in Brisbane in particular, by reducing the number of dailies from three to one. The most senior investigative journalist interviewed said that he quite often found it easier to get Brisbane stories published in the quality (non-Murdoch) Sydney newspapers, particularly in the more challenging areas of the environment, consumerism, and politics. Freelancers not only found it hard to get published, but their rates had gone down. "What could have commanded a few years ago $2000, you'll be offered now about $700—take it or leave it. We don't care how good the story is. We don't care how relevant it is. You know, this is the product."

So, when the group of senior journalists was asked about their news priorities, they invariably cited the story-behind-the-story, or the insider story, but at the same time expressed frustration that publishing outlets were not always easy to find. Their view was that the Internet had become an increasingly important source of news for journalists, with the emphasis on source. Again, their view was that not enough journalists read other newspapers on the web. For example, the *New York Times* was cited as an important agenda-setter for journalists worldwide, particularly since the 9/11 tragedy, but equally, new websites such as the *Drudge Report* and an Australian equivalent, *crikey.com*, were also given emphasis as important agenda-setters for journalists—not necessarily publishing outlets in themselves.

It should be stressed that the media agendas were the main concern of journalists. In Brisbane that "agenda" meant no stories that would rock the boat, mainly in terms of the political and executive culture. Murdoch's *Courier Mail* does not even provide email for its journalists—just for the executive team. One freelancer remarked, "It used to be a lot easier when there were more outlets. Journalists used to move[1] round a lot." Another member of the same group, a former commercial television news producer, remarked, "They [local journalists] may be entirely great at uncovering stories, great foot soldiers, at finding a story—and they are rewarded for that. They are promoted for that, and they stay within the industry instead of thinking, 'This is all too hard; I'm going to get out.' You come in with some critical analysis, start asking some questions and making decisions based on 'Well that's crap anyway; I'm not interested.' ... Let me tell you that commercial television has a complete bloody dearth of people who are going to ask questions behind."

The ABC radio journalists were much the same; however, they too had come under criticism from the investigative group and so were more supportive of the "mainstream" news agenda that did not rock the boat.

It was the television journalists who stood apart. They seemed unconcerned with the ruminations of covert politics and agendas and seemed more caught up in the overt world, where politics did not play a major part. At the time of the interview the most significant story for the television journalists was a scandal involving a Brisbane-based national swimming coach in court over sexual abuse allegations from some of his former child swim stars. Clearly it was an important national story, but it was the television journalists' lack of interest in politics and questions of social structure and accountability that stood them apart. Their world was dominated by "pictorial significance" and individual, rather than social, accountability.

Comparing People's News Preferences with What Is in the Newspaper

Using Spearman rank order correlation coefficients, the relationship between the ranking of news items by journalists, public relations practitioners, and audience groups were examined in addition to the ranking of each group with the prominence given to the items in the newspapers analyzed. Table 8.6 presents the Spearman rank coefficients for the Sydney and Brisbane groups.

There was weak agreement among the journalists, public relations practitioners, and audience groups with the newspaper ranking. In Sydney, all the correlations between the rankings by group participants and the newspaper ranking were small and not significant. This was also true for rankings in the smaller city of Brisbane with only a moderate but significant correlation between public relations practitioners and newspaper ranking (rho = .50) and for the low SES participants (rho = .39). The *Courier Mail* produces relatively stronger, but still weak, relationships in terms of news priorities with its constituent workers and readers. Brisbane is a much smaller city, and the newspaper is more parochial and thus more connected with its community.

TABLE 8.6 Spearman Rank-Order Correlation Coefficients between Newspaper Item Prominence and Focus Group Rankings

	Journalists	Public Relations	High SES Audience	Low SES Audience	Newspaper
Journalists	–	.71[c]	.64[c]	.64[c]	.29
PR	.86[c]	–	.80[c]	.82[c]	.50[b]
High SES Audience	.61[c]	.65[c]	–	.75[c]	.35
Low SES Audience	.63[c]	.65[c]	.78[c]	–	.39[a]
Newspaper	.18	.17	.11	.19	–

Note: Sydney coefficients in upper triangle, Brisbane in the lower. SES = socioeconomic status.
[a]$p < .05$; [b]$p < .01$; [c]$p < .001$.

Sydney has more of a national focus and is therefore less communitarian. But the agreement among journalists, public relations professionals, and audiences seems to fit with the generally highly personal response to "what's news?" and the general distancing of all groups from the news media agenda.

In contrast, in Brisbane and Sydney there was significant agreement with moderate to high coefficients among journalists, public relations professionals, and audiences about the relative importance of the news items, as reflected by the card-sorting task. In Sydney, the range was from .64 to .71, and in Brisbane the range was from .61 to .86. Thus, using item headlines, journalists, public relations practitioners, and audience groups rank news items in a similar order of importance.

Discussion

The high correlation among all media on sports seems consistent with the large amount of space devoted to this news category. And while high levels of sports are expected in radio and television news bulletins, it is clearly not the place for cultural coverage; newspapers scored this second.

Otherwise, particular media had fairly predictable interests: business, commerce, and industry for newspapers; disasters, accidents, and human interest for television; and internal politics and internal order for radio. Newspapers, being media of record and heavily text based, carry the kind of news that requires detailed explanation; commercial television operates heavily in the emotional sphere. Public broadcasting system ABC radio is strongly tied into the political process.

In broad terms, Australian newspapers do not necessarily give prominence to deviance whereas radio and television tend to do so. However, there is no question that the *Courier Mail*, for instance, has high levels of deviant news to which it gives prominence. But there is also socially significant competing prominent news: whole supplements devoted to healthy living, fashion, computers, employment, sports, food, entertainment, and so on. These affirming prominent topics tend to dilute the effect of hard news that often deals in deviance.

In general terms, in all three media there was correlation between news prominence and social significance. This result would tend to mollify tabloid extremes. Indeed, while the *Courier Mail* and *Channel Nine* news tend to be tabloid, they are moderate, suggesting there is a balance between deviance and social significance. In both deviance and social significance, while prominence was affected, the size of story was not.

Although the audience groups were not recruited on the basis of SES there seemed to be some kind of rough alignment of socioeconomic status; whether this resulted by chance or group dynamics is a moot point. There appeared to be two broad characteristics of the low SES groups: (1) They generally liked the kind of stories where public figures were being brought to task; and (2) they felt angered at stories that were purely about political administration, policy, and structure—"just a lot of nonsense"—and usually turned it off.

It was surprising how many respondents from the audience and professional groups valued *ABC Radio National* as an enjoyable, informative, and positive source of knowledge—in contrast to most of the news services they either worked for or were being surveyed about. *ABC-RN* is rather like *BBC2*: informative radio talks, interviews, documentaries, new music, and so on. Another general refuge from the media was the World Wide Web. For public relations workers, it was a medium beyond their control because it was not corporately owned to any great extent; for journalists too, the lack of corporate presence was a plus, as it meant there were no agendas: an approximate public sphere.

Another common thread was the lack of interest by commercial television journalists in politics or facts. They were highly visual and preferred an emotional rather than factual basis for their stories. There was a sense in which television acted in the same way as a digest of quotations—it enabled audiences to perform socially the next day by being "across" issues in the news but not necessarily to know much about them. That role was for newspaper journalists, presumably.

In general across all groups there was a concern that while "deviance" and "social significance" appeared as extremes on a single news/social continuum, in reality they were not mutually exclusive. Stories that had intense deviance, such as a swimming coach sexually molesting his child swim stars, also had potentially intense social significance. Putting such a person through the news process was akin to villains in a medieval village being placed in the stocks: they were pilloried by journalists, the judiciary, and audience and thereby reinforced social morals through their own deviant or negative example. Valence operated similarly. Many focus group members thought that although a story may have initial negative valence, the consequences of reading it, viewing it, or hearing it might have positive social outcomes.

The most important event in people's lifetimes tended to produce "generational" responses. For older people, the first moon landing was important, along with John F. Kennedy's assassination. However there were some near universals: the death of Princess Diana, the fall of the Berlin Wall, the 1991 Gulf War, and the Vietnam War. Some Australian political and social events competed with global issues, and some personal issues competed effectively against worldly, public occurrences.

Overall, it is somewhat puzzling that no correlation exists between people's assessments of newsworthiness (journalists, public relations practitioners, and audience groups) and newspaper prominence but that significant (moderate to strong) correlations among people can be found for ranking of news items. Thus, newspapers seem to be out of step not only with their readers, but with their constituent workers—public relations professionals and journalists—as well public relations professionals and journalists.

9
What's News in Chile?

SOLEDAD PUENTE AND CONSTANZA MUJICA

The Media System in Chile

Chile is a long and isolated country. Bordered by the Pacific Ocean on one side and the Andes Mountains on the other, and despite its length of 2,653 miles (4,270 kilometers), it is highly centralized. Various initiatives, especially by the government, have sought to change this situation, but none have been able to alter the fact that the capital, Santiago, is very powerful. The 2002 census showed that a third of the Chilean population lives in the capital (INE 2003), despite the government's efforts to change the situation.

And yet Chile is also a small country with 15.1 million inhabitants (INE 2003). The population has only increased by about 1.8 million (INE 2003) during the past ten years, indicating a decline in the birth rate (INE 2003). However, its small size and isolation have not been obstacles to the developing media industry, especially in some regions where it has been quite successful.

All in all, Chile currently has eight national and forty-four regional newspapers—most of them run by businesses tied to the respective regions—and an additional three free newspapers (ANP 2003) distributed in Santiago.

The media are generally in the hands of large journalistic conglomerates, whether national or through the involvement of transnational companies that have entered the Chilean media scene in recent years. However, this does not detract from the fact that initiatives of independent groups have been able to generate business models of their own (Dermota 2002; Krohne 2002).

Very strong print media companies with national coverage exist, such as the newspapers *El Mercurio* and *La Tercera*. The former, based in Santiago, was founded in 1900 (ANP 2003) by the Edwards family, who had a strong publishing business. The family also owns *La Segunda*, an afternoon paper, and *Las Ultimas Noticias*. The three newspapers have national circulations (ANP 2003). Another important media conglomerate is *Copesa*—Consorcio Periodístico de Chile (The Journalistic Consortium of Chile). This company owns two newspapers, *La Tercera* and *La Cuarta*, as well as the weekly news magazine *Qué Pasa* and others (Dermota 2002; Krohne 2002). All of these media have their own websites, as online journalism has been rapidly developing in Chile and is linked with the prestigious media.

Television is another area of national influence with rather unique characteristics. Chile has seven very high frequency (VHF) television channels (CNTV 2003), the most important of which are Televisión Nacional de Chile and Canal 13, both owned by the Corporación de Televisión de la Pontificia Universidad Católica de Chile. In third place is Megavisión, a private station. These three channels cover the entire country (Krohne 2002).

Television was traditionally considered a nonprofit business in the hands of the universities. The first authorized stations were projects headed by the large Chilean universities: one belonged to the Universidad de Chile and the other to the Pontificia Universidad Católica de Chile. In 1968 a third station, Televisión Nacional de Chile, entered as a state-owned channel. Toward the end of the military government, this structure began to change when private organizations were also authorized to enter the industry.

Today, only the Universidad Católica's station continues in university hands, and a Chilean entrepreneur operates the channel originally directed by the Universidad de Chile. As is the case with the rest of the audiovisual media, the state-owned channel is economically self-sufficient and is organized as a public channel. As for the penetration of cable television, 2005 data showed that 35.9% of all Chilean homes were subscribers (CNTV 2003). Television is regulated by the National Television Council, which oversees programming quality, particularly with respect to violence and obscenity (Godoy 1999).

The current strength in radio is in FM broadcasting, and the business has been organized almost entirely around large consortiums. Radio has undergone two major changes since the mid-1990s: the Santiago chains have expanded to cover the entire nation, and foreign operators have acquired local stations (Godoy 1999). In terms of news radio, Radio Cooperativa is a significant reference: Cooperativa. However, it must be pointed out that the so-called *companionship*, or *talk*, radio is currently a very important programming area, especially during peak hours.

The work of journalists in the 150 or so Chilean media outlets is regulated by the May 2001 Press Law, which took eight years to be legislated and which overturned the former Publicity Abuse Law (Krohne 2002). Chilean journalists and editors recognize the value of the media in controlling authorities and thus in denouncing major cases of societal imbalances. On the other hand, they complain that journalists—as colleagues and editors—do a poor job of really working for citizens' rights to information, and their conviction that through journalism they contribute to the consolidation of democracy and working for a country with stable and respected institutions (Gronemeyer 2002). This represents a challenge for the new generations of professionals who must fight for autonomy and independence that has weakened among Chilean journalists (Gronemeyer 2002).

The Study Sample

Although national characteristics made the choice of a large city easy, the requirement for a location that allows for a sample of all media types, ideally with a local character, made the choice of a small city more difficult.

Santiago, the country's capital and largest city, was selected along with Concepción, which, although an important city within the Chilean context, is classified as being small vis-à-vis the objectives of this study. Concepción has its own newspapers as well as television and radio stations, all of which are distinct from those in the large city.

Greater Santiago, including all of its municipalities, has nearly 4.8 million inhabitants (INE 2003) and, as previously mentioned, is the central reference for the country, mainly due to the centralized power structure. This has been enhanced by recent technological advances in the field of communication, which have led to reduced costs and shorter distances. While recognizing the importance of the capital for the rest of the country, this situation is not always reflected in the news agenda because a significant percentage of the news focuses on Santiago and its unique concerns.

Given its importance, the newspaper selected in Santiago was *El Mercurio*. The corpus consisted of 3,540 items. The television news program selected was *Teletrece*, aired between 9:00 and 10:00 p.m. on the Pontificia Universidad Católica de Chile channel. Despite its highs and lows, it is considered to be the most independent medium. The analysis consisted of 221 items. The radio news program selected was *Diario de Cooperativa*, aired between 1:00 and 2:00 p.m., which gives the day's first version of events. The analysis yielded 233 items.

Concepción is 318 miles from Santiago and has over 900,000 inhabitants (INE 2003). It is in the Bío Bío Region and is considered to be very independent of Santiago, owing to its citizens' fight to maintain their autonomy. The newspaper of choice was *El Sur*, founded in 1882 and owned by the *El Sur* newspaper company (ANP 2003). The analysis consisted of 1,207 items. Radio Bio Bio was founded in 1966 by Nibaldo Mosciatti, an Italian immigrant, and was at the time of the study the region's most important radio station. The company at the time of the study owns thirty-three stations from Arica to Punta Arenas (Mosciatti 2003). Radio Bio Bio's *Radiograma* news program is aired at 1:00 p.m. and runs between fifty and seventy minutes. The analysis consisted of 334 items. Finally, the television news program selected was the 50-minute Canal 9 *Regional de Concepción* television news program, which airs after the Santiago news, beginning at 10:45 p.m., with 154 items analyzed.

Topics in the News

Although the topics of the news in the two cities seem similar at first—the main topics are cultural events, sports, and internal politics—a detailed analysis reveals significant differences between the two cities and their media.

Intuition suggested that the study would reveal more similarities between the topics most addressed by each medium and its editorial line and its public perception, in accordance with the characteristics of what was considered news. It was therefore surprising to discover that the newspaper *El Mercurio* placed such a high value on cultural events when it is perceived as a conservative newspaper focused on politics and economics.

Another notable point concerns local issues with regard to internal order, which were addressed by all the media with the exception of the Santiago newspaper. Despite the fact that Santiago is the nation's capital and its media present national coverage, the treatment of regional news is quite limited, and the consequences of the events are analyzed according to their possible importance to the Santiago metropolitan region. This fact is always mentioned by people of the region and appeared frequently in the focus group discussions.

As can be seen in Table 9.1, the newspapers in the study proved to be more generalist than their broadcast counterparts. In *El Mercurio*, the most important topics were cultural events (16.2%), sports (14.0%), and internal politics (13.4%). However, in *El Sur* internal politics was the most important (20.4%), followed by international politics (9.9%) and cultural events (8.1%). The data show the value that the regions attribute to local occurrences, indicating the policy of independence with respect to the capital and illustrating the value a city as far away as Concepción gives to foreign events.

With respect to television, Santiago news on *Teletrece* placed the highest values on internal politics (20.1%), sports (10.9%), and health/welfare/social services (7.8%). Concepción's *Canal 9* regional news program placed more emphasis on internal politics (32.0%), followed by internal order (18.5%) and labor relations and trade unions (6.7%). Examined together, the three values show that the sum of these topics, which have a high identity value and demonstrate an effort to be independent of Santiago, is 57.2.

In Santiago's *Diario de Cooperativa* radio news program, the three most important topics were internal politics (35.9%), health/welfare/social services (8.5%), and international politics (7.0%). The regional program *Radiograma de Concepción* showed internal politics (30.4%) as well as sports (15.7%) and internal order (13.3%) as part of the news. It is striking that neither the Concepción newspaper *El Sur* nor the Canal 9 news program placed very high values on sports (7.6% and 1.7%, respectively), although it was one of the top three topics for *Radiograma* (15.7%).

With the exception of *El Mercurio* in Santiago, the highest value by far was placed on internal politics. It was particularly dominant in television in Concepción (32.0%) and on radio in both cities (35.9% and 30.4%). This seems to be in line with the close attention Chileans give to politics, especially among people over 30 who lived through the political turmoil of the 1970s and 1980s and the transition from a socialist government to a military government and then to democracy.

Although there were specific differences in the data of the different media, an analysis of the Spearman rank correlations shows more similarities between them than is evident at first glance (see Table 9.2). The highest Spearman rank

TABLE 9.1 Distribution of General Topics of News Items by City and Medium

Topics	Newspaper		Television		Radio	
	Santiago	Concepción	Santiago	Concepción	Santiago	Concepción
Cultural Events	16.2	8.1	5.8	2.2	1.1	.3
Sports	14.0	7.6	10.9	1.7	3.5	15.7
Internal Politics	13.4	20.4	20.1	32.0	35.9	30.4
Business/Commerce/ Industry	8.0	7.8	5.1	3.9	2.1	2.4
Economy	6.9	5.5	4.4	2.2	6.7	2.1
Human Interest Stories	4.9	3.0	5.5	1.1	1.8	.6
International Politics	4.1	9.9	6.1	3.4	7.0	5.1
Internal Order	4.0	5.3	5.5	18.5	3.5	13.3
Health/Welfare/ Social Services	3.4	3.7	7.8	3.9	8.5	3.0
Transportation	3.1	3.4	5.1	3.4	6.3	1.5
Communication	3.0	2.5	1.7	1.1	.0	.6
Education	2.5	4.6	2.4	5.6	.4	2.1
Ceremonies	2.2	3.2	6.5	3.4	5.6	1.2
Entertainment	1.9	2.2	.7	.0	.0	.0
Housing	1.7	.9	1.4	2.2	.0	1.5
Other	1.6	.0	.0	.0	.0	.0
Social Relations	1.6	1.5	4.1	1.1	2.5	3.3
Disasters/Accidents/ Epidemics	1.4	1.2	.7	3.9	3.2	3.9
Military and Defense	1.2	1.8	1.4	.0	2.1	.9
Labor Relations and Trade Unions	1.1	2.5	3.1	6.7	6.3	6.9
Science/Technology	1.1	1.6	1.0	1.1	.0	.6
Weather	1.1	.9	.0	.6	2.5	2.1
Environment	.6	.8	.3	1.7	.4	1.8
Fashion/Beauty	.6	.4	.0	.0	.0	.0
Energy	.2	1.2	.3	.0	.0	.3
Population	.1	.2	.0	.0	.7	.3
Total[a]	100.0	100.0	100.0	100.0	100.0	100.0
	(n = 3415)	(n = 1713)	(n = 293)	(n = 178)	(n = 284)	(n = 332)

Note: Distributions given in percent.

[a]Total percentage may not actually be 100.0 due to rounding error.

TABLE 9.2 Spearman Rank-Order Correlation Coefficients between Rankings of News Topics in Various Media

	Santiago Newspaper	Concepción Newspaper	Santiago Television	Concepción Television	Santiago Radio	Concepción Radio
Santiago Newspaper		.88[c]	.84[c]	.52[b]	.44[a]	.40[a]
Concepción Newspaper			.89[c]	.66[b]	.57[b]	.54[b]
Santiago Television				.67[c]	.69[c]	.58[b]
Concepción Television					.60[b]	.77[c]
Santiago Radio						.77[c]
Concepción Radio						

[a] $p < .05$; [b] $p < .01$; [c] $p < .001$.

correlation occurred between a newspaper and a television station (.89). Concepción's *El Sur* newspaper and Santiago's *Teletrece* news program appear very similar in the variety of topics addressed as well as the nearly identical value placed on the topic of internal order. Due to the characteristics of the media and given the importance of the national level, the Santiago television news has been defined as national and at the service of all Chileans, although many consider it to have a very strong primary emphasis on what happens in the capital. For the Concepción newspaper, it is important to be considered a local medium, but with national-level quality. In fact, it is one of the most prestigious regional media in the country, and many of Chile's great journalists have begun as reporters for *El Sur de Concepción.*

Spearman rank correlation values greater than .80 were found regarding the two newspapers analyzed (.88) and the Santiago television station with the Santiago newspaper (.84). These numbers comply with the intuition that print media should have a similar agenda given their characteristics as references for readers and demonstrating the importance attached to values such as objectivity and balance in the coverage of their reports. This fact merits further analysis beyond this study in order to more thoroughly discuss the work of the agenda setting among the media and how they influence each other.

Some interesting Spearman rank correlations were obtained among the broadcast media. The correlations between television and radio and between Santiago's *Diario de Cooperativa* and Concepción's *Radiograma* were both .77, probably due to the great importance given to internal order. For the three outlets—Santiago radio and Concepción radio and television—internal order was the main topic and comprised more than 30% of their items. Although there were large differences among the other topics addressed, there was another interesting similarity in the area of labor relations and trade unions, where each of the three broadcast media devoted more than 6% of their news to the topic. It should be noted, however, that during the composite week analyzed, a strike by governmental employees occurred, followed by a strike in the hospitals.

The Chilean correlations are both surprising and not surprising. The data seem to suggest that the media communicated with each other, and indeed to some extent they did. Some of the greatest problems with Chilean journalism—despite efforts to the contrary—are that (1) the different media are very similar in the topics addressed; (2) they tend to set their agendas with an eye on each other; and (3) the topics selected closely follow the United Press International (UPI) and Orbe (Chilean) news services, which the media subscribe to and which guide the possible journalistic actions of the day. Research on the values and habits of journalists in Chile illustrates how decision-making in terms of selecting and prioritizing the news results in the need to train autonomous journalists who value their freedom and its consequences (Gronemeyer 2002).

Deviance in the News

In Chile, deviating from the norm generally results in societal sanctions. The country is quite conservative and strongly resists abrupt changes; therefore, it imposes social sanctions on extremes of all kinds. Even today, politics is what most attracts older generations as a result of the transitions of the past thirty years. Although the salience of politics has decreased in the agenda, especially among younger people, it is still the national objective to avoid extreme confrontation and to seek unity rather than division. This tendency explains the low levels of deviance in the Chilean media agenda. For something to appear in the media agenda, it needs to be in the extreme levels of deviance, especially in the more conservative media such as those analyzed in this study.

In general, the empirical findings of the content analysis regarding the deviance measures present a mixed bag (see Table 9.3). Of the thirty analyses of variance (ANOVAs), nine were significant, but only a few indicated a linear relationship—that is, the more unusual events had higher prominence scores. Furthermore, there was a tendency for prominence scores to be highest in the mid-range level of the three dimensions of deviance (statistical, social change, and normative), which was particularly evident for the Santiago and Concepción newspapers and for the Santiago television news.

Specifically regarding statistical deviance, the findings for the *El Mercurio* (Santiago) newspaper were significant for both the verbal and verbal plus visual data, while *El Sur* (Concepción) yielded significant results only for the verbal data. The data also showed that Chilean television tended particularly toward news that was somewhat unusual and quite unusual.

TABLE 9.3 Mean Verbal and Visual Prominence Scores for Intensity of Deviance

Intensity of Deviance	Santiago					Concepción				
	Newspaper		Television		Radio	Newspaper		Television		Radio
	Verbal only (n = 2792)	Verbal plus Visual (n = 1196)	Verbal (n = 179)	Visual (n = 179)	Verbal (n = 204)	Verbal only (n = 935)	Verbal plus Visual (n = 473)	Verbal (n = 155)	Visual (n = 155)	Verbal (n = 288)
Statistical Deviance										
(1) common	290.4[b]	516.2[b]	93.7	89.2[a]	167.5	341.2[b]	570.3	125.7	171.8	185.6
(2) somewhat unusual	355.6	545.0	121.3	160.8	149.8	455.2	693.0	120.8	166.1	134.7
(3) quite unusual	382.3	696.9	176.9	165.3	263.5	319.7	700.3	164.5	130.5	144.6
(4) extremely unusual	319.0	646.3	27.3	39.0	102.3	346.5	–	117.0	171.3	144.3
Social Change Deviance										
(1) not threatening to status quo	334.0	567.2	129.2	120.5	149.7	383.1	597.5	126.5[a]	156.6	124.7[a]
(2) minimal threat	329.0	721.4	153.0	156.2	192.4	406.8	795.6	161.9	137.6	173.7
(3) moderate threat	385.9	780.5	60.1	157.0	169.4	445.6	1099.7	169.9	97.2	147.1
(4) major threat	347.5	524.6	103.3	16.0	–	549.6	–	218.6	168.0	183.3

Normative Deviance

(1) does not violate any norms	339.2	576.2[a]	135.7	132.2	173.6[a]	403.8[b]	604.9	144.6	148.3	63.5
(2) minimal violation	300.6	496.0	97.2	126.4	356.9	548.1	587.7	169.2	218.0	122.5
(3) moderate violation	274.5	269.3	126.8	67.5	104.7	265.2	737.5	168.6	169.3	83.0
(4) major violation	329.1	1429.2	90.9	186.8	–	372.0	384.5	159.9	58.0	186.3

[a]$p < .05$; [b]$p < .01$; [c]$p < .001$

With respect to social change deviance, the ANOVAs were significant and in a linear direction regarding the verbal aspects of both television and radio in Concepción, suggesting that these news items had a potential to threaten or change the status quo in the city, region, country, and even in the international system.

As for normative deviance, the most poignant finding was that the verbal plus visual news items in *El Mercurio* have extremely high prominence when high in deviance. The ANOVA for the Concepción newspaper, while significant, did not show any clear pattern. Finally, neither the television nor the radio news programs indicated any clear picture.

Social Significance in the News

The empirical data for the four dimensions of social significance (political, economic, cultural, and public), as presented in Table 9.4, also present a mixed picture. Overall, thirteen of the forty ANOVAs were significant, with only several—mainly in the Concepción newspaper—indicating a linear relationship.

In terms of political significance—the extent to which the content of a news item has potential or actual impact on the relationship between people and government or between governments—the ANOVAs for both Concepción's television and radio news programs yielded statistically significant differences regarding the verbal component. Furthermore, as was the case regarding deviance (Table 9.3), there was no linear relationship but rather higher prominence scores for the mid-range values of political significance. None of the newspapers or the Santiago television news program had significant values in prominence for political significance.

With respect to economic significance—the extent to which the content of the news item has potential or actual impact on the exchange of goods and services, including the monetary system, business, tariffs, labor, transportation, jobs, markets, resources, and infrastructure—three of the ten ANOVAs were significant but with no linear relationship. Once again, there was a tendency for the mid-range values to be highest in prominence, especially in the Santiago newspaper and the verbal component of the Concepción television news program. In the Concepción newspaper and the visual component of its television news, a linear trend was observed, although it was not statistically significant.

Of the ten ANOVAs calculated for the prominence of cultural significance—the extent to which the content of the news item has potential or actual impact on a social system's traditions, institutions, and norms, such as religion, ethnicity, or arts—four were statistically significant. What is more important, however, is that in three of these cases, linear trends indicating higher prominence positively correlated to the level of cultural significance. This was the case for both the verbal and visual components for *El Sur* (Concepción) and for the verbal plus visual component of *El Mercurio* (Santiago). A partial linear trend for the verbal component of *El Mercurio* was also was

TABLE 9.4 Mean Verbal and Visual Prominence Scores for Intensity of Social Significance

Intensity of Social Significance	Santiago							Concepción						
	Newspaper		Television		Radio			Newspaper		Television		Radio		
	Verbal only ($n=2792$)	Verbal plus Visual ($n=1196$)	Verbal ($n=179$)	Visual ($n=179$)	Verbal ($n=204$)			Verbal only ($n=935$)	Verbal plus Visual ($n=473$)	Verbal ($n=155$)	Visual ($n=155$)	Verbal ($n=155$)	Visual ($n=155$)	Verbal ($n=288$)
Political Significance														
(1) not significant	332.0[c]	561.6	117.0	131.3	154.9			400.0	605.4	122.4[a]	136.2			85.3[c]
(2) minimal	360.1	662.7	188.5	115.2	194.7			349.4	482.9	164.7	193.4			162.1
(3) moderate	344.2	728.5	145.7	153.3	210.1			407.5	811.5	180.4	149.9			189.4
(4) major	323.0	854.0	7.0	–	–			739.3	–	114.5	189.3			153.8
Economic Significance														
(1) not significant	314.6[c]	573.4	129.0	124.1	168.7			395.0	592.2	133.4[a]	149.4			141.0[a]
(2) minimal	381.8	526.0	61.1	222.1	217.7			425.0	846.3	162.9	156.0			96.3
(3) moderate	475.0	833.4	171.0	189.0	237.0			378.8	1026.5	206.2	177.0			185.2
(4) major	308.6	259.9	–	–	–			998.0	–	162.8	–			182.7

[a] $p < .05$; [b] $p < .01$; [c] $p < .001$.

(Continued)

TABLE 9.4 Mean Verbal and Visual Prominence Scores for Intensity of Social Significance (*Continued*)

Intensity of Social Significance	Santiago					Concepción				
	Newspaper		Television		Radio	Newspaper		Television		Radio
	Verbal only (*n* = 2792)	Verbal plus Visual (*n* = 1196)	Verbal (*n* = 179)	Visual (*n* = 179)	Verbal (*n* = 204)	Verbal only (*n* = 935)	Verbal plus Visual (*n* = 473)	Verbal (*n* = 155)	Visual (*n* = 155)	Verbal (*n* = 288)
Cultural Significance										
(1) not significant	337.9[c]	574.9[c]	119.6	127.6	180.1	370.8[b]	587.7[c]	166.8	155.6	145.2
(2) minimal	269.7	507.7	216.1	139.5	166.3	378.1	563.8	142.5	131.9	131.7
(3) moderate	403.9	1080.2	137.0	152.6	–	521.6	1141.3	113.1	124.6	175.7
(4) major	617.5	3162.0	–	250.0	–	–	–	149.2	–	162.8
Public Significance										
(1) not significant	323.1	566.3[b]	67.2	109.5	81.6	316.0[c]	564.6[b]	143.3	149.6[a]	121.7
(2) minimal	362.3	834.0	150.7	131.4	180.6	354.6	652.6	148.1	112.7	156.3
(3) moderate	403.2	981.6	120.4	159.8	238.9	500.5	853.4	135.0	243.1	164.0
(4) major	358.4	400.5	–	–	–	1830.0	–	170.2	145.2	164.8

[a]$p < .05$; [b]$p < .01$; [c]$p < .001$.

noted. No discernable patterns were found regarding the broadcast media in the two cities.

Finally, with respect to the prominence for public significance—the enhancement of or threats a news item represents for the public's well being—four of the ten ANOVAs were statistically significant. The only clear pattern—that of linear trends—was found in the two components of *El Sur*'s news items.

Deviance and Social Significance as Predictors of News Prominence

The final statistical analyses of the data were the eight stepwise regression analyses examining the combined impact of all the deviance and social significance variables in predicting the prominence of the news items.

The results in Table 9.5 show that when the variable is analyzed solely from the verbal perspective, the Santiago media gave the same results (.03), except in the case of *El Mercurio*. While the amount of explained variance of prominence was not very high in any of the cases, it is clear that deviance and social significance in Concepción were higher than in Santiago. In fact, deviance and social significance in Concepción explained 16% of the variance of prominence for television and 12% for radio. Also, the combined analysis for the visual and verbal components for the Concepción newspaper explained 8% of the variance in the prominence. As for the beta weights of the variables, in Santiago there were very few significant coefficients, whereas in Concepción there were more significant variables in the regressions of each of the media.

These findings might be due to the high value placed on the Santiago newspaper as a national medium, whereas the focus of the Concepción newspaper is more regional. Moreover, the Concepción media seemingly establish their respective agendas with the knowledge that they represent secondary media. This is especially the case with television, since the regional newscasts are aired following the newscasts from the capital city. A similar situation exists with regard to radio news.

People Defining News

Eight focus groups were organized in late July and early August 2001, with four each in Santiago and Concepción. The conversations in both cities followed the same methodology. In both cases focus groups were organized for journalists, public relations agents, and audience representatives from high and low socioeconomic status (SES). The two groups of journalists included editors and represented the various media organizations whose products were analyzed in the content analysis. Different companies were included in the public relations groups, some exclusively dedicated to the field and others as part of communications departments within organizations. Special care was taken to include male and female participants as well as a variety of ages in the audience groups.

TABLE 9.5 Stepwise Regression Analyses of Intensity of Deviance and Social Significance on News Prominence

	Santiago								Concepción							
	Newspaper Prominence Verbal only $R^2=.01^a$ (n = 2787)		Newspaper Prominence Visual and Verbal $R^2=.03^a$ (n = 1128)		Television Prominence $R^2=.03^a$ (n = 179)		Radio Prominence $R^2=.03^a$ (n = 204)		Newspaper Prominence Verbal only $R^2=.04^c$ (n = 935)		Newspaper Prominence Visual and Verbal $R^2=.08^c$ (n = 439)		Television Prominence $R^2=.16^c$ (n = 155)		Radio Prominence $R^2=.12^c$ (n = 287)	
Independent variables	r	Std. Beta	r	Std. Beta	r	Std. Beta	r	Std. Beta	r	Std. Beta	r	Std. Beta	r	Std. Beta	r	Std. Beta
Deviance																
– Statistical, verbal content	$.05^b$	$.05^a$	$.16^c$	$.17^c$	.14	ns	.11	ns	.03	ns	.06	ns	.09	ns	–.08	$–.1^a$
– Statistical, visual content	–	–	$.08^b$	ns	$.17^a$	$.17^a$	–	–	–	–	.09	ns	–.14	ns	–	–
– Social Change, verbal content	.01	ns	.00	$–.06^a$	–.06	ns	.04	ns	$.07^a$	$.07^a$	.09	$.10^a$	$.23^b$	$.34^c$	$.13^a$	ns
– Social Change, visual content	–	–	.03	ns	.07	ns	–	–	–	–	.08	ns	–.11	$–.39^c$	–	–
– Normative, verbal content	–.01	ns	.02	ns	–.06	ns	.00	ns	–.05	ns	–.04	ns	.07	ns	$.16^b$	$.19^b$
– Normative, visual content	–	–	$.06^a$	ns	–.01	ns	–	–	–	–	–.01	ns	.02	ns	–	–

Social Significance

Social Significance															
– Political, verbal content	.00	ns	.04	ns	.09	ns	.09	ns	.01	ns	-.05	ns	.07	ns	.26[c]
															.22[c]
– Political, visual content	–	–	.05	ns	-.01	ns	–	–	–	–	.03	ns	.13	ns	–
															–
– Economic, verbal content	.06[b]	ns	.03	ns	.05	ns	.09	ns	.00	ns	.02	ns	.22[b]	ns	.13[a]
															.16[a]
– Economic, visual content	–	–	.01	ns	.12	ns	–	–	–	–	.12[a]	.11[a]	.02	ns	–
															–
– Cultural, verbal content	.04[a]	ns	-.01	ns	.09	ns	-.01	ns	.11[c]	.10[b]	.15[b]	.17[c]	-.18[a]	-.16[a]	.07
															ns
– Cultural, visual content	–	–	.03	ns	.06	ns	–	–	–	–	.14[b]	ns	-.10	ns	–
															–
– Public, verbal content	.04	ns	.05	ns	.04	ns	.17[a]	.17[a]	.17[c]	.15[c]	.17[c]	ns	.08	ns	.13[a]
															ns
– Public, visual content	–	–	.08[a]	.08[a]	.11	ns	–	–	–	–	.19[c]	.17[c]	.05	.20[a]	–
															–

[a] $p < .05$; [b] $p < .01$; [c] $p < .001$. ns = not part of final stepwise regression equation.

The Santiago groups met at the School of Communications of the Escuela de Periodismo, Pontificia Universidad Católica de Chile. In Concepción, Claudia Tapia of the School of Journalism of the Universidad Santísima Concepción recruited the members of the focus groups. The meetings were held in a salon of a local hotel during the researchers' two visits to the city.

The different focus groups provided support for the theoretical notions of the current study, particularly among the audience groups for whom the most significant news events included high deviance with mixed levels of social significance.

All of the groups were able to define the concept of *news*. Even without knowing the traditional criteria for news, the majority of the members of the focus groups chose related concepts in presenting their definitions. In the case of journalists, some listed the criteria of proximity, consequence, prominence, sex, and oddity, and others included academic definitions. However, the majority of the responses were restricted to words related to novelty, impact, or social relevance and significance.

The first concept in the discussions was novelty, perceived as something previously unknown—because it was hidden, rarely happens, or happened for the first time. As a high SES person from Santiago put it, "News is the spectacular, the novel, what is worth reporting for the rarity of its occurrence." A low SES woman in her mid-30s, a cleaning lady from the capital city, said, "For me news is information. Information is knowing what is happening at the world level, because there is news everywhere. Or also what is happening within my own family, for example, because sometimes in the family things happen and you don't know, and when you find out, it's news." And a public relations practitioner from Santiago defined news this way: "For me news is something new, different, and that holds a certain interest for everyone else."

Social relevance, which received the greatest number of definitions, was generally understood to be that which affects the life of a community. In fact, among three of the four groups in Concepción, the tendency was to consider it as synonymous with the local, the regional, and even the personal. A high SES male university professor in Concepción stated, "Now that I live here, I am much more affected by the regional and local news and the way that things in our daily lives must necessarily pass through the control, supervision, and authorization from Santiago." A low SES person from the same city said, "[It is] anything that affects me or the people I love or the people near me, people I interact with, and in that sense, that which affects me as a citizen."

Santiago dwellers, however, had a more difficult time discussing community. The journalists did not even consider the idea of "local"; they included the national and world levels within society. A public relations person in Santiago said that it was "something that is important for most people," while a low SES person in the same city explained that "news is what happens during the day or night, at any moment, that is important for the country." And a journalist put it this way: "I think that it's not enough for something novel to be considered information or news just because it's novel; to be news it has to be novel and have general interest as well, and as a consequence it must have some repercussion on the community."

Significance—something with long-term consequences, where news is not only an event but also a complete process, or a characteristic of the event that remains current over time—also was discussed in the groups, although less frequently. For example, a Santiago resident, a high SES engineer in his late 60s, described it as "something that is going to last, that I can continue thinking about later," while a low SES man in his early 40s in the same city said, "There are things that have happened and that have been forgotten and have not been resolved; the most important are those that are not resolved."

Through conversations came the idea that, for an event or process to be news, it necessarily had to appear in the media. In some cases, it was suggested that other elements were not even necessary. A high SES Santiago resident said, "Images are very important today. There are things that should not be news, but which are very strong as images, such as an accident that has no importance." As a male journalist in Concepción put it: "Everything is news. It's a matter of discovering it, transforming it, and delivering it: the close-at-hand, the magnificent, or anything out there that is extraordinary can be made significant."

Among the group of Santiago journalists, the day of the focus group produced a conversation that clearly demonstrated the thesis of the study, especially in relation to deviance. Just the night before, a Canal 13 news camera operator had filmed the death of a resident who drowned in a canal surrounding Santiago's poorest neighborhoods, Zanjón de la Aguada. The footage was shown on the nightly news, and the event shocked journalists from that station who were participating in the focus group. They and other professionals spent a large part of the time in the group discussing the incident.

Interesting responses were given when group participants were asked to name the three most significant events in their lives. Across all eight focus groups, the two events most often mentioned related to Chilean politics: the 1988 plebiscite when the winning "No" option put an end to the government of Augusto Pinochet (twenty-eight mentions) and Pinochet's detention in London (twenty mentions). The Catholic dominance of Chilean society prompted the mentioning of the Pope's visit to Chile (fifteen mentions). Two world events received much attention as well: the first landing on the moon and the fall of the Berlin wall (thirteen mentions each). Finally, the 1985 Chilean earthquake and the military coup there received twelve mentions each.

The proximity of the events was another feature. One of the most notable cases was when an older woman in the high SES group mentioned the discovery of penicillin as one of the greatest events in her life, because it saved the life of a cousin who had lived with her family and who was about to die—thus an example of news with deviance and social significance.

An analysis and evaluation of the events mentioned made it possible to group the responses into five categories: the unexpected, social change, proximity, directly experiencing the event, and striking images.

The unexpected involved events that seemed unforeseen to participants and were therefore classified as surprising. Here are several examples:

- "Pinochet's detention in England ... something we never expected to happen"—a low SES person in Concepción.
- "The military coup in 1973 ... the situation had an impact on me because it produced something that had seemed like it wasn't going to happen, and suddenly it did"—a journalist in Concepción.
- "The fall of the Berlin Wall and the end of real socialism ... The most impressive thing is that nobody really predicted it ... it was almost like the taking of Bastille. No one understood its magnitude when the changes started in the Soviet Union, that it was going to happen so quickly and that was going to have the consequences that it had"—a high SES woman in her mid-40s in Santiago.

Social change refers to situations that put an end to long-term social and political processes and at the same time initiate new and unknown situations. The participants associated this criterion, as well as the prior one, with uncertainty. The following are examples of comments from participants:

- "The 'No' campaign was impressive; the support, the people who were mobilized, the social and spiritual ideals, the collective spirit, the spirit that united the campaign in social terms"—a public relations professional in Santiago.
- "The detention of Pinochet in London; I was in my first year, so I realized that all of Chilean public opinion got involved, it was full of conflicts and produced a lot of uncertainty about what was going to happen"—a high SES Argentinean woman in her late 30s, living in Concepción for more than eleven years.
- "The fall of the Berlin Wall ... I guess that seeing that the order established for years that you grew up with could fall apart and from that moment the world was different for everyone also had a strong impact on me"—a female Santiago journalist in her mid-40s.

The notion of proximity appeared several times when participants of the focus groups referred to acts that seemed close to them, having lived through an experience similar to the one involved in the event. A high SES Concepción woman in her late 30s and mother of two, put it this way: "Just two weeks ago I had miraculously escaped an accident and then I see what happens to someone as well known as she was (Lady Di), so it made me rethink a lot of things." Or having been connected with an event in her own life, a high SES female history teacher in Concepción in her mid-30s said, "The possible war with

Argentina in 1978 affected me greatly; I remember that I was very afraid because my dad could have gone to war; it was very close."

Many of the participants, and especially the journalists and public relations professionals, mentioned the importance of events in relation to directly experiencing them. Participants said,

- "The presidential campaign because I participated in it; I traveled as a journalist; it was a very stressful and interesting period because never in my life had I thought I would be working on a presidential campaign"—a public relations practitioner in Santiago.
- "I was working in radio during the plebiscite, and I think that that time was very special for all of us who covered it on the radio"—a Santiago journalist.
- "The first thing I mentioned was the earthquake of 1985 because I was caught in the street and it was really impressive to see how the earth moved like a snake. This really made an impression on me and that image of the way the earth moved has stayed in my mind"—a low SES Santiago resident in the audience groups commenting on earthquakes, such as the one in 1985 in Santiago and in 1960 in the south.

The fifth concept that appeared was striking images. Events occurring after the arrival of television appeared in four cases: the first man landing on the moon, the assassination of Egyptian president Anwar Sadat, the bombing of the Moneda during the coup d'état in 1973, and the explosion of the Challenger space shuttle in 1987. Referring to the Sadat story, a Santiago public relations practitioner in her late 20s said, "I must have been very little, but it was the first time I ever saw a terrorist act. I remember seeing the bloody hand in the middle of the platform and all that really had a great impact on me." A high SES professional translator in her 40s from Santiago had this to say: "The bombing within the city limits really affected me because you always hear bombs outside; in other countries, it's somewhat distant, but when you see that they are bombing the Moneda, which is a governmental symbol, it was incredible."

Focus group participants were asked whether they preferred positive or negative news. The general response was that although they wished positive news would be more frequent, they recognized that this was impossible. The majority responded that they looked for both types of news in the media, although one mentioned that "as a culture we are always more concerned with the bad news." But one of the low SES men in his 40s who works as a technical assistant summed it up most poignantly: "News is like a person; it has both a good side and a bad side."

- A common perception among all the groups seemed to be that although positive news is preferable, in reality positive and negative events occur, and it is inappropriate to deny any of them. "We'd all love it if all the news were positive, but unfortunately the majority of the news is more negative than positive," said a low SES person in Concepción. A journalist in Concepción remarked, "I just prefer to have the information, whether it's positive or negative will be determined by the public. I don't know— sometimes I hear something negative, and after all is said and done, it could turn out to be something positive." Finally, two high SES partici- pants in Santiago had this to say: "I like positive news, but above all else, I like the truth"—a male in his early 30s. "You can't turn your back and say that nothing negative is happening today—to just think positive."

In any event, there is the perception that there is more emphasis on nega- tive news in Chile. A low SES person in Concepción put it this way: "For example, if you look at all the times that you turn on the TV or read a newspa- per, it's all tragedies." In contrast, a Santiago public relations practitioner said, "Definitely the positive. I hate all this dark stuff that we have a culture where we all look for defects, for the disastrous, for the negative."

Along these lines, many of the participants expressed their desire for solu- tions or positive perspectives to the problems presented. A high SES econo- mist in his 40s in Santiago said, "I think that news is neither positive nor negative. It's all in the treatment that it's given, in the focus. I think, as some- one else said, that it's terrible news that everything is flooding, but it's great news that all Chileans organize to help the people who need it."

Despite the general preference, these groups perceived the negative news as a necessity to get around in the world. A high SES male professor of econo- mics from Concepción in his early 30s said, "I see a headline, and I'm inter- ested in the topic, and whether it's positive or negative, I buy it. If, for example, it says 'Unemployment Dropped Two Points,' great, because I con- sider that to be something good that is happening. But maybe if the headline says 'Unemployment Up Three Points' maybe this will interest me more to see what the problems are." And a female freelance public relations worker from Santiago in her early 30s remarked, "I take a look at the negative news the same as the good, to know about it, because that way you can pay more attention.... If the little girl reacts in such-and-such a way, maybe it's because the babysitter is punishing her..."

Knowing someone involved in the news event, or having participated in it directly, related to the value of proximity at the time of selecting the news. This allows people to relate to the media and their peers. It is interesting to note that the function of deviance in the news is not only useful for better participation in society but also as a topic of conversation in social gatherings, such as with the storms in Santiago the day before one of the focus groups met.

In any event, the groups in Concepción maintained the idea that beyond other criteria, such as positive or negative, what is essential when turning to the news is proximity, defined both as that which is "local" and "useful to me." As a high SES agronomist, a father of six from Concepción put it: "What interests me is that which affects my daily life, my family… if there is a drop in taxes … I'm going to look into it … and obviously I'm going to put my fingers to the calculator…. Now, in the opposite case, if the interest rate goes up, I also do a follow-up to see what the effect will be to prevent problems."

Comparing People's News Preferences with What's in the Newspaper

The previous focus group member comments provide the essence of the sentiments expressed by participants in the two cities. The final question was what relationship, if any, exists between the actual portrayal of the news in the newspapers of the two cities and the perceptions of the group members. The findings are presented in Table 9.6.

Based on the gatekeeping exercise in Santiago, there were moderately high Spearman rank correlations between the perceptions of the various groups, except for the correlation between the two audience groups of low and high SES, which was lower—albeit statistically significant—as well. However, the relationships between the actual prominence of the items in El Mercurio and the perceptions of their newsworthiness by the group members were rather weak and negative, with only one correlation being statistically significant and one being close to zero. In other words, the perception of the four focus group participants were hardly related to the agenda presented by the main city newspaper, and if they were, the relationship was negative—that is, items presented as prominent were perceived as less newsworthy and vice versa.

The situation in Concepción, on the other hand, was rather different. First, there was considerable agreement among the members of the four groups, even higher than in Santiago, with all Spearman rank correlations being .80 or higher (the highest correlation was between the journalists and the high SES participants.) Second, the Spearman rank correlations between the actual prominence of the El Sur items and the perceptions of the four groups were quite low but all positive, thus indicating a slight relationship between the

TABLE 9.6 Spearman Rank-Order Correlation Coefficients between Newspaper Item Prominence and Focus Group Rankings

	Journalists	Public Relations	High SES Audience	Low SES Audience	Newspaper
Journalists	–	.67[c]	.63[c]	.62[c]	−.20
Public Relations	.89[c]	–	.76[c]	.65[c]	−.34
High SES Audience	.93[c]	.80[c]	–	.38[a]	−.37[a]
Low SES Audience	.89[c]	.81[c]	.82[c]	–	−.14
Newspaper	.34	.33	.24	.26	–

Note: Santiago coefficients in upper triangle; Concepción in the lower; SES = socioeconomic status.
[a] $p < .05$; [b] $p < .01$; [c] $p < .001$.

agenda of the newspaper and the perception of the journalists, public relations practitioners, and the public.

Discussion

More than the question of whether Chilean media provide the best journalism, this study allows us to begin to reflect on the equilibrium between the media agenda and public's perceptions. The media analysis and the discussion in the focus groups tend to confirm the research hypotheses, although the statistical data do not always reflect this.

The Santiago media indicated the greatest distance between the commentaries of the audiences and the agenda, in accordance with assumptions of the study. However, in Concepción more variance in the prominence of the news items was explained by deviance and social significance.

One possible explanation for the lack of explanatory power of deviance and social significance is that the Santiago media have a national rather than local perspective, despite the opinions to the contrary expressed by people in the periphery of the country. On the other hand, in the various regions—especially in Concepción, the regional city selected for the study—greater emphasis is placed on people's identity through their media.

National characteristics suggest that the reasoning behind the statistical data is cultural. Chile is an isolated country with a powerful elite that does not allow great excesses. Extremes are highly censored by the majority, regardless of one's political position; thus, many people feel that their possibilities are restricted, despite the fact that they live in a democratic society.

The Chilean characteristic of rejecting conflicts and extreme positions is also reflected in the analysis of the different media, especially with respect to those of Santiago: *El Mercurio*, *Teletrece*, and *Diario de Cooperativa*. For example, none of the ANOVAs of prominence were significant for social change deviance. In Concepción, on the other hand, the data were more significant. Interestingly—and this might be material for future studies—the findings for *Teletrece*, one of the country's two most important news programs, did not yield a single significant finding among all the social significance variables with regard to prominence of the news items.

It was also surprising to note the strong attachment to the traditional definitions of news exhibited by the focus group participants. Although at times the conversations achieved the goal of creating a distance from the expected responses, as soon as the reference turned to journalism the participants returned to the most traditional concepts of news known by everyone.

In conclusion, judging from the hypotheses of the study the Chilean data did not provide strong support. However, when considering the cultural characteristics of the country and the current state of the press, the data do support the basic thesis. And, as was pointed out by the news audiences, interest in the news was not always reflected in the news agenda, especially in the Santiago media.

10
What's News in China?

GUOLIANG ZHANG AND JONATHAN J. H. ZHU, WITH BING HONG,
SHENGQING LIAO, GANG HAN, WEI DING, YUN LONG, AND YE LU

The Media System in China

Although relatively low in terms of media consumption per capita, China nevertheless has the largest media system in the world. In 2001, there were 2,000 newspapers, 9,000 magazines, 2,000 radio channels, and 3,600 television channels across the country. Newspapers publish nearly 200 million copies per issue, which translates to about 500 million readers nationwide. The broadcast media enjoy an even larger audience, with 700 million listeners for radio and 1 billion viewers for television. The national rate of cable subscription is 40%, with the rate going over 70 to 80% in most cities. Members of cable households have access to thirty or more television channels, whereas noncable viewers can typically receive five to eight terrestrial channels. Beyond these print and broadcast outlets, the Chinese media operate 2,000 online editions serving a total of 60 million Internet users domestically.

Despite the unmatched size, virtually all Chinese media are owned by the state, which has direct impact on editorial policy and personnel. At the center of the system is the propaganda department of the ruling Chinese Communication Party (CCP) that issues, on a regular basis, specific guidelines for the media on what to report, how to report, and what to avoid. Departure from the directives typically results in the involved media executives and/or journalists being removed from their posts. On the other hand, working for Chinese media brings both social prestige and financial benefits, which compares favorably with business managers, medical doctors, and other professionals (Chen, Zhu, and Wu 1997). Therefore, strong deterrents and incentives exist for the editors and reporters of Chinese media to follow the party line closely in their daily work. Consequently, crackdowns on undisciplined media personnel have happened infrequently.

Given the massive size of the media outlets and the centralized control by the CCP, news coverage in Chinese media is highly ubiquitous and uniform. These qualities are most noticeable in the coverage of all international events and major national events (e.g., state affairs, national meetings, state and CCP decisions), in which cases all local media are required to run stories originated from a few designated national media (e.g., Xinhua News Agency, the *People's Daily*, or China Central TV). Of the less important or local events, those that fit the national propaganda agenda specified in CCP guidelines are more likely to be reported. While varying in jargon over time, the national agenda for propaganda invariably includes such themes as promotion of Communist heroes as role models for the public, demonstration of economic achievements to legitimize government policies, and censure of dissident views to ensure political stability. The expectation of the news is to showcase the themes with localized angles.

However, the impact of state ownership on media content has been offset by the ongoing commercialization process (Zhao 1998). Although state ownership remains unchanged, since the 1990s Chinese media have been self-financed, with revenues from advertising and subscriptions instead of state subsidies. Fortunately, the booming economy in China, which itself has been transformed from central planning to market driven, has provided the needed advertising revenue, which amounts to more than US$10 billion a year, a figure unmatched anywhere in Asia-Pacific. However, as with commercial media systems elsewhere, the competition for audiences (and ultimately for advertising revenue) has become increasingly intense for Chinese media. Driven by competition, Chinese media organizations have begun to offer news with increasingly diverse topics and sometimes with sensationalized presentation styles. In other words, Chinese media organizations are now assigned a dual role: to be propaganda mouthpieces of the state on the one hand, and profit-seeking businesses on the other. Commercially driven practices are tolerated when they do not deviate from CCP's guidelines or compete with the state's agenda. However, whenever a clash arises, the latter always prevails.

Chinese audiences, the largest in size in the world, are both attentive to and critical of the news provided by the domestic media. The audiences, especially those living in cities where greater social change has taken place, pay close attention to the news. For example, audience surveys have repeatedly found that, among the 400 million urban residents, almost everyone watches television, more than 80% read newspapers, and nearly half listen to radio. However, they are generally not satisfied with the news in the domestic media, which is either less interesting than or inconsistent with what they observe directly or learn from personal or alternative media sources, such as foreign broadcasts and the Internet (Zhu and He 2002).

In short, given the unique structure and functions of the Chinese media system, it is reasonable to expect Chinese media to act idiosyncratically in the selection and presentation of news events when compared with their counterparts elsewhere. Meanwhile, the recent commercialization has introduced inconsistency and irregularity to news content that is otherwise unaccountable. Discrepancies may also arise between the media and the audience, which demands more from the former.

The Study Sample

As in other countries under study, one larger and one smaller city in China were chosen: Beijing in the north and Jinhua in the east. Beijing, the capital city of the country, is where all national media organizations are located, including more than 200 national newspapers, the only national television network (CCTV), the only national radio network (CNR), and two news agencies (Xinhua and China News). With an urban population of 9 million, Beijing is also one of the largest cities in both China and the world. As such, Beijing is the largest media market in the country. In addition to the national media outlets mentioned already, residents in Beijing have access to more than thirty local newspapers, forty television channels (a quarter of which are locally programmed), and more than a dozen local radio channels.

By Chinese standard, Jinhua is smaller in size, with an urban population of 260,000, or about 3% of that of Beijing. Located inside Zhejiang Province in East China, Jinhua ranks number 220 in terms of population size among 660 cities in China. Residents in this small city have the same access to national print and broadcast media outlets as their counterparts in Beijing. However, they prefer local media that consist of two daily newspapers (the *Jinhua Daily* and the *Jinhua Evening Post*), one radio station (Jinhua Radio) and one television station (Jinhua TV), each operating three channels. The number of local media outlets is typical for other cities of similar size in China.

The three media outlets chosen from Beijing for content analysis were the *People's Daily*, Channel 1 of Chinese Central Television (CCTV 1), and Channel 1 of China National Radio (CNR 1). Unlike most other countries in the study that involve only local media, the three media from Beijing are not local but are instead national, available to audiences in both the large and small cities under study here. The *People's Daily* is the official newspaper of the CCP, with a nationwide circulation of 2 million, which makes it the second largest newspaper in the nation. CCTV is the only national television network, operating eleven channels (at the time of the study). CCTV 1 is a general-purpose channel for a wide range of programs, including news, entertainment, and sports. All provincial channels, as required by the state, broadcast CCTV's thirty-minute evening news program, aired between 7:00 and 7:30 p.m. As such, the channel enjoys the largest viewership among the eleven channels of CCTV (and in fact any other channel). CNR is the only national radio network, with the general-purpose channel CNR 1 being the most popular among its eight channels.

In the small city of Jinhua, the *Jinhua Evening Post*, Jinhua TV, and Jinhua Radio were selected for the study. As noted previously, there are two daily newspapers in the city, both operated by the same publisher and having the same editorial staff, with a clear-cut division of labor between the two. The older, larger *Jinhua Daily* primarily targets institutional subscribers such as local government officials, business managers, and village leaders, whereas the younger *Jinhua Evening Post* (with seven years of history) aims at 80,000 individual subscribers living in the city and satellite towns. As such, the *Jinhua Daily* resembles the *People's Daily*. The *Evening Post* has therefore been chosen to maximize variation of media content in our sample. At the time of the study, the *Evening Post* published eight pages a day on weekdays and four pages on weekends. As aforementioned, Jinhua TV and Jinhua Radio are the only local television and radio station, respectively, in the city. Each operates three channels, including a comprehensive channel, an economic channel, and a miscellaneous channel. Included in the sample are the comprehensive channels of Jinhua TV and Jinhua Radio, each resembling CCTV 1 and CNR 1, respectively.

Topics in the News

How much news is in Chinese media? Our content analysis found that, on a daily basis, the *People's Daily* publishes 127 news stories (with an average size of 16 cm²) and twenty-two news pictures (with 14 cm² per picture), and the *Jinhua Evening Post* runs sixty-one stories (with 8 cm² per story) and sixteen pictures (with 8 cm² per picture). During their respective prime-time news programs, CCTV 1 broadcasts twenty-eight stories (with an average length of 62 seconds), Jinhua TV, seventeen stories (with 48 seconds per story), CNR, fifty stories (with 31 seconds per story), and Jinhua Radio, twenty-nine stories (with 36 seconds per story).

As shown in Table 10.1, the top three topics on the six Chinese media all fall into such hard news categories as internal politics, business/commerce/industry, and international politics. On the whole, internal politics account for 20% of the news space or airtime on the six media, followed by business/commerce/industry (13%) and international politics (10%). Altogether, the three topics account for 40–50% of the news hole. In fact, the share would be even larger if similar categories such as economy (5%), transportation (4%), and telecommunication (3%) were merged with business/commerce/industry. The emphasis on political and business news is clearly consistent with the designated functions of the Chinese media as discussed earlier.

Several noticeable patterns emerge from Table 10.1. Of the six media, the *Jinhua Evening Post* stands out as an exception, with the top three topics all devoted to such soft news topics as human interest stories, education, and sports, in that order. This is due to the division of labor between the *Post* and its big brother paper, *Jinhua Daily*, which is responsible for hard news. This type of division of labor is quite common for most cities in China. It is also informative to note that the three national media carry far more international politics than their local counterparts do (with an average ratio of 14% versus 5%). In particular, of the three local media, Jinhua TV does not carry any story on international politics. In fact, Jinhua TV does not report any international news—political or otherwise—which is, again, a general practice for all local television stations in China that lack both resources and authorization to cover international news. Local newspapers, such as *Jinhua Evening Post*, and

TABLE 10.1 Distribution of General Topics of News Items by City and Medium

Topics	Newspaper		Television		Radio	
	Beijing	Jinhua	Beijing	Jinhua	Beijing	Jinhua
Business/Commerce/ Industry	15.5	6.7	10.6	15.6	8.7	21.4
Internal Politics	15.2	5.8	27.4	35.9	25.5	10.0
International Politics	13.7	7.6	14.4	.0	13.4	8.6
Cultural Events	7.1	7.0	9.1	4.7	2.5	7.1
Human Interest Stories	5.8	15.5	3.4	.8	2.8	3.8
Sports	4.9	10.1	1.0	.0	.8	2.4
Health/Welfare/ Social Services	4.7	4.9	4.3	10.2	5.0	3.8
Environment	4.5	1.6	1.4	6.3	3.1	2.9
Economy	4.4	1.1	8.2	2.3	7.6	8.6
Internal Order	4.3	3.1	.5	8.6	4.8	11.4
Science/Technology	3.8	2.7	3.8	2.3	2.5	2.4
Education	3.7	13.2	1.4	.0	1.4	3.8
Communication	2.4	1.3	.0	3.1	2.0	.5
Transportation	1.8	2.2	3.8	8.6	2.5	2.4
Ceremonies	1.7	.4	3.8	.0	3.6	2.4
Energy	1.2	.2	1.9	.0	.0	1.0
Military and Defense	1.2	.7	1.0	.0	3.1	.0
Weather	1.2	1.8	1.0	.0	5.9	.5
Housing	.8	3.1	1.4	.8	1.4	1.9
Social Relations	.8	.9	.5	.0	.0	1.0
Disasters/Accidents/ Epidemics	.5	2.0	1.0	.0	1.7	.5
Population	.4	.2	.0	.0	1.4	.0
Entertainment	.3	5.2	.0	.0	.0	.0
Fashion/Beauty	.0	2.0	.0	.0	.0	.0
Labor Relations and Trade Unions	.0	.4	.0	.8	.3	3.8
Other	.0	.0	.0	.0	.0	.0
Total[a]	100.0	100.0	100.0	100.0	100.0	100.0
	(n = 952)	(n = 446)	(n = 208)	(n = 128)	(n = 357)	(n = 210)

Note: Distributions given in percent.
[a]Total percentage may not actually be 100.0 due to rounding error.

local radio stations, such as Jinhua Radio, equally do not have the resources but could adopt international stories from Xinhua News Agency and other national media sources.

Consonance of News Agenda To formally test the level of homogeneity or consonance among the six media in their news topics, we analyzed Spearman rank-order correlation coefficients for each pair of the six media. As shown in Table 10.2, thirteen of the fifteen pairwise rank-order correlations are statistically significant at the .05 level or above, suggesting a high degree of consonance across the six media, which is predicted by our earlier review of the media system in China.

It is perhaps more interesting to compare and contrast the consonance in news topics between the media dimension (i.e., *intramedia consonance*) and the geographic dimension (*intralocality consonance*). While the former may indicate the impact of media organizational, technological, and professional factors, the latter is more likely a reflection of political or local cultural forces. We can contrast the relative strength between the two by calculating the average correlations within each of the media sectors and within each of the localities against the baseline consonance (i.e., the average of all fifteen correlations in Table 10.2). By averaging the correlations between the two newspapers (.66), between the two television stations (.48), and between the two radio stations (.65), we arrive at an intramedia consonance score (.60) that is virtually the same as the baseline consonance (.59). In other words, the news topics between two print or broadcast media are just as similar as between any other pair of media selected completely at random. The intra-Beijing consonance (.74), as measured by the average of the correlations among the three national

TABLE 10.2 Spearman Rank-Order Correlation Coefficients between Rankings of News Topics in Various Media

	Beijing Newspaper	Jinhua Newspaper	Beijing Television	Jinhua Television	Beijing Radio	Jinhua Radio
Beijing Newspaper		.66[c]	.79[c]	.59[b]	.73[c]	.80[c]
Jinhua Newspaper			.47[a]	.30	.31	.54[b]
Beijing Television				.48[a]	.70[c]	.70[c]
Jinhua Television					.54[b]	.66[c]
Beijing Radio						.65[c]
Jinhua Radio						

[a] $p < .05$; [b] $p < .01$; [c] $p < .001$.

media (.70, .73, and .79), is much stronger than the baseline, whereas the intra-Jinhua consonance (.50), which is the average of the correlations among the three local media (.30, .54, and .66), is much weaker than the baseline. In short, there appears to be an average intramedia consonance, strong consonance among the national media, but weak consonance among the local media.

Diversity of News Agenda The data in Table 10.1 also permit us to analyze the diversity of news topics across the six media. Following our earlier work on issue diversity in a media agenda-setting context, we calculated the H-statistic to measure the diversity for news agenda for each of the six media (McCombs and Zhu 1995, 502). As it turns out, the resulting H-statistic is the highest for the two newspapers (3.87 for the *People's Daily* and 3.98 for the *Jinhua Evening Post*) and the lowest for the two television stations (3.48 for CCTV and 2.93 for Jinhua TV), with the two radio stations falling in between (3.73 for CNR and 3.78 for Jinhua Radio). The findings are probably due to a combination of two factors: the newspapers carry a lot more news items than the broadcast media, which has a positive impact on the diversity measure, and television news is the most regulated in China, which bears a negative impact.

Where Is Sports News?

Sports news takes a much lower place in Chinese media than it does in the media of other countries under study, where it often takes first or second position. As a whole, sports news accounts for only 5% of the news hole and is ranked tenth on a list of twenty-six topics in the six media (seventeenth on CCTV, twenty-first on CNR, eighteenth on both Jinhua TV and Jinhua Radio, but relatively higher in the two newspapers—sixth in the *People's Daily* and third in the *Jinhua Evening Post*). To be sure, there is no lack of sports in the Chinese media. For example, there is a sports channel on CCTV, a national *Sports Daily*, and countless sports magazines across the nation. However, sports are typically treated as entertainment and are kept outside of mainstream news. Sports come to the news section only when married with a political theme (e.g., national pride).

Deviance in the News

Deviance was measured in this study by three variables: statistical deviance, social change deviance, and normative deviance, each on a 4-point scale. Our content analysis shows that the overwhelming majority (74–91%) of the news stories in the six Chinese media fall into the "common" or "minimally unusual" categories of the deviance scale. Only less than 3% of the stories were classified by coders as "extremely unusual." Judged by the three measures of deviance, the news stories in the Chinese media are relatively more statistically deviant, less normatively deviant, and least socially deviant. In general, soft news is slightly more likely to be unusual than hard news; international news is more likely to be deviant than national news, which is in turn likely to be more unusual than local news.

Table 10.3 shows the results of thirty analyses of variance (ANOVA) tests of the deviance hypothesis, which posits that a news event's deviance determines how prominently the media cover the event. The hypothesis is largely unsupported by the data because significant differences are found only in six (or 20%) of the thirty tests. Between the two cities, the six significant tests are all found in the small city, Jinhua. That is, the deviance of an event carries no impact on how much space or airtime any of the three national media in Beijing allocates to the event.

Four of the six significant comparisons are found in the *Jinhua Evening Post*. In fact, the prominence of news events in the verbal content ("verbal prominence") on the newspaper is significantly related to all three measures of deviance (i.e., statistical, social change, and normative deviance). However, the verbal plus visual prominence on the newspaper is only related to normative deviance. Furthermore, in the four tests where deviance significantly affects news prominence, the direction of the relationships is often inconsistent with the hypothesis that predicts that the more deviant a news event is, the more prominent the event is treated by the media. The predicted pattern emerges only in the case of social change deviance, where extremely unusual events receive the highest level of verbal prominence in the newspaper. In three other comparisons—statistical deviance in verbal significance, and normative deviance in both verbal and visual significance in the *Jinhua Evening Post*—the most deviant events are actually the least prominent. However, if we ignore those most deviant events (which account for less than 3% of the sample, as mentioned already), the pattern of the relationships would be in the predicted direction: the "quite unusual" events receive more prominent coverage than the "somewhat unusual" events, which are in turn covered more prominently than the "common" events.

The remaining two statistically significant tests involve visual prominence on Jinhua TV. As predicted by the deviance hypothesis, those "somewhat unusual" events appear on the station more prominently than those "common" events. On the other hand, none of the three measures of deviance has anything to do with the verbal prominence on Jinhua TV or Jinhua Radio.

It should be noted that the four broadcast media outlets in the two cities are less likely to report highly deviant events than the newspapers in the two cities. Especially on Jinhua TV, the coders judged not a single story as "quite unusual" or "extremely unusual," which is partly due to the lack of international news—which is often more deviant than domestic news—on the local television. These findings present no surprise at all given the concerns among media regulators in China that television news carries the strongest effect on audiences and therefore should avoid sensationalism as much as possible. In fact, much less negative news is shown on the two television channels under

TABLE 10.3 Mean Verbal and Visual Prominence Scores for Intensity of Deviance

Intensity of Deviance	Beijing						Jinhua					
	Newspaper		Television		Radio		Newspaper		Television		Radio	
	Verbal only ($n=889$)	Verbal plus Visual ($n=158$)	Verbal ($n=199$)	Visual ($n=199$)	Radio Verbal ($n=352$)		Verbal only ($n=425$)	Verbal plus Visual ($n=113$)	Verbal ($n=116$)	Visual ($n=116$)	Verbal ($n=204$)	
Statistical Deviance												
(1) common	158.8	286.8	96.7	88.5	49.7		74.5[a]	130.8	–	43.7[c]	69.6	
(2) somewhat unusual	163.0	218.8	139.2	146.9	63.7		91.2	95.8	102.4	146.9	78.1	
(3) quite unusual	178.4	214.0	145.4	143.0	75.5		125.2	194.3	–	–	69.7	
(4) extremely unusual	169.2	98.4	42.1	72.0	53.6		54.0	46.5	–	–	–	
Social Change Deviance												
(1) not threatening to status quo	160.5	254.4	137.9	140.1	68.8		76.7[c]	122.5	102.9	99.9	74.9	
(2) minimal threat	168.8	197.0	117.5	91.6	66.3		134.3	163.0	99.2	128.8	75.6	
(3) moderate threat	170.1	136.5	136.7	150.3	62.8		96.1	87.3	–	–	19.7	
(4) major threat	210.6	106.6	89.0	–	–		306.0	–	–	–	–	

[a]$p < .05$; [b]$p < .01$; [c]$p < .001$.

(*Continued*)

TABLE 10.3 Mean Verbal and Visual Prominence Scores for Intensity of Deviance (*Continued*)

	Beijing					Jinhua				
	Newspaper		Television		Radio	Newspaper		Television		Radio
Intensity of Deviance	Verbal only ($n=889$)	Verbal plus Visual ($n=158$)	Verbal ($n=199$)	Visual ($n=199$)	Verbal ($n=352$)	Verbal only ($n=425$)	Verbal plus Visual ($n=113$)	Verbal ($n=116$)	Visual ($n=116$)	Verbal ($n=204$)
Normative Deviance										
(1) does not violate any norms	160.1	254.4	158.5	142.7	80.8	76.9[c]	105.2[c]	85.3	69.0[b]	80.0
(2) minimal violation	188.7	172.0	94.6	95.3	61.5	97.1	146.5	110.7	137.0	51.2
(3) moderate violation	142.0	162.6	108.7	135.1	50.9	178.5	411.7	–	–	39.5
(4) major violation	116.5	111.4	72.0	72.0	57.0	38.5	116.0	–	–	–

[a]$p < .05$; [b]$p < .01$; [c]$p < .001$.

study (only 2% being "mostly negative or bad news") than on the two radio channels (9%) or the two newspapers (11%).

Social Significance in the News

Social significance was measured using four components: political significance, economic significance, cultural significance, and public significance, each on a 4-point scale ranging from "no significance" to "major significance." The six Chinese media under study appear to focus on the low end of the significance scale, with 76–95% of the stories falling into the "no significance" or "minimal significance" categories, and 3% or less into "major significance." As in deviance, the skewed distribution of significance affects the subsequent bivariate and multivariate analyses. Of the four measures of significance, the news stories under study score the highest on public significance, followed by economic significance, political significance, and cultural significance, in that order.

Table 10.4 summarizes forty ANOVA tests of the significance hypothesis that the more socially significant a news event, the more prominently the media are likely to cover the event. Slightly better than in the previous tests of the deviance hypothesis, where only one out of five tests is statistically significant, ten (or 25%) of the forty tests are significant here, with two for the *People's Daily* in Beijing and the remaining eight for the three local media in Jinhua.

The two significant tests for the *People's Daily* involve political and public significance. Consistent with the significance hypothesis, those events of "major significance"—either politically or publicly—are given the highest level of verbal prominence in the national newspaper; conversely, those events of little political or public significance receive the lowest prominence. Economic and cultural significance bear no relationship to the verbal prominence in the paper. Likewise, none of the measures of social significance affects the verbal plus visual prominence in the newspaper, verbal and visual prominence on CCTV, or verbal prominence on CNR.

Verbal prominence in the *Jinhua Evening Post* is significantly influenced not only by all three measures of deviance (as shown in Table 10.3) but also by all four measures of social significance (as shown in Table 10.4). In contrast, none of the four measures of social significance is related to verbal plus visual prominence in the newspaper. However, similar to deviance, the direction of the relationship between social significance and verbal prominence in the paper is often inconsistent with the hypothesis that predicts a unidirectional and linear relationship between social significance and news prominence on the media. For example, while those events of little political or economic significance are indeed the least prominent in the paper, those events of major cultural or significance are treated the least prominently. In fact, there appears to be a curvilinear relationship between social significance of news events and

TABLE 10.4 Mean Verbal and Visual Prominence Scores for Intensity of Social Significance

Intensity of Social Significance	Beijing					Jinhua				
	Newspaper		Television		Radio	Newspaper		Television		Radio
	Verbal only (n = 889)	Verbal plus Visual (n = 158)	Verbal (n = 199)	Visual (n = 199)	Verbal (n = 352)	Verbal only (n = 425)	Verbal plus Visual (n = 113)	Verbal (n = 116)	Visual (n = 116)	Verbal (n = 204)
Political Significance										
(1) not significant	166.3[a]	219.2	123.3	122.1	52.3	83.1[a]	122.2	108.2	101.1	76.8
(2) minimal	154.5	330.9	118.5	135.4	70.5	137.5	186.4	89.9	106.4	68.1
(3) moderate	151.5	257.3	169.5	159.8	96.6	109.2	68.0	220.0	–	31.1
(4) major	274.8	116.8	83.3	72.0	17.8	107.7	53.5	–	–	–
Economic Significance										
(1) not significant	150.4	240.8	123.2	123.8	65.5	78.8[b]	108.3	71.4[b]	65.4[c]	60.2[b]
(2) minimal	177.1	226.5	129.3	175.0	59.2	103.6	177.6	127.6	163.0	71.7
(3) moderate	191.1	199.9	169.3	112.7	79.5	158.2	253.7	–	–	156.0
(4) major	173.5	108.7	132.1	–	75.8	149.0	–	–	–	–
Cultural Significance										
(1) not significant	168.1	258.0	137.3	134.5	62.0	76.5[b]	117.2	98.8	94.9	77.4
(2) minimal	143.7	175.6	81.9	127.1	99.2	124.9	154.3	119.6	156.6	54.0

(3) moderate	164.7	254.4	140.7	100.9	71.4	119.1	121.4	–	–	56.6
(4) major significance	231.0	134.5	–	–	65.7	60.0	–	–	–	–
Public Significance										
(1) not significant	136.0[b]	220.7	138.2	122.4	61.7	71.0[a]	117.5	44.6[c]	60.7[b]	64.2
(2) minimal	175.8	257.6	121.9	140.2	78.9	96.1	141.0	119.5	136.7	97.4
(3) moderate	170.2	276.2	149.6	143.0	56.1	127.8	102.5	147.8	147.8	64.5
(4) major	252.5	114.6	59.5	–	52.7	60.0	116.0	–	–	–

[a]$p < .05$; [b]$p < .01$; [c]$p < .001$.

verbal prominence in the *Jinhua Event Post*, as those events in the middle—"minimally" or "moderately significant"—of the significance spectrum almost always appear most prominently in the newspaper.

On Jinhua TV, verbal prominence is affected only by public significance, whereas visual prominence is related to economic and public significance. These relationships are all in the predicted direction. However, as in the case of deviance, the small-city television station hardly carries any news of moderate or major significance. Instead, it concentrates on events of non- or minimal significance. Likewise, of the four social significance measures, only economic significance has something to do with news prominence on Jinhua Radio in the direction as predicted by the significance-driven hypothesis. Once again, none of the stories aired on the radio falls into "major significance." The lack of resources and authorization to report international news and major national news largely explains the triviality of news coverage on local Chinese broadcast stations such as Jinhua TV and Jinhua Radio.

Deviance and Social Significance as Predictors of News Prominence

The previous two sections examined separately the impact of deviance and social significance on news prominence. Since deviance and significance are related, it is necessary to examine all specific measures of deviance and significance simultaneously. Table 10.5 shows the results of eight regression analyses in which the impact of all six measures of deviance and eight measures of social significance on news prominence are estimated stepwise.

As indicated by the size of R^2 in the heading of Table 10.5, the fourteen measures of deviance and social significance as a whole explain a small amount of variance in news prominence in the six media under study. The average variance explained is about 10%, with a minimum of zero and a maximum of 31%. Between the two cities, the three local media outlets in Jinhua are relatively more predictable by deviance and social significance, with an average of 16% of the variance in news prominence accounted for, as compared with 4% in the three national media in Beijing. Of the three national media, only the *People's Daily* is somewhat predictable (with 14% of the variance explained). The news prominence on the two national electronic media (i.e., CCTV and CNR) is completely unrelated to deviance and social significance.

Comparing measures of deviance and social significance, the latter seems to have a slightly stronger impact on news prominence. For example, verbal prominence in the *People's Daily* and on Jinhua Radio is significantly predicted only by public significance and economic significance, respectively. In two other regression analyses (i.e., the combined verbal and visual prominence in the *People's Daily* and the verbal prominence in the *Jinhua Evening Post*), two measures of social significance are statistically significant, whereas only one measure of deviance is significant.

TABLE 10.5 Stepwise Regression Analyses of Intensity of Deviance and Social Significance on News Prominence

	Beijing								Jinhua							
	Newspaper Prominence Verbal only — Total $R^2 = .01^b$ ($n = 889$)		Newspaper Prominence Visual and Verbal — Total $R^2 = .14^c$ ($n = 150$)		Television Prominence — Total $R^2 = $ ns ($n = 199$)		Radio Prominence — Total $R^2 = $ ns ($n = 352$)		Newspaper Prominence Verbal only — Total $R^2 = .07^c$ ($n = 425$)		Newspaper Prominence Visual and Verbal — Total $R^2 = .21^c$ ($n = 96$)		Television Prominence — Total $R^2 = .31^c$ ($n = 116$)		Radio Prominence — Total $R^2 = .04^b$ ($n = 204$)	
Independent variables	r	Std. Beta	r	Std. Beta	r	Std. Beta	r	Std. Beta	r	Std. Beta	r	Std. Beta	r	Std. Beta	r	Std. Beta
Deviance																
– Statistical, verbal content	.02	ns	–.04	ns	.00	ns	.04	ns	.11ᵃ	ns	.01	ns	ns	ns	.01	ns
– Statistical, visual content	–	–	–.16	ns	.09	ns	–	–	–	–	.00	ns	.46ᶜ	.37ᶜ	–	–
– Social change, verbal content	.04	ns	–.08	ns	–.02	ns	–.02	ns	.15ᵇ	ns	.12	ns	–.01	ns	–.06	ns
– Social change, verbal content	–	–	–.18ᵃ	–.65ᶜ	–.02	ns	–	–	–	–	.02	ns	.07	ns	–	–
– Normative, verbal content	–.01	ns	–.09	ns	–.12	ns	–.08	ns	.19ᶜ	.16ᵇ	.41ᶜ	.39ᶜ	.11	ns	–.11	ns
– Normative, visual content	–	–	–.17ᵃ	ns	–.05	ns	· –	–	–	–	.35ᶜ	ns	.31ᵇ	ns	–	–

ᵃp < .05; ᵇp < .01; ᶜp < .001; ns = not part of final stepwise regression equation.

(Continued)

TABLE 10.5 Stepwise Regression Analyses of Intensity of Deviance and Social Significance on News Prominence (Continued)

| | Beijing | | | | | | | | Jinhua | | | | | | | |
| | Newspaper Prominence Verbal only Total R^2 = .01[b] (n = 889) | | Newspaper Prominence Visual and Verbal Total R^2 = .14[c] (n = 150) | | Television Prominence Total R^2 = ns (n = 199) | | Radio Prominence Total R^2 = ns (n = 352) | | Newspaper Prominence Verbal only Total R^2 = .07[c] (n = 425) | | Newspaper Prominence Visual and Verbal Total R^2 = .21[c] (n = 96) | | Television Prominence Total R^2 = .31[c] (n = 116) | | Radio Prominence Total R^2 = .04[b] (n = 204) | |
Independent variables	r	Std. Beta	r	Std. Beta	r	Std. Beta	r	Std. Beta	r	Std. Beta	r	Std. Beta	r	Std. Beta	r	Std. Beta
Social Significance																
– Political, verbal content	.04	ns	–.01	ns	.07	ns	.08	ns	.10[a]	ns	–.00	ns	–.05	ns	–.09	ns
– Political, visual content	–	–	–.01	.28[a]	.07	ns	–	–	–	–	.01	ns	.02	ns	–	–
– Economic, verbal content	.06	ns	–.00	ns	.08	ns	.03	ns	.16[b]	.14[b]	.23[a]	ns	.25[b]	ns	.19[b]	.19[b]
– Economic, visual content	–	–	–.08	ns	.03	ns	–	–	–	–	.23[a]	.20[a]	.43[c]	.32[c]	–	–
– Cultural, verbal content	.02	ns	–.03	ns	–.03	ns	.03	ns	.16[b]	.12[a]	.03	ns	.07	ns	–.07	ns
– Cultural, visual content	–	–	–.09	ns	–.04	ns	–	–	–	–	.07	ns	.18	ns	–	–
– Public, verbal content	.10[b]	.10[b]	.07	.36[b]	–.01	ns	–.02	ns	.15[b]	ns	.08	ns	.29[b]	ns	.04	ns
– Public, visual content	–	–	–.02	ns	.04	ns	–	–	–	–	.03	ns	.33[c]	ns	–	–

[a]p < .05; [b]p < .01; [c]p < .001; ns = not part of final stepwise regression equation.

More interestingly, contrary to the deviance hypothesis, one of the significant deviance predictors ("visual social deviance") has a negative impact ($\beta = -.65$) on verbal–visual prominence in the *People's Daily*. That is, the more threatening to the status quo a news event appears visually, the less prominently the event will be reported in the newspaper. While the finding is contrary to the theoretical framework of the current study, it is highly plausible within the context of Chinese media system. As noted earlier, the primary function of the media in China, especially the national media, is to showcase the superiority, achievements, and stability of the status quo, all of which are hardly deviant. A 35-year-old journalist from Jinhua who participated in our focus groups provided an illuminating illustration: "If following my own instinct, I could easily put those negative stories, such as a bomb explosion somewhere [that] everyone is eager to know [about], on the top of the front pages. However, in practice, we ran such stories that are truly the most important of the day only in a corner of an inside page."

People Defining News

Eight focus group discussions were held in the two research sites, Beijing and Jinhua. A total of seventy-five people participated and were evenly divided into groups of journalists, public relations professionals, audiences with higher socioeconomic status (SES), and audiences with lower SES. Both deviance and social significance came up during the focus group discussions, although the two concepts were mentioned infrequently and often implicitly. Between the two concepts, deviance seems to be relatively more salient than social significance. However, most participants nominated, on a voluntary basis with far more frequency, personal relevance as the most salient ingredient in their conception of news. For example, when asked to recall what news they encountered on the day of the focus group meetings, participants cited information based on personal relevance forty-one times, as compared with nineteen times for deviance and seventeen times for significance. Likewise, when asked to identify which news would attract their attention, personal relevance came up thirty-six times, as compared with fourteen times for deviance or twenty-three times for social significance. Similarly, when asked to define what *news* is, personal relevance was referred to nineteen times, as compared with fifteen times for deviance and thirteen times for significance. In fact, when defining news, the most frequently mentioned term was "objectivity" or "truth" (twenty-five mentions). "Timeliness" also came up frequently (seventeen times).

Deviance Throughout the focus group discussions, descriptors such as "deviant," "abnormal," and "unusual" did not come up regularly. However, participants often made references to news as "unexpected," "surprising," "unique," "mysterious," or "counterintuitive," including such examples as a 72-year-old man applying for study at a medical school in Shanghai, a 13-year-old boy swimming across the Qiongzhou strait, poor children in rural areas going to school without shoes, a thief returning a stolen wallet to the owner, a 17-year-old boy killing his mother, and the unusually hot weather during that summer. One focus group participant, a 50-year-old university professor in Jinhua, said, "I am attracted to things that create a sense of suspension in your mind."

Of the seventy-five participants, one person, a public relations practitioner in Beijing, explicitly used the probability concept (i.e., statistical deviance) to interpret deviance: "News contains information that is negatively related to probability. The crash between [Chinese and U.S.] planes seldom happens. The less likely to occur, the more informative an event is." A member of the low SES group in Beijing echoed the view that "the shocking nature of the news is more intriguing."

Interest in deviant news could also be detected from participants' preference for negative news over positive news. When asked to indicate their preference between positive and negative news, 29% of participants preferred only negative news, as compared with 25% exclusively for positive news and another 45% with no preference one way or the other. One reason given by those who prefer negative news is that "positive news is always predictable, whereas negative is not."

A small number of participants explicitly stated they were interested in deviant news, but not in socially significant news. A 32-year-old male professional in Beijing claimed, "I'm curious about international news or tragic events. Domestic news such as meeting reports is just rubbish. I'm only aroused by international news or accidents. That's my mentality."

Social Significance When discussing news with social significance, participants used terms such as "major impact," "with impact," "a lot of people affected," and "the world affected." In other words, significance is interpreted either as an abstract scale of magnitude or as a concrete size of people who are affected. A 32-year-old male professional worker in Beijing said that "[news] is just those trivial things that may turn into big events. These are unplanned incidents but could affect a lot of people, affect the world." The view underscores "trivial things" and "unplanned incidents," as opposed to planned events of a grand scale, such as official meetings or commercial ceremonies that dominate national and local news in China.

It is also interesting to note that the notion of major impact (i.e., significance) was often brought up in the context of international conflicts (as opposed to domestic affairs), corruption cases (as opposed to communist hero models), or disasters (as opposed to achievements). A 32-year-old male public relations practitioner in Beijing commented, "I am attracted mostly to the news that bears major impact on foreign relations."

Perhaps more revealing is an examination of the relative importance Chinese audiences attach to significance in relation to other considerations. When asked to define what news is, a male member of the high SES group in

Beijing listed four dimensions in this order of descending importance: "[News] should be first of all timely, true, relevant to the public, and of major impact." Here, significance came up but was ranked number four after timeliness, truth, and personal relevance. Many other participants put personal relevance above everything else, but the position of significance remained unchanged.

Social significance was the most important factor to participants when they were asked to recall three most memorable "lifetime events." A total of fifty-four events were nominated. The top seven events, each nominated by ten or more participants, included the following:

1. The Tiananmen Square incident in 1989 (mentioned thirty-four times)
2. The Cultural Revolution from 1966 to 1976 (fifteen times)
3. The arrest of the "Gang of Four" in 1976 (fourteen times)
4. The death of Chairman Mao Zedong and/or Premier Zhou Enlai in 1976 (thirteen times)
5. The launch of economic reforms and open-door policies in 1978 (eleven times)
6. Entrance examinations for high schools or universities from 1978 to the present (eleven times)
7. The U.S. bombing of the Chinese Embassy in Belgrade in 2000 (ten times)

Altogether, these seven events accounted for 58% of the 187 events nominated. While all seven events appear to have high social significance, two of them (reforms and open-door policies and entrance exams) do not seem to be deviant. In fact, significance and deviance play an equal role in 70% of the 187 nominations, with 59% of the nominations for events both significant and deviant and 11% for events neither significant nor deviant. Of those "non-significant and nondeviant" events, all but one are personally relevant. The news that the Chinese soccer team lost the 1998 World Cup games is the only event that seems to be low on all three factors of social significance, deviance, and relevance.

What distinguishes the two factors is the comparison between the remaining events, 26% of which are significant but not deviant and 4% of which are deviant but not significant, showing that social significance carries more weight than deviance in the reconstruction of collective memory of lifetime events among Chinese audiences.

Personal Relevance Personal relevance refers to events that bear direct implications to an individual's job, family, or life, as explained by a member of the high SES audience group in Beijing. "I'm only interested in information about things I have to deal with [e.g., purchasing a house], nothing else." Across all focus group sessions, the participants spontaneously and invariably offered personal relevance as the most important consideration in their definition of news.

Even for events that are not deviant or significant per se, Chinese audiences often interpret the news from a personal angle. Nine participants mentioned China's entry to the World Trade Organization because, as they explained, it would bring down import tariffs so that they could afford imported cars or other previously expensive products. When learning of a bomb explosion in Shijiangzhuang city, a highly deviant event, audiences in Beijing immediately became concerned about the safety in Beijing. A journalist working for CCTV was very interested in Beijing's bid for hosting the Olympic Games in 2008 because if Beijing succeeded CCTV would have to be relocated.

To some Chinese audiences, there seems to be an inherent conflict between social significance and personal relevance. When asked whether social significance or personal relevance was more important, a 26-year-old female professional worker from Beijing commented that "personal relevance [i.e., related to life and job] is more important to me. … [News] is what I am concerned about. If I am not concerned about [an event or issue], it is then not news, even if it is reported on TV." Another 37-year-old man in the low SES audience group in Beijing noted that "events of national significance are not necessarily news whereas events about ordinary citizens can be news. … [Our media] keep reporting so-called major issues that are not news at all but become silent on many things that are directly related to the fundamental interests of the public."

Comparing People's News Preferences with What's in the Newspaper

We have so far reported media content data and audience data separately. A divergent pattern seems to have risen from the two types of data: while the Chinese media give more weight to social significance, the audiences—some of whom are journalists—prefer deviance and personal relevance. The gatekeeping exercise performed in this study provides data for us to formally test the (dis)similarity between the two sides.

As shown in the first four columns and rows of Table 10.6, the Spearman rank-order correlation coefficients between the four groups of participants vary from .58 to .76 in Beijing and .56 to .75 in Jinhua. All are statistically significant at the .01 level or above. Given the similar size across the coefficients, no subdivision seems to exist among any of the audience groups. In other words, the data suggest a consistent and substantial level of congruence among Chinese audiences, including ordinary residents with high or low SES, professional journalists, and public relations practitioners, in their ranking of which news items are important and unimportant. A closer look at the focus group discussions reveals that the congruence is mainly achieved by the journalists and public relations practitioners who acted more as ordinary audiences than as media professionals. Such a tendency was quite apparent during the focus group sessions in which most of the journalists and public relations workers referred to the media as "them" (i.e., governmental regulators) instead of "us."

TABLE 10.6 Spearman Rank-Order Correlation Coefficients between Newspaper Item Prominence and Focus Group Rankings

	Journalists	Public Relations	High SES Audience	Low SES Audience	Newspaper
Journalists	–	.60[c]	.70[c]	.58[b]	–.20
Public Relations	.56[b]	–	.76[c]	.71[c]	–.31
High SES Audience	.64[c]	.73[c]	–	.70[c]	–.46[a]
Low SES Audience	.71[c]	.75[c]	.63[c]	–	–.37[a]
Newspaper	.01	.13	.04	.12	–

Note: Beijing coefficients are in the upper triangle; Jinhua coefficients in the lower. SES = socio-economic status.
[a]$p < .05$; [b]$p < .01$; [c]$p < .001$.

Despite the high consonance among the media (see the section on news topics) and the high consistency among the audiences as already reported, there seems to be a gap between the two. As shown in the last row of Table 10.6, the coverage prominence of the thirty news items in the *Jinhua Evening Post* is not correlated with the ranking of the same news stories by any of the four groups in Jinhua. Shown in the last column of Table 10.6 is the relationship between the *People's Daily* and all four groups in Beijing. It is also informative to note the ascending order in which the four groups depart from the newspaper: journalists (.20), public relations practitioners (.31), audiences with low SES (.37), and audiences with high SES (.46). Although the first two correlations are not statistically significant, they are quite substantial in size.

To obtain a better understanding of how and why the two sides diverge so much, we compared the ranking of the audiences and the newspapers on an item-by-item basis. In Beijing, members of the four focus groups agreed with the *People's Daily* on the newsworthiness of only two stories. There was slight disagreement on twelve stories moderate disagreement in seven stories and high disagreement on nine stories. Likewise, participants from the four focus groups in Jinhua ranked three stories the same as the *Jinhua Evening Post* did, but differed somewhat on seven stories and differed a lot on another eight stories.

It is perhaps more revealing to look at the "extreme" items ranked as most prominent by one party but least prominent by another. The most prominent items in the *People's Daily*, all of which appeared on the front page, included (1) a story with a size of 353 cm² on an education commissioner in Jiaxing who held a high moral standard, (2) a story sized 633 cm² on the Chunland Corporation's rapid growth, and (3) a story sized 524 cm² on President Jiang Zemin's attendance at a Peking opera performance. The four groups of participants in Beijing ranked these stories as being among the least newsworthy.

In contrast, the least prominent stories in the *People's Daily* included: (1) an article with a size of 37 cm² on U.S. Defense Secretary Cohen's speech on sanctions against Iraq, (2) a story with a size of 25 cm² on the State Quality Control Bureau's ban on sales of fake cotton, and (3) a story with a size of 48 cm² on ten senior citizens 100 years old winning the title of "Healthiest Men of the Century." All three of these stories were ranked as more newsworthy by people in the focus groups. The Iraq and fake cotton stories were ranked as being most newsworthy by many people and even the senior citizen story was ranked as more newsworthy than the *People's Daily's* most prominent stories.

The discrepancy between the *Jinhua Evening Post* and the audiences in Jinhua is less drastic but still noticeable. For example, one of the stories most prominent in the *Post*—on the stock market in Shanghai with a size of 317 cm² on the front page—was ranked as only seventh in newsworthiness by people in the focus groups. On the other hand, two of the least prominent news items, one about road construction (25cm² on an inside page) and another (a letter to the editor) about electric bills (17 cm² on an inside page) were ranked in the focus groups as near the top in newsworthiness.

The item-by-item comparisons confirm the patterns uncovered in previous sections. On the one hand, Chinese media devote a lot of space or lengthy air-time and prominent positions to significant individuals (e.g., the president or Communist heroes) or events (e.g., a fast-growing company or stock market) to showcase achievements or ideological values. On the other hand, members of the audience rank events that are deviant (e.g., international conflicts) or relevant (e.g., sales of fake cotton, road construction, and electric bills) as being much more newsworthy.

Discussion

What is meant exactly by the term *news* in China? The current study reveals two operational versions: an official version observable from media content, and a public version detectable through focus groups and other methods. The two versions diverge considerably. On the media side, there are consistent and pervasive routines that emphasize positive (as opposed to negative), normal (as opposed to deviant), and regular (as opposed to significant) political and economic events. In the words of a participant of our focus groups, "The news on our media only showcases peace and prosperity." Such an approach to news is highly consistent with the structure, functions, and tradition of the media system in the country. Of course, there are variations across the media, such as the focus on soft news by the local *Jinhua Evening Post* newspaper, and the shortage of foreign news on the local television station Jinhua TV. The variations reflect in part the impact of recent commercialization within the media sector and in part the concerted division of labor among different types of media under the central planning system.

There also appears to be a congruent definition of news among Chinese audiences. In operational terms, the public version of news underscores truth, timeliness, and, probably most of all, personal relevance, which is—as documented in the study—directly at odds with the media's definition of news. To borrow the language from media agenda-setting research, the current study shows that the government is successful in setting the agenda for the media, but the media are not effective in setting the agenda for the public.

The existence of two detached definitions of news is a typical feature of socialist societies where public ideology and private opinion coexist (see, for example, Shlapentokh 1986, 1989). A full exploration of the latter goes beyond the scope of the study, but a participant in our focus groups helps to explain why prominence in the media does not necessarily translate to salience among the audiences: "We usually don't read front pages, especially the headline stories that are invariably and repetitively about state leaders. ... We are forced to read these but are not interested at all." Another participant revealed the underlying mechanisms such as reading between the lines and boomerang: "When the *People's Daily* ran a story on the unity of ethnic groups in Xingjiang [in the northwest of China] without any context, we knew right away that there must be something wrong over there; when Christmas was coming, we knew that the *People's Daily* was going to publish articles on orphans begging on streets in the U.S."

Despite the differences, the two versions of news in China share one thing in common: News prominence is largely not predicted by deviance and social significance, the two central theoretical concepts of the current study. If deviance and social significance offer limited explanation of news prominence on the Chinese media, then what else does? One of the missing variables here probably is propaganda values. A cursory reading of the Chinese media could easily reveal many stories about official meetings, model individuals, exemplary institutions, and successful accomplishments, with large space or lengthy time and prominent placement. While most of these stories do not score high on deviance and social significance (how politically, culturally, or publicly significant is the attendance of the state president to a Chinese New Year performance?), they do have important missions to serve—to set the national agenda, to correlate public opinion, to confer social status, and to maintain the status quo (Lazarsfeld and Merton 1948). News of that type is in fact highly predictive. A broadcast journalist said in a focus group session that "every day we cover nothing but news on official meetings. All you need to do is to replace the location, time, and people's names of the old news." As another journalist put it, "News is about leaders. The higher the rank, the more lavish the activity looks like, and the more likely it hits the news."

Why are Chinese audiences more interested in things that are true (i.e., consistent with what they have directly experienced or learned from alternative sources) or relevant to their personal lives than things that are deviant and significant? Based on the focus group discussions, it seems that Chinese audiences' preference for trustworthy and relevant news is not necessarily a manifestation of something like "Chineseness" but rather is more a direct response to the dissatisfactory information environment around them. As found elsewhere, audiences tend to approach the media for information instrumental to fulfill their immediate goals (Ball-Rokeach 1985), and their focus on short-term problems is constrained by limited emotional resources (Hilgartner and Bosk 1988). For Chinese audiences who are in the midst of drastic social transformations, trustworthy and personally relevant news is more useful, whereas deviant and significant news is probably something too luxurious to possess at the moment. If that conjecture is plausible, it then follows that as the supplies of news become more reliable and relevant the news taste of Chinese audiences will change to catch up to the standard flavor of deviance and significance.

11
What's News in Germany?

CARSTEN REINEMANN AND MARTIN EICHHOLZ

The German Media System

There are approximately 82 million inhabitants of Germany. The German media system is based upon the *Basic Law* (the German constitution) that guarantees freedom of expression and the press. The regulation of the media falls within the jurisdiction of the sixteen German states, all of which have passed additional laws that ensure the existence of a free and democratic media system. Since the deregulation of broadcasting in the mid-1980s, the German media system has seen immense expansion and differentiation, leading to vastly increased competition (Hans-Bredow-Institut 2000; Kleinsteuber 1997).

The German press is characterized by a strong position of local and regional newspapers. More than 90% of the German subscription papers are local or regional, and their readership and circulation are significantly larger than that of the national papers. About 70% of the German population 14 years and older read a local or regional paper every day; 22% read a (national or regional) tabloid paper; and less than 6% read a national broadsheet (Schulz 1999). Nevertheless, the national dailies and weeklies have a high impact on national politics and other media. Another characteristic feature of the German press is its high degree of economic concentration, with the leading five publishers owning a market share of more than 40%. As a consequence, about 40% of the German population has access to only one daily paper with local information (Kleinsteuber 1997; Reinemann 2003; Roeper 2000; Schuetz 2000).

Germany also has very competitive television and radio markets, which are largely decentralized due to the individual states' responsibility for media laws and the licensing of new broadcasting companies. Nearly two-thirds of the German population 14 and older watch television on an average weekday, and about 80% listen to a radio program. Since its deregulation, broadcasting in Germany has consisted of a mixed system of commercial and public service networks and stations. Today, public and private broadcasting entities possess about equal shares of the television and radio markets. Public service channels are required by law to present integrated, comprehensive, and politically balanced content. Programming requirements for the private media are less strict, and political balance must be achieved across the entire private media spectrum rather than within each individual broadcaster.

The market share of the leading national television channels has not exceeded 15% in recent years, mostly because the average German household can receive about thirty free television channels. Most channels are available nationwide or statewide; those on the state level are usually public broadcasters. The few local television stations have miniscule audiences and are of very minor importance within the media system. Radio in Germany is of a predominantly regional nature, and the average household can receive about thirty different stations. Most of the statewide public radio broadcasting companies offer several stations for different audiences, while private broadcasters usually run only one station that reaches local audiences in some states and statewide audiences in others (Darschin and Kayser 2001; Hans-Bredow-Institut 2000; Kleinsteuber 1997; Klingler and Mueller 2001).

Besides the classical media, online media have become more and more important in Germany. Since 2000, the vast majority of German newspapers, radio stations, and television channels have had an Internet presence, and several radio stations are available online. The online audience is considerable; about 30% of the German population 14 years and older had Internet access in 2000, and half of this online audience used the Internet for news at least once a week (Neuberger 2000; Eimeren and Gerhard 2000).

News is a very important part of German media content. For example, national news can be found in national and regional newspapers, on national television channels and regional radio stations, and on the Internet. Most of the five leading national television networks broadcast news throughout the day and a 15- to 30-minute main news bulletin in the evening. In addition, there are two 24-hour television news channels.

Radio stations usually offer brief 2 1/2–minute newscasts on the hour (Schoenbach and Goertz 1995), and most public broadcasters offer at least one 24-hour news station (called Info-Radio). Television and radio news on commercial and public service media entities differ in their thematic structures: The relative amount of information provided is greater on public television channels, and public radio stations have more political and hard news than their commercial counterparts (Schoenbach and Goertz 1995).

An average German daily newspaper is thirty-eight pages long, including twenty-six pages of editorial content. Local information has the biggest share with an average of seven pages, followed by four pages of sports (Schoenbach 1997). More than 80% of newspaper readers read local news every day, 60% national news, 40% foreign news, and 30% the economic and cultural sections (Schulz 1999). Content analyses have shown that an event's proximity, negativity, reference to elite people or nations, and political partisanship are important news values for German journalists (Eilders 1997). International comparative studies have shown that the journalistic culture in Germany is unique with respect to role perceptions, editorial control, and other aspects. Research shows that, whereas a plurality of German journalists value their role of promoting ideas and values, only one in five U.S. journalists in that study gave the same reply (Donsbach 1999; Patterson 1998; Pfetsch 2001). On the other hand, there is some recent evidence that German newsroom practices are becoming more similar to those of other Western countries. This is in part traced back to the increasing competition that reduces the relevance of political motives in journalism (Donsbach 1999).

The Study Sample

Berlin and Mainz were the cities chosen for the study. Berlin is Germany's capital and is outstanding in many ways. By far the country's largest city—with 3.3 million residents—it boasts a unique cultural scene and ethnic diversity and the nation's most competitive media market, including two national and three local television channels, ten daily newspapers, and more than twenty local radio stations. The Berlin newspaper selected for the content analysis was the *Berliner Zeitung*, which is published daily except Sunday. In 1999, it had a circulation of 250,000, the highest of all subscription papers in Berlin. Although a regional paper, the *Berliner Zeitung* has become a paper of national importance since the German government moved to Berlin in 1999. The television newscast analyzed was the daily 30-minute *Abendschau*, the main news program of the B1 regional public service channel, aired at 7:30 p.m., which has on average 300,000 viewers and a market share of 30%. Finally, the radio newscast in the analysis was the 9:00 a.m. edition of the commercial Berliner Rundfunk, the most successful Berlin radio station in 1999, with 500,000 listeners and a market share of 15%.

Mainz, with 200,000 residents, is the capital of the state of Rhineland-Palatinate, the sixth largest German state, with a population of about 4 million. The media market in Mainz is far less competitive than in Berlin, but the city is known as the home of two national television channels (ZDF and 3sat). In addition, Mainz has a regional public television channel and radio station and two local newspapers. The newspaper in the analysis was the *Allgemeine Zeitung Mainz*, the local market leader. It too is published daily except on Sunday, with a circulation of about 70,000 in the Mainz region. The television newscast analyzed was the *Rheinland-Pfalz aktuell*, the main daily news edition of the public channel SWR and the only regional television newscast of importance. It is aired from 7:45 to 8:00 p.m. and has approximately 170,000 viewers, representing a market share of 14%. Finally, the radio program selected was the 9 a.m. morning news on the private Radio RPR station, the most successful radio station in Rhineland-Palatinate, with 900,000 listeners and a market share of 30%.

The six German news outlets yielded a total of 3,933 news items during the study period. The two newspapers account for the vast majority of news items (1,929 in the *Berliner Zeitung* and 1,640 in the *Allgemeine Zeitung of*

Mainz), followed by television news items (154 in *Abendschau* and 95 in *Rheinland-Pfalz aktuell*) and radio items (70 on Radio RPR and 45 on *Berliner Rundfunk*).

Topics in the News

Which topics did the German news media deal with? Overall, 15% of all the items dealt with internal politics, the subject the media most frequently covered, usually by reporting on statements or activities of individual politicians and judicial or executive decisions. Following internal politics, the three most prevalent topics were business/commerce/industry (12%), cultural events (12%), and sports (11%). Two other topics, internal order (e.g., crime, police) and human interest stories accounted for an additional 6% each. Due to the considerable demographic, economic, and cultural differences between Berlin and Mainz and the various differences among newspaper, television, and radio reporting, differences were expected in the distributions of topics in the two cities and the different media. Table 11.1 presents the news topics by medium and by city. Several points emerge from the table.

First, the range of topics is more diverse among newspaper items than among television or radio items. Based on the more comprehensive and detailed list of topics (which was later condensed into broader categories), newspaper items made reference to 213 different topics, compared to 126 covered by television items and only seventy-four covered by radio items. This finding is not only due to the fact that newspapers can carry more information than the other two media but also based on the nature of television, and especially radio, newscasts being more structured in their topical composition.

Second, the share and relative importance of the topics vary considerably among the media. For example, internal politics was the most prominent topic in four of the six media, having a share of about 20% in the radio news and the Mainz television newscast and a share of 16% in the Berlin newspaper. This does not mean, however, that radio news is especially "political" in Germany. Many of the "internal politics" news items on radio deal with judicial matters, meaning that these events deal first and foremost with crime. The fact that internal politics is only the second most prominent topic in the Berlin television newscast (14%) and that it is only in third place in the Mainz newspaper (12%) should thus not be overemphasized. Interestingly though, the topic most intensively covered in the Mainz newspaper is sports (15%), which comes second in the Berlin radio news (16%). Sports are not as prominent in the other media.

The importance of other topics also differs among the media. The share of business/commerce/industry, for example, varies between 7% in the Berlin radio news and 14% in the Berlin newspaper. Cultural events are quite prominent in the German newspapers (14% in Berlin and 11% in Mainz) and in television news (8% in both cities) but are nearly nonexistent in radio news (1% in both cities). This is mostly due to the higher frequency of listings and

TABLE 11.1 Distribution of General Topics of News Items by City and Medium

Topics	Newspaper		Television		Radio	
	Berlin	Mainz	Berlin	Mainz	Berlin	Mainz
Internal Politics	16.4	11.8	14.2	19.9	19.8	20.5
Business/Commerce/ Industry	14.9	9.4	11.2	12.2	7.2	10.9
Cultural Events	11.2	14.1	8.3	8.1	.9	1.3
Sports	8.5	15.3	3.5	9.0	16.2	5.1
Human Interest Stories	5.3	6.7	6.2	3.6	3.6	1.3
Internal Order	5.3	5.6	16.5	6.8	15.3	14.7
Communication	3.9	2.2	3.2	1.8	.9	3.2
Economy	3.9	2.7	3.8	4.1	.9	1.3
International Politics	3.6	1.9	1.8	1.4	2.7	3.2
Transportation	3.3	3.6	5.3	1.4	1.8	8.3
Ceremonies	3.2	6.1	2.7	.9	3.6	2.6
Disasters/Accidents/ Epidemics	3.0	2.3	4.4	6.3	6.3	10.3
Social Relations	2.9	3.2	2.4	3.2	1.8	1.3
Education	2.5	1.7	2.1	1.8	1.8	.6
Housing	2.5	2.1	5.6	3.6	2.7	.6
Health/Welfare/ Social Services	1.9	3.2	2.9	5.9	1.8	1.9
Science/Technology	1.6	1.5	.6	2.7	1.8	1.3
Entertainment	.9	2.1	.0	.0	.9	.0
Other	.9	.4	.0	.0	.0	.0
Environment	.8	1.1	1.2	.0	.0	.6
Military and Defense	.8	.4	.6	.0	2.7	1.3
Weather	.7	.7	2.1	5.0	6.3	9.0
Energy	.6	.5	.9	.5	.0	.6
Labor Relations and Trade Unions	.5	.6	.0	.9	.0	.0
Population	.5	.7	.6	.9	.9	.0
Fashion/Beauty	.2	.2	.0	.0	.0	.0
Total[a]	100.0	100.0	100.0	100.0	100.0	100.0
	($n = 3864$)	($n = 3379$)	($n = 339$)	($n = 221$)	($n = 111$)	($n = 156$)

Note: Distributions given in percent.

[a]Total percentage may not actually be 100.0 due to rounding error.

brief announcements concerning cultural events in local newspaper sections and on public television news. In contrast, internal order is much more commonly addressed on radio (15% in both cities) and on television (17% in Berlin and 7% in Mainz) than in the newspapers (5% in Berlin and 6% in Mainz). This makes internal order the most intensively covered subject in the Berlin television newscast, the second most prominent on the Mainz radio station, and the third most prominent on Berlin radio. Finally, disasters/accidents/epidemics are among the five most frequently covered topics in both radio stations (6% in Berlin and 10% in Mainz), as are reports on the weather (6% in Berlin and 9% in Mainz).

Despite these differences, the statistically significant Spearman rank-order correlations in Table 11.2 indicate that the topical structure of the media in both cities is quite similar. A careful analysis reveals that the topical structures are more similar between the similar media types than among the media of the same city. In both newspapers, for example, internal politics, business/commerce/industry, cultural events, sports, and human interest stories are among the five topics most often covered. Accordingly, the Spearman rank correlation between the two newspapers is very high (.90). The topical structures of the television and radio newscasts are also very similar, although their correspondence is not as high as for the newspapers (for television, .83 and for radio, .81). On the other hand, the average Spearman correlation coefficient for the topical structure in Mainz was only .75 and .71 for the Berlin media. This means that the specific ways of news making in different types of media have a slightly greater effect on the topical structure of news than the common circumstances of the city in which the news medium is located.

TABLE 11.2 Spearman Rank-Order Correlation Coefficients between Rankings of News Topics in Various Media

	Berlin Newspaper	Mainz Newspaper	Berlin Television	Mainz Television	Berlin Radio	Mainz Radio
Berlin Newspaper		.90[c]	.85[c]	.75[c]	.62[b]	.71[c]
Mainz Newspaper			.84[c]	.77[c]	.62[b]	.65[c]
Berlin Television				.83[c]	.66[c]	.74[c]
Mainz Television					.70[c]	.73[c]
Berlin Radio						.81[c]
Mainz Radio						

[a]$p < .05$; [b]$p < .01$; [c]$p < .001$.

Thus, some of the differences seem to reflect general trends in news-making among different types of media. For example, journalists working for television and radio generally must select fewer news items and must present them in shorter formats than their newspaper colleagues. These hurried news-making conditions lead to a focus on outstanding and unambiguous events that can easily be determined as newsworthy. The present data suggest that this is especially true for news about internal order and disasters, which is quite often presented by most of the electronic media. In addition, some topical differences between radio and television might be rooted in organizational structure and ownership (both sampled television newscasts are produced by public channels, while the two radio stations are both privately owned) and the neglect of cultural news might be indicative of the character of private radio stations in Germany. The differences in the topical structure of the newspapers in Mainz and Berlin might be explained differently in that the Berlin newspaper reflects its geographic proximity to the center of German politics and its character as a quality paper. Thus, since the mid-1990s, the *Berliner Zeitung* has tried to compete with the large national broadsheets and has gained an excellent reputation for the quality of its political, economic, and cultural reporting (Reinemann 2003). Therefore, the *Berliner Zeitung* allocates more space to political and economic stories. In addition, the great importance of sports in the *Allgemeine Zeitung* reflects a characteristic feature of most German regional newspapers given the fact that local and regional sports are very important for readers.

Deviance in the News

The basic assumption in the study concerning deviance is that deviant news items are longer and more prominently placed compared with news items that are not deviant. In order to test this hypothesis, average prominence scores were calculated for news in the different media for each level of statistical deviance, social change deviance, and normative deviance. In addition, one-way analyses of variance (ANOVAs) were computed in order to test the statistical significance of the differences between the prominence scores. Table 11.3 presents the mean scores with those statistically significant based on the analyses of variance.

In eighteen of the thirty analyses conducted, significant differences between the levels of deviance were obtained. This does not necessarily mean, however, that all the significant differences indicate a linear relationship between deviance and prominence. Sometimes prominence scores were higher for a lower level of deviance than for a higher level. Looking at the two extreme levels of deviance among the significant ANOVAs, it was found that the lowest level of deviance corresponded with the lowest prominence scores in fourteen out of eighteen cases and that the highest level of deviance corresponded with the highest prominence scores in nine out of eighteen cases. This indicates that

TABLE 11.3 Mean Verbal and Visual Prominence Scores for Intensity of Deviance

| | Berlin | | | | | Mainz | | | | |
| | Newspaper | | Television | | Radio | Newspaper | | Television | | Radio |
Intensity of Deviance	Verbal only ($n = 1917$)	Verbal plus Visual ($n = 563$)	Verbal ($n = 154$)	Visual ($n = 154$)	Verbal ($n = 45$)	Verbal only ($n = 1627$)	Verbal plus Visual ($n = 431$)	Verbal ($n = 95$)	Visual ($n = 95$)	Verbal ($n = 70$)
Statistical Deviance										
(1) common	245.2[c]	500.9[b]	125.3	116.0[b]	18.8[c]	198.8[c]	374.5[c]	102.2	103.6[a]	17.4[b]
(2) somewhat unusual	237.2	695.8	138.3	254.6	55.1	183.2	498.3	114.1	250.5	18.1
(3) quite unusual	325.3	570.9	195.5	189.8	74.6	231.7	622.1	126.6	223.9	35.3
(4) extremely unusual	384.7	848.3	178.4	–	69.6	309.1	528.8	157.4	164.2	72.4
Social Change Deviance										
(1) not threatening to status quo	253.0[c]	518.3[b]	126.0	128.7[b]	50.6	204.2[b]	417.4	114.2	118.2	31.6
(2) minimal threat	266.4	783.0	171.0	309.9	74.3	238.4	577.1	117.4	204.0	46.0
(3) moderate threat	323.6	187.6	199.2	133.0	–	264.5	542.7	105.5	–	13.7
(4) major threat	466.3	994.8	256.0	–	89.5	307.9	585.3	205.0	–	45.1

Normative Deviance

(1) does not violate any norms	253.2[a]	512.6[c]	106.0[a]	124.5[b]	42.0[a]	215.3	424.0	112.1	121.0	22.7[a]
(2) minimal violation	334.6	645.5	211.7	273.1	78.1	224.0	385.2	169.5	145.5	55.9
(3) moderate violation	300.6	801.8	125.5	–	58.6	198.0	682.3	127.0	49.0	35.3
(4) major violation	279.7	1093.0	137.4	324.0	72.6	218.4	630.7	72.7	–	39.7

[a] $p < .05$; [b] $p < .01$; [c] $p < .001$.

deviance very often had an effect on the size and/or placement of news items, in line with the hypotheses of the study.

On the other hand, the relevance of deviance differs considerably among the media. Effects of deviance were most often found for the *Berliner Zeitung*: The differences between prominence scores were statistically significant for verbal and visual content on all three dimensions of deviance, and the direction of the influence was generally in correspondence with the assumptions of the study. As for the Berlin and Mainz radio newscasts, two out of three analyses produced significant results. Although there is no linear relationship, statistical and normative deviance do affect the placement and size of news items in the newscasts of both radio stations. Regarding the Berlin television newscast, four of the six comparisons were significant: Prominence scores for the visuals varied systematically along all the dimensions of deviance, while the prominence of verbal content was only affected by the level of normative deviance. Again, there was no linear relationship, but in all cases the lowest levels of prominence corresponded with the lowest levels of deviance.

In contrast, only half of the analyses yielded significant findings for the Mainz newspaper *Allgemeine Zeitung*, but the directions of the significant influences are in correspondence with the assumptions of the study. Thus, statistical deviance had an effect on both the verbal and visual content; social change deviance affected the verbal content, although no significant effect appeared for normative deviance. Finally, in just one of six comparisons, effects of deviance were found for the reporting of the Mainz television newscasts, with the level of statistical deviance being significantly related to the size and placement of the news items.

In sum, of the three dimensions of deviance, statistical deviance was most often related to the mean prominence of the news items. In eight of the ten analyses, the level of statistical deviance influenced the size and placement of news stories according to the hypothesis of the study. Normative deviance had an effect in six of the ten analyses, including a clear difference between the two cities. In Berlin, the prominence of at least one type of content was influenced in all three media. In Mainz, however, the only significant effect was for the radio newscast. This might suggest that journalists in the big city of Berlin tend to give more prominence to normatively deviant news than their colleagues in the smaller, quieter city of Mainz.

Finally, social change deviance had a significant effect in only four of the ten analyses. Interestingly, a pattern fully in line with the assumptions of the study was found for verbal reporting in both newspapers: the higher the level of social change deviance, the more prominent the news items. In addition, there were effects for the verbal plus visual content of the *Berliner Zeitung* and the visual content of the Berlin television newscast. The reason for the more frequent relationships with newspaper content was that the potential shift of power is normally visible or even discussed only in longer or background reports rarely found in the electronic media.

Social Significance in the News

Just as for deviance, the basic hypothesis is that socially significant news items are more prominent in the media than news items that lack social significance. To test this hypothesis, mean prominence scores for the news items in the different media were calculated, as were the forty possible one-way ANOVAs. Of these analyses, twenty-three produced significant effects (see Table 11.4). In these cases, the level of social significance was related to, and was presumed to influence, the average prominence of the news items. Also, as in the case of the deviance measures, the direction of influence was not always strictly linear, although the lowest level of significance corresponded with the lowest prominence scores in seventeen of the cases and the highest level of significance corresponded with the highest prominence scores in nineteen of the cases. It can thus be concluded that social significance very often affects the prominence of news items in accordance with the assumptions of the study, yet the patterns of influence were quite different from those found for deviance.

In contrast to the results obtained in the analysis of deviance, the content of the Mainz newspaper was most often influenced by social significance. In five of the six analyses, statistically significant results were obtained, with only one nonsignificant finding regarding the relationship between verbal plus visual content and political significance. For the Berlin radio newscast, three of the four comparisons yielded significant differences, with only the cultural dimension failing to produce significant findings. Half of the analyses produced significant differences among the prominence scores for the Berlin newspaper, the Berlin television newscast, and the Mainz television newscast. However, each medium presented a specific pattern of influences. Finally, only one of the four analyses provided interpretable data for the Berlin radio station, with only the influence of public significance being in the predicted direction.

What about the four dimensions of social significance? Political and economic significance affected the prominence of news items most often, for instance, in seven of the ten analyses. Thus, the relevance of a news item for political or economic processes and decisions seems to be a major criterion of newsworthiness for most of the German media. Only verbal content in the Berlin television newscast, visual content in the Mainz television newscast, and the reporting of the Mainz radio station were not influenced by political significance. The lessened prominence of politically significant news items in the electronic media is in line with other studies about regional and local news television and radio programs (Brosius and Fahr 1996; Schoenbach and Goertz 1995). Economic significance, on the other hand, neither affected the

TABLE 11.4 Mean Verbal and Visual Prominence Scores for Intensity of Social Significance

Intensity of Social Significance	Berlin					Mainz				
	Newspaper		Television		Radio	Newspaper		Television		Radio
	Verbal only (n = 1917)	Verbal plus Visual (n = 563)	Verbal (n = 154)	Visual (n = 154)	Verbal (n = 45)	Verbal only (n = 1627)	Verbal plus Visual (n = 431)	Verbal (n = 95)	Visual (n = 95)	Verbal (n = 70)
Political Significance										
(1) not significant	261.4[c]	502.5[c]	125.2	123.3[c]	43.7[a]	206.0[b]	414.4	92.2[a]	100.5[a]	24.9
(2) minimal	228.3	595.4	156.4	298.4	71.0	203.5	477.3	120.4	128.0	34.0
(3) moderate	260.7	722.7	198.0	374.0	75.3	246.3	571.9	143.7	227.3	53.5
(4) major	427.3	1009.0	212.4	–	89.5	333.6	567.0	238.2	238.2	47.4
Economic Significance										
(1) not significant	266.9	530.8	126.6[a]	130.2[b]	44.1[b]	204.3[b]	408.5[b]	99.5[a]	101.3[b]	30.6
(2) minimal	304.0	548.2	135.1	297.1	108.0	210.0	441.7	104.6	162.1	47.8
(3) moderate	249.5	549.0	208.1	145.0	74.8	234.1	684.8	161.8	275.8	13.2
(4) major	335.2	356.0	454.7	–	79.1	328.5	602.4	245.0	–	47.8
Cultural Significance										
(1) not significant	219.4[c]	509.9[a]	150.5	140.9	57.2	202.9[b]	430.1[a]	110.7	115.7	33.3
(2) minimal	371.5	606.0	172.6	180.6	30.5	219.4	339.6	136.3	128.5	38.4

(3) moderate	359.6	681.1	99.0	279.3	53.5	234.7	551.1	108.6	82.2	42.5
(4) major	392.5	727.1	194.7	290.0	—	302.0	527.7	202.8	210.3	—

Public Significance

(1) not significant	238.1[c]	518.0	132.0[a]	142.4	47.5[a]	192.3[c]	399.1[c]	93.8	120.0	22.7[b]
(2) minimal	334.1	720.4	131.3	231.1	66.6	243.4	588.0	154.8	—	48.5
(3) moderate	299.7	430.8	227.3	—	72.8	272.4	523.2	159.2	—	58.4
(4) major	442.4	662.0	331.4	—	138.0	366.2	733.3	83.0	—	66.8

[a]$p < .05$; [b]$p < .01$; [c]$p < .001$

verbal and verbal–visual contents of the Berlin newspaper nor, once again, the stories in the Mainz radio newscast.

Public significance affected the size and placement of news items in six of the ten analyses. The only nonsignificant results appeared for the relationship between public significance and the verbal plus visual content of the Berlin newspaper and the Berlin and Mainz television newscasts, as well as for the verbal content of the Mainz television newscast. It should be noted that the findings regarding the visual content of the television newscasts are difficult to interpret because there were no items with visual content of "moderate" or "major" public significance in the Berlin and the Mainz television newscasts and no items with "minimal" public significance in the Mainz television newscasts. Thus, the fact that three of the four nonsignificant results were found for visual content is even more notable and suggests that the public significance of events—perhaps with the exception of wars or disasters—may not be visually presented.

Finally, only three of the ten analyses yielded significant influences of cultural significance on newspaper content—in Mainz regarding both verbal and visual contents and in Berlin only regarding verbal content. What is probably responsible for these effects is the large number of listings of local and regional cultural events that appear in the newspapers on a daily basis, while cultural events play a much smaller role in the electronic media.

Deviance and Social Significance as Predictors of News Prominence

In order to simultaneously test the influence of all dimensions of deviance and social significance on news prominence, eight stepwise regression analyses were calculated. For each of the two cities, one analysis was calculated for verbal newspaper prominence, verbal and visual newspaper prominence, as well as for television and radio prominence.

As can be seen in Table 11.5, all the regression analyses produce significant effects. In several cases, however—especially for newspapers—the amount of the variance explained was rather small, with only 3% of verbal content prominence for the *Allgemeine Zeitung* and 5% for the *Berliner Zeitung*. The predictive power of the models is slightly higher when visual prominence and verbal prominence are considered together, with 8% explained variance in Mainz and 16% in Berlin. All in all, deviance and social significance explain a relatively small part of the prominence of newspaper items. The findings for the electronic media are more promising, with 22% explained variance for television news in Mainz, 24% for radio in Mainz, and 26% for television news in Berlin. Finally, the predictive power of the regression model was much higher for the Berlin radio newscast, with 47% of the variance explained by only two factors: statistical deviance and economic significance.

Comparing the different dimensions of deviance and social significance between verbal and visual content suggests that the selection and placement of visual content follows different rules than the selection of verbal content.

TABLE 11.5 Stepwise Regression Analyses of Intensity of Deviance and Social Significance on News Prominence

	Berlin								Mainz							
	Newspaper Prominence Verbal only Total $R^2 = .05^c$ (n = 1917)		Newspaper Prominence Visual and Verbal Total $R^2 = .16^c$ (n = 551)		Television Prominence Total $R^2 = .26^c$ (n = 154)		Radio Prominence Total $R^2 = .47^c$ (n = 45)		Newspaper Prominence Verbal only Total $R^2 = .03^c$ (n = 1627)		Newspaper Prominence Visual and Verbal Total $R^2 = .08^c$ (n = 418)		Television Prominence Total $R^2 = .22^c$ (n = 95)		Radio Prominence Total $R^2 = .24^c$ (n = 70)	
Independent variables	r	Std. Beta	r	Std. Beta	r	Std. Beta	r	Std. Beta	r	Std. Beta	r	Std. Beta	r	Std. Beta	r	Std. Beta
Deviance																
– Statistical, verbal content	.11[c]	.06[b]	.26[c]	.19[c]	.20[a]	ns	.61[c]	.56[c]	.12[c]	.09[b]	.13[a]	ns	.16	ns	.32[b]	.28[a]
– Statistical, visual content	–	–	.15[b]	ns	.26[b]	ns	–	–	–	–	.23[c]	.19[c]	.38[c]	ns	–	–
– Social change, verbal content	.11[c]	.08[b]	.24[c]	.15[c]	.12	ns	.37[a]	ns	.09[c]	ns	.10[a]	ns	.06	ns	.09	ns
– Social change, visual content	–	–	.12[b]	ns	.31[c]	ns	–	–	–	–	.12[a]	ns	.03	ns	–	–
– Normative, verbal content	.04	ns	.16[c]	ns	–.04	ns	.18	ns	–.01	ns	.04	ns	–.02	ns	.17	ns
– Normative, visual content	–	–	.18[c]	.09[a]	.20[a]	.26[c]	–	–	–	–	.09	ns	–.11	ns	–	–

[a]$p < .05$; [b]$p < .01$; [c]$p < .001$; ns = not part of final stepwise regression equation.

(Continued)

TABLE 11.5 Stepwise Regression Analyses of Intensity of Deviance and Social Significance on News Prominence (*Continued*)

	Berlin								Mainz							
	Newspaper Prominence (Verbal only) Total R^2 = .05[c] (n = 1917)		Newspaper Prominence (Visual and Verbal) Total R^2 = .16[c] (n = 551)		Television Prominence Total R^2 = .26[c] (n = 154)		Radio Prominence Total R^2 = .47[c] (n = 45)		Newspaper Prominence (Verbal only) Total R^2 = .03[c] (n = 1627)		Newspaper Prominence (Visual and Verbal) Total R^2 = .08[c] (n = 418)		Television Prominence Total R^2 = .22[c] (n = 95)		Radio Prominence Total R^2 = .24[c] (n = 70)	
Independent variables	r	Std. Beta	r	Std. Beta	r	Std. Beta	r	Std. Beta	r	Std. Beta	r	Std. Beta	r	Std. Beta	r	Std. Beta
Social Significance																
– Political, verbal content	.09[c]	ns	.19[c]	ns	.13	ns	.34[a]	ns	.09[b]	ns	.09	ns	.07	ns	.24	ns
– Political, visual content	–	–	.20[c]	.11[b]	.37[c]	.27[c]	–	–	–	–	.13[b]	ns	.15	.27[b]	–	–
– Economic, verbal content	.02	ns	.05	ns	.25[b]	.19[a]	.31[a]	.27[a]	.08[b]	ns	.14[b]	ns	.20	ns	.08	ns
– Economic, visual content	–	–	–.02	ns	.32[c]	ns	–	–	–	–	.16[b]	ns	.35[c]	.34[c]	–	–
– Cultural, verbal content	.16[c]	.15[c]	.19[c]	.18[c]	.09	–.16[a]	–.17	ns	.08[b]	.07[b]	.00	ns	.28[b]	ns	.09	ns
– Cultural, visual content	–	–	.13[b]	ns	.19[a]	ns	–	–	–	–	.06	ns	.21[a]	ns	–	–
– Public, verbal content	.12[c]	.08[b]	.21[c]	ns	.22[b]	.24[b]	.41[b]	ns	.14[c]	.12[c]	.13[b]	ns	.18	ns	.35[b]	.31[b]
– Public, visual content	–	–	.08	ns	.14	ns	–	–	–	–	.20[c]	.16[b]	ns	ns	–	–

[a] $p < .05$; [b] $p < .01$; [c] $p < .001$; ns = not part of final stepwise regression equation.

Thus, verbal content was significantly influenced by statistical deviance in five of the six cases. In fact, one of these analyses produced the largest effect of all, with a standardized beta weight of .56 on the prominence of news items in the Berlin radio newscast.

In half of the analyses, public and cultural significance produced statistically significant effects. Public significance influenced the prominence of verbal newspaper content in the Mainz and Berlin newspapers as well as for Berlin television and Mainz radio news. Cultural significance positively affected verbal and visual–verbal prominence in the Berlin newspaper and the prominence of verbal content in the Mainz newspaper. This finding is in line with the ANOVAs reported in the previous section, namely that the large listings of cultural events were mainly responsible for this effect. Only these listings were found to have high cultural significance. In contrast, cultural significance had a negative effect on the prominence of verbal content in the Berlin television newscast (–.16). This means that an item's prominence was reduced when it was culturally significant. This result can be explained by the fact that culturally relevant items normally appear at the end of a television newscast, if at all.

Only two statistically significant influences were found for economic significance and social change deviance. Economic significance affected the prominence of news items in the Berlin radio and television news. Social change deviance influenced the size and placement of verbal content only in the *Berliner Zeitung*. This finding comes as no surprise when looking more closely at the reporting of the *Berliner Zeitung*. Journalists there tend to follow a model of quality journalism much more closely than their colleagues in the other media investigated. The newspaper's editorial staff provides much more background information on political processes and social problems. Finally, normative deviance and political significance had no significant effect on news prominence. This suggests that on the verbal level, statistical deviance and cultural and public significance often influenced news prominence, social change deviance and economic significance had some effect on some of the media, while normative deviance and political significance had no effect at all.

For visual content, the findings were different. Here the political dimension of social significance had the greatest effect on news prominence, being significant for three of the four analyses. Only the combination of visual and verbal content in the Mainz newspaper was not influenced by political significance. Statistical deviance in the Mainz newspaper, economic significance on Mainz television, and public significance in the Mainz newspaper affected prominence in one medium each. Finally, social change deviance and cultural significance had no significant effect at all on visuals. This suggests that events with political significance—in which politicians can be seen—have the best chance of being covered with large and prominently placed photos or long video sequences at the beginnings of newscasts. Given the typical structure of newspaper photos and the importance of talking heads in television newscasts, this finding makes sense intuitively.

People Defining News

The eight German focus groups were conducted in April 2000. Following is a summary of the main findings of the discussions as they relate to the key constructs of the study.

Deviance

Many German focus group participants across the professions and cities referred to the construct of deviance and its various dimensions when describing and explaining their interest in news. Among the concepts related to deviance, statistical deviance and being unusual were most popular. For example, one Mainz high socioeconomic status (SES) participant argued that what is newsworthy is something that "you have never seen before, never heard before, it is especially horrible, especially far, especially high, or especially sad." Many Mainz and Berlin journalists also made statements referring to deviance, and some Berlin public relations practitioners mentioned specific examples of how deviance is used commercially to attract consumers, such as a very popular German television show in which common people can win money by doing extraordinary things.

Some difficulties in the definition of deviance also came up: mainly the fact that deviance can fade over time. Participants described this fading process through examples of desensitization, whereby news consumers become accustomed to a certain type of deviant story. Many of the participants also linked deviance to news valence, often using negative news items as examples for deviance. However, when probed, many participants revised their position to say that positive news could be just as deviant as negative news, thereby supporting the theoretical notion that deviance is a bidirectional construct in terms of news valence.

When asked for the reasons behind their interest in deviant news, several respondents brought up biological evolution and related issues. The most basic reference spoke of the human need to monitor the environment. Interestingly, low SES participants were somewhat more likely to make such remarks. A second type of comment referred to the inherent ability of deviant news to demand attention. Some focus group participants felt helpless when trying to explain the attraction of deviance, commenting that they are forced to look at deviant news. As one female high SES participant in Berlin said, "Imagine a headline reading '1000 dead bodies discovered in mass grave.' You can't ignore such a message; it just draws you in." A third group of remarks spoke directly to the concepts of survival and biological evolution, and some participants also connected the negativity of news and its relevance for survival. One female public relations practitioner described the influence of biological evolution on news interest in near perfect congruence with the theory that "you only survive if you inform yourself, are curious, and look for reasons behind things so that you can develop your own ideas and strategies to survive. ... How can I get through this life? ... Anything is possible if you leave the cave."

Social Significance

The construct of social significance did not play a major role in the German focus groups and was hardly ever mentioned. Although the participants often used news stories of high social significance as examples to illustrate their definitions of newsworthiness, they did not discuss the concept itself in any depth. In Mainz the high SES participants discussed whether people have a responsibility to be informed and to follow the news. It started with one participant pointing out that he has ignored mass media news since 1989. As a result, several other participants immediately accused him of not fulfilling his civic duty of being informed. Said one male participant, "There is a responsibility to be informed. ... Democracy is not a gift; we had to fight for it, and this forces us to make sure that what happened at the beginning of the twentieth century won't happen again."

News Valence

The concept of news valence accounted for a significant amount of conversation and sometimes led to heated discussions among the participants. However, the news consumers often agreed that the concept is difficult if not impossible to define. The German journalists and public relations practitioners were less likely than the German news consumers to regard the definition of news valence as problematic. Overall, most of the participants were convinced that people are more interested in negative than in positive news. Interestingly, many journalists in Mainz became defensive when asked whether people prefer bad news or good news. It appeared that these news people felt accused and perceived a need for justifying their profession. Some Mainz journalists argued that they do indeed publish a great amount of positive news. Others maintained that they do not report good news because the audience is not interested in it or because their reporting would not have any effect.

The participants mentioned a variety of reasons for people's heightened interest in negative news. On the most basic level, some participants simply referred to human nature. However, more prominent were explanations that described how negative news could make the news consumer feel better and also could evoke empathy for the victims portrayed in such news stories. But negative news stories do not only make people feel better; they can also be newsworthy because they evoke empathy among news consumers. Said a male Berlin television journalist, "Negative news stories usually cause greater empathy because people are mentally taking part in what is going on. This is especially the case when a story is about the fate of a specific person." Another string of explanatory comments pointed at the mass media. One Mainz high SES participant claimed that the news diet provided by the traditional mass media is responsible for peoples' interest in negative news. Finally, several high SES participants in Mainz described concepts that have the power to override or intervene with news valence as a determinant of newsworthiness. For example, some participants argued that a news item's topic or individual factors such as mood or stage of life are more important than news valence in determining their news interest.

Sociocommunicative Function of News

Particularly among the German low SES participants, another issue came up, namely how following the news provides them with topics for daily conversations with friends and colleagues. This sociocommunicative function of news appeared to be among the key drivers of news interest among low SES participants, and participants assigned it the potential to significantly increase or decrease one's social standing. Furthermore, it appeared that the "un-informed" are often excluded from basic communication processes and that the threat of this stigma exerts enormous social pressure. "You need to know about the news so you can talk to other people and don't appear to be living on a desert island." Several low SES participants illustrated the social benefit of knowing the news of the day with the increased attention that a storyteller receives.

At the end of each group's general discussion, the participants were asked about the three most important news events in their lives. A wide array of news stories came up, ranging from well known international events to obscure regional or local occurrences. Nearly two-thirds of these events were negative in nature. The largest group of events addressed wars, crimes, assinations, major political or cultural events, catastrophes, accidents, and scientific achievements. Based on the participants' explanations, several key factors emerged that could explain why these particular events were remembered. First and foremost was the fear and threat associated with remembering certain news events (such as the Gulf War, the Balkan crisis, and the Chernobyl nuclear disaster). In contrast, happiness and joy could serve to remember events (such as the fall of the Berlin Wall). A third major factor was personal impact or involvement. Finally, many news events were mentioned because they seemed to trigger feelings of empathy.

Comparing People's News Preferences with What's in the Newspaper

One of the hypotheses in the study is that there would be a positive relationship between the rankings of newsworthiness among journalists, public relations practitioners, and audience members. It was also hypothesized that there would be a positive relationship between the rankings of the three groups of people and the actual prominence of the newspaper items. These two hypotheses were tested by means of the gatekeeping task described earlier in the book. Using Spearman rank-order correlation coefficients, it was possible to determine the relationships between the actual rankings of the prominence of the newspaper items and the subjective rankings of the three groups of people. Table 11.6 presents the twenty Spearman rank-order correlation coefficients for the two cities. A cursory examination revealed that nineteen of the twenty correlations were statistically significant at various levels ranging from .05 to .001. Thus, these findings suggest that people in Germany have similar definitions of news across gender, age, profession, and socioeconomic factors, yet a closer inspection revealed some interesting differences.

Among the Mainz groups, eight of the ten group intercorrelations were significant. The strongest relationships exist between the public relations practitioners and the journalists as well as between the public relations practitioners and the high SES audience participants (in both cases, .85). The relationship between journalists and high SES participants was also quite significant. The similarity in the news definitions between the journalists and the PR practitioners group comes as no surprise given the two groups' comparable professional backgrounds and the common phenomenon of moving from the journalism profession to the public relations profession and back. The strong similarities between high SES audience participants and public relations practitioners and journalists, respectively, are more surprising but could be related to these groups' similar socioeconomic backgrounds. The outlier among the Mainz participants was the low SES audience group, with correlations averaging .59 with the other groups.

In Berlin, all correlations were significant. The strongest relationships were between the public relations practitioners and both groups of citizens (high

TABLE 11.6 Spearman Rank-Order Correlation Coefficients between Newspaper Item Prominence and Focus Group Rankings

	Journalists	Public Relations	High SES Audience	Low SES Audience	Newspaper
Journalists	–	.74[c]	.78[c]	.82[c]	.48[b]
Public Relations	.85[c]	–	.91[c]	.87[c]	.49[b]
High SES Audience	.74[c]	.85[c]	–	.86[c]	.47[b]
Low SES Audience	.52[b]	.59[b]	.67[c]	–	.48[b]
Newspaper	.58[b]	.38[a]	.35	.38[a]	–

Note: Berlin coefficients are in the upper triangle; Mainz coefficients in the lower. SES = socioeconomic status.
[a]$p < .05$; [b]$p < .01$; [c]$p < .001$.

and low SES). The other coefficients were also high, however, so the low SES audience group was not an outlier in Berlin.

The data relating to the relationships between the actual media coverage (in the newspapers) and the various groups of people yielded seven significant correlations among the eight correlations calculated. And yet while there was clearly a relationship between the definition of newsworthiness by the group members and the actual prominence of the newspaper items, these correlations were generally lower than those among the various focus groups.

In Mainz, the relationship with the actual coverage was the highest for the journalists' rankings. Since journalists were the actual producers of the local media content presented, these findings make intuitive sense. The correlations with the other focus groups' rankings were much lower, with a mean rho of only .37. Moreover, the only nonsignificant coefficient was between the high SES audience and the actual newspaper coverage. In Berlin, on the other hand, all four correlations including the one for the journalists ranged from .47 to .49. All in all, the implicit definitions of newsworthiness of various news stories provided by the focus group participants are more similar to each other than with the actual prominence given to these stories in the local newspapers.

Discussion

The topical structure found in German newspapers as well as in radio and television news seems to be quite representative of German media in general. In addition, the topical structures of the media are very similar. This becomes evident in the high and statistically significant rank-order correlations. However, the topical structures are more similar between media of the same type than among media located in the same city.

The differences among the three types of media are clearly visible. German newspapers give a more comprehensive picture of current events and have much more culturally relevant news than the electronic media. The five topics most often covered are internal politics, business/commerce/industry, cultural events, sports, and human interest stories. The five topics most often reported about on German regional television are nearly the same, but internal order replaces sports in Berlin and human interest stories in Mainz. On both radio stations, internal order is also one of the five most important topics. In addition, disasters/accidents/epidemics and weather appear as very important topics, replacing cultural events and human interest stories in Berlin and also sports in Mainz. These findings correspond with other analyses of German news media that show the electronic media allocating more editorial time and space to negative and sensational stories.

With respect to the analyses of variance, the influence of deviance and social significance were statistically significant in just over half of the tests. Overall, the direction of these effects was generally in line with the hypotheses of the study. On the other hand, the patterns of influence were different among the media. This supports the assumption that criteria of news selection and presentation are specific for different types of media and even for single media outlets. These medium-specific criteria limit the influence of biologically and culturally derived concepts. In addition, the individual dimensions of deviance and significance vary in their importance. Most often, statistical deviance and political and economic significance affected the average prominence scores in the ANOVAs. Public significance and normative deviance were significant factors nearly as often, but social change deviance and cultural significance only affected the average prominence scores half as often or less than statistical deviance. Thus, the dimension most deeply rooted in biological evolution has an effect on news prominence most often.

This important finding is supported in the multivariate regression analysis. In nearly all of the media, verbal news content was significantly influenced by statistical deviance. In half of the analyses, cultural and public significance were of importance for verbal news prominence. In contrast, economic significance and social change deviance affected verbal news prominence only seldom. Political significance and normative deviance—which produced significant effects in the bivariate analyses—provide proof that there was no genuine influence on verbal prominence when all other dimensions of deviance and significance were considered. This picture changes when looking at visual news content, where political significance most often affects prominence.

The multivariate analyses also show that the predictive power of deviance and social significance differs among the media. While the amount of variance explained is rather small for the newspapers, there is a better fit for the electronic media. Statistical deviance and economic significance explain nearly 50% of the variance for the Berlin radio news. This makes clear that the explanation of news selection is much more difficult for newspapers than for electronic media. Keeping in mind the great variety of content in regional newspapers, ranging from hard political news to the listing of cultural events, regional sports, comic strips, and local beat reporting, this finding is understandable. Thus, attempting to validate a general theory of news selection for the entire spectrum of news content is clearly a very difficult thing to do. Perhaps better findings for newspapers would emerge if their content was divided by topics or news sections. Since different departments that might have special criteria of news selection produce different sections of newspapers, this might be a promising idea for future analysis and research.

Some of the basic concepts forming the core of the study were well represented in the minds of the German people producing and consuming news. This is especially true for statistical deviance. The fact that statistical deviance was even perceived to be connected to the concept of biological evolution by some of the focus group members underlines the importance of this factor, as shown in the content analysis. On the other hand, significance as a concept was not overly present from the start, possibly because the cultural dependence of ordinary everyday activities, such as consuming news, is very difficult to recognize without comparing it within different cultural contexts. Finally, the gatekeeping task suggests that there was a basic consensus among the German focus group participants about the newsworthiness of news items. The correlations between groups of people were usually higher than those between people and their newspapers, pointing once again to the fact that media news selection is not only influenced by biological and cultural factors but also by numerous other factors at the institutional and personal levels.

What can be concluded from the German part of the study? Dimensions of deviance or social significance do affect the prominence of news items in all the German media investigated. Moreover, these concepts are relevant for both German news producers and news consumers. The predictive power of deviance and social significance is much higher, however, for the electronic media than for newspapers. In addition, the patterns of influences do seem to depend to a certain degree on the specific characteristics of the single media outlets. But in spite of the restrictions and problems mentioned, the German part of the study supports the view that deviance and social significance are valuable concepts for the investigation of news-making.

12
What's News in India?

KAVITA KARAN

The Media System in India

Geographically, India is the seventh largest country in the world, with a population of over 1 billion, comprising 16% of the world's population. Only China has a larger population. The literacy level is comparatively low at 65%, with men being more literate than women. India has a parliamentary system of government and takes pride in its unity amidst diversity. There is a large urban–rural divide, with 70% of the population living in rural and semi-urban areas. This diversity is manifest in the multiplicity of religions, castes, and languages. India presents a picture of interesting coexistence: high technology coexists with primitive technology, affluence with poverty, development with underdevelopment, very high literacy with illiteracy; and so on.

Low levels of literacy and high poverty are major barriers to uniform modes of communication. The most important is language: the country's twenty-six states and six union territories share eighteen scheduled languages and 844 dialects. Hindi is the official language and is spoken by about 40% of the population. English is generally used in combination with either Hindi or with the state's official language in government, business, and commercial transactions.

The print medium developed through private initiative and has operated in a virtually control-free environment for more than fifty years in post-independent India, except for a short while in the mid-1970s during a national emergency (Mankekar 1978). Newspapers and magazines have consistently provided readers with top-class news and editorial comment both in English and regional languages. Readership of newspapers had increased consistently, with about 85 million newspaper readers in the country when last reported (NRS 2002a). Early newspapers started by Indians alongside the British in the early nineteenth century were aimed principally at arousing awareness and involving the masses in the freedom struggle. After independence, the press actively reported India's march toward political and economic progress and freely criticized people, programs, and policies that came in the way of progress of the nation (Karan and Mathur 2003).

189

A consistent increase has occurred in the number of publications every year (largely in the regional language segment) despite threats from the electronic media, especially private news channels. The Registrar of Newspapers in India reported that newspapers were published in 101 languages and dialects in 2001. These include 5,364 dailies and 47,296 periodicals. Of this, the majority are published in Hindi, including the Bengali daily *Anand Bazar Patrika,* published in Kolkata with a circulation of nearly 900,000, and two English dailies, the *Hindu,* published in Chennai, and the *Hindustan Times,* published in Delhi, both with circulation over 900,000 (Press in India 2001). A majority of Indian newspapers are owned by individuals, with the remainder owned by corporations.

While the Press Council of India regulates the press, the government could potentially limit its autonomy by virtue of several laws that govern the press, including the Official Secrets Act of 1923, the National Security Act of 1980, and the Parliamentary Proceedings (Protection) Act of 1977. Furthermore, India's constitution does not directly guarantee freedom of the press, although this has been recommended by a constitution review committee. Controls are also exercised by way of a requirement to register newspapers and a policy of price control (Iyer 2000; Venkateshwaran 1993).

Newspapers are considered a credible source of information, and readership is high among the literate. Newspapers are also heard through community readings in rural areas. An average newspaper ranges from between fifteen and twenty-five pages. In recent years, the Indian press, as elsewhere, has undergone a dramatic transformation. New technology has enabled smartly presented editorial content (e.g., new columns, features, and supplements on careers, lifestyle, fashion, films, and investments) that has evoked tremendous interest in all walks of Indian life. Innovative pricing and marketing strategies are being evolved to increase market share. Most newspapers are also available on the Internet. Despite newspapers being owned by corporations, in the past advertising and circulation departments rarely dared to influence editorial content. Today, however, publications are systematically dismantling the walls between their editorial and business departments to cater to new demands of business and readers—ideas that would be considered scandalous by die-hard journalists even a decade ago, as content and presentation were considered and acknowledged as being purely their turf.

Radio and television are the popular mass media, with national and regional channels competing for media space and viewership. The government of India took upon itself the task of expanding the frontiers of broadcasting with its twin arms of broadcast (All India Radio [AIR]) and television (Doordarshan) to serve the objectives of national development and focused on information, education, and entertainment for the rural and the semiliterate audiences. After decades of debate on autonomy for the electronic media, the parliament passed an act in 1990 that took effect in 1997, forming the Prasar Bharathi (the Broadcasting Corporation of India) to oversee the functioning of radio and television. Despite privatization, both AIR and Doordarshan continue to remain under government control.

Given its potential of overcoming literacy barriers, the development of radio as the frontline medium of mass communication has been a major thrust on the government's agenda. Its network of 210 broadcasting centers covers 90% of the area of the country and serves 99% of the population. All India Radio broadcasts in sixty-eight languages and dialects in its home services, while its external services broadcast in sixteen foreign and ten Indian languages, making it perhaps the largest broadcasting network in the world. The transistor revolution with the availability of low-cost sets in the mid-1960s increased the popularity of radio listening even in the most remote areas of the country.

Television has transformed the information entertainment landscape of the country. Here again, the concerted efforts of the government have led to Doordarshan emerging as the largest terrestrial television network. Television is claimed to reach 79 million homes in India.

The most dramatic change since the mid-1990s was the appearance of private cable television. India had virtually control-free private television channels to complement its control-free print media. This development is significant despite aberrations related to distribution of channels by cable operators, tariffs to consumers, and regulation for control and up-linking. State-owned television channels combined with cable-delivered channels provide a wide variety of news, educational, and entertainment channels, operating virtually around the clock and keeping Indians glued to their sets.

What is also significant is that since the mid-1970s India has had a fairly well-developed program production infrastructure that does not need to depend on foreign sourcing. Doordarshan and AIR have built impressive facilities over the years. Media conglomerates have diversified into program production, supplying programs to both public and private channels.

The Study Sample

The Indian cities chosen for the study were New Delhi and Hyderabad. New Delhi is India's capital, with a population of 13.8 million, and is the nerve center of political, cultural, and media activities. Most of the nationally circulated English newspapers are either published in or have an edition from New Delhi. The newspaper selected for the study was the *Hindustan Times*, a multi-edition national daily published in New Delhi and nine other cities. It is one of the oldest newspapers launched during India's freedom struggle. Its New Delhi edition has the highest circulation among its competitors. As for the broadcast media, the daily 9:00 p.m. national English news broadcasts of Doordarshan and All India Radio, which reach about 90% of the population, were recorded for the study.

The second city selected was Hyderabad, the capital of Andhra Pradesh, with a population of 6.4 million. Located in south India, Andhra Pradesh is emerging as a technologically savvy state in the country. The media market is competitive with a number of English, Telugu (state language), and Urdu newspapers as well as television channels.

The newspaper selected was *Eenadu* [Today], the largest circulated Telugu daily published in Hyderabad and seven centers in the state, with a circulation of 1.2 million and a readership of 1.7 million. The paper is published daily and has special sections for women, youth, and children. The urge to read this paper has also resulted in people registering at adult literacy centers. The regional Telugu newscasts at 7:00 p.m. on All India Radio and 7:30 p.m. on television were considered for the study.

The news items collected on the selected seven days were independently coded by a team of ten people who were trained in three sessions before starting on the exercise. Intercoder reliability was established before the newspapers and the news on television and radio were coded.

Topics in the News

Table 12.1 presents the news topics by medium and city. Several interesting points emerge from this table. The topics covered were similar across the three media analyzed, although differences occurred in the priority given to them.

Internal politics occupy a prominent place in the Indian media. One-fourth of the radio news in both of the cities covered internal politics. Political news on Hyderabad television (43.5%) was twice that in national news bulletins (20.3%). Interestingly, newspapers had the lowest coverage of internal politics—18.2% in the *Hindustan Times* and 15.6% in *Eenadu.*

Despite the similarly high coverage of internal politics in national and local media, there are vast differences in coverage of international politics. International politics had sizeable coverage at the national level on New Delhi television (22.1%), ranking highest and even surpassing internal politics, and on radio (17.1%), ranking third. While the *Hindustan Times*, a national newspaper, had 6.9% coverage of international news, the regional daily, *Eenadu*, published roughly half of that (3.9%). As an overall trend, the regional media of Hyderabad seemed indifferent to international politics, giving it less than 5% coverage in all the three media.

Indians are interested in sports, especially cricket. Coverage of sports news was the second most extensively covered topic after politics in the Hindustan Times (15.3%). However, its regional counterpart Eenadu devoted less than half of that percentage of space to sports (7.7%). Sports coverage was higher in New Delhi radio (17.8%) and television (10%, ranked fourth) broadcasts, which was rather surprising. Conversely, the regional media in Hyderabad devoted between nearly 5% and 8% to sports across all three media.

Given India's long history of communal strife and caste wars and in more recent times insurgency, terrorism, white-collar crime, and corruption, it is no surprise that internal order news—or more specifically law and order

TABLE 12.1 Distribution of General Topics of News Items by City and Medium

Topics	Newspaper		Television		Radio	
	New Delhi	Hyderabad	New Delhi	Hyderabad	New Delhi	Hyderabad
Internal Politics	18.2	15.6	20.3	43.5	26.3	24.5
Sports	15.3	7.7	10.0	6.1	17.8	4.8
Internal Order	10.0	12.1	12.1	13.3	13.2	10.3
Business/ Commerce/ Industry	8.4	9.5	6.4	3.1	2.8	5.5
Human Interest Stories	8.3	10.2	2.1	.0	1.1	.6
International Politics	6.9	3.9	22.1	4.8	17.1	.3
Cultural Events	5.3	12.0	2.8	.7	.7	8.2
Education	4.1	4.6	.0	.3	.4	3.9
Communication	3.0	1.4	1.1	.0	2.5	.9
Social Relations	2.9	1.7	2.8	2.7	1.4	6.7
Entertainment	2.4	2.1	.0	.0	.0	.0
Science/Technology	2.1	1.1	.0	1.7	.0	.9
Economy	1.9	3.0	2.1	2.0	.7	8.2
Military and Defense	1.8	.9	2.8	.3	1.8	.0
Disasters/Accidents/ Epidemics	1.7	1.8	1.1	1.7	1.8	2.1
Health/Welfare/ Social Services	1.6	2.5	.7	2.7	.4	4.8
Ceremonies	1.4	1.1	1.8	.3	2.5	4.5
Environment	1.4	1.1	2.5	.0	2.1	1.5
Labor Relations and Trade Unions	.8	1.1	1.8	2.7	4.6	3.0
Transportation	.8	1.3	.0	.0	.7	.0
Fashion/Beauty	.6	.5	.0	.3	.0	.0
Housing	.4	1.3	.0	.7	.0	.6
Energy	.3	1.4	.0	4.4	.0	7.6
Population	.2	.6	.4	.3	.4	.0
Weather	.2	.6	7.1	8.2	1.8	.9
Other	.1	.6	.0	.0	.0	.0
Total[a]	100.0	100.0	100.0	100.0	100.0	100.0
	(n = 1894)	(n = 2712)	(n = 281)	(n = 294)	(n = 281)	(n = 330)

Note: Distributions given in percentage.

[a] Total percentage may not actually be 100.0 due to rounding error.

news—reflects the crime profile in the country. Between 10 and 13% of the news hole in all media was devoted to such news.

In the last two decades, phenomenal growth has taken place in the investor public in India. To cater to the growing interest in business/commerce/industry, newspapers have separate business sections and pages, be they national or regional. This is why the coverage of business/commerce/industry news occupies between 8.4% and 9.5% in the newspapers. There was less coverage in the news on television and radio, but business does feature in other television and radio programming.

Human interest stories got significant coverage in the newspapers (8.3% in the *Hindustan Times*; 10.2% in *Eenadu*) but virtually none in the broadcast media. Cultural events were less reported in the national media but were better covered in the Hyderabad regional newspapers (12%) and on radio (8.2%), with only 0.7% on television. Weather appeared to take up a large chunk of daily television news, with 7 to 8% coverage, and was negligible in radio and in newspapers.

Taking all the media at the national and regional levels, the trends on reporting are similar in terms of subject identification but with varying degrees of coverage, as seen in Table 12.2.

Using Spearman rank-order correlation coefficients, the rankings of the various topics were highly similar between the two newspapers (.86), less correlated between the two television newscasts (.60), and even weaker between the two radio newscasts (.41). In addition, on average, the intercorrelations among the three New Delhi media were higher (.67) than among the three media in Hyderabad (.56), thus indicating more structural similarity in the larger city. Thus, despite the private ownership of the two newspapers, they

TABLE 12.2 Spearman Rank-Order Correlation Coefficients between Rankings of News Topics in Various Media

	New Delhi Newspaper	Hyderabad Newspaper	New Delhi Television	Hyderabad Television	New Delhi Radio	Hyderabad Radio
New Delhi Newspaper		.86[c]	.59[b]	.31	.57[b]	.49[a]
Hyderabad Newspaper			.47[a]	.39	.42[a]	.65[c]
New Delhi Television				.60[b]	.85[c]	.47[a]
Hyderabad Television					.46[a]	.65[c]
New Delhi Radio						.41[a]
Hyderabad Radio						

[a]$p < .05$; [b]$p < .01$; [c]$p < .001$.

were relatively more similar to each other than the two pairs of television and radio newscasts operated by the public and private channels. These differences seem to reflect the general trends in news reporting in different types of media, editorial policies, and regional influences that impact news coverage. Thus, the similarity in the topical structure in both newspapers reflects the national trends where journalists tend to share similar news values of reporting.

Furthermore, reporting of the news tends to be similar in each region given the values of proximity and importance attached to issues and consumer interest. Newspapers with national circulation like *Hindustan Times* give wider coverage to national and international issues and less coverage of local issues specific to the city, while regional papers like *Eenadu* give wider coverage to regional issues, internal order, human interest stories, and cultural events. Finally, the relative similarity between radio and televisions news is predictable, as news bulletins emanate from a common source: the Broadcasting Corporation of India. Moreover, the distinctive differences in the topics covered on radio and television news in both cities indicate the regional differences and efforts made to localize the news.

Deviance in the News

In order to examine the relationship between deviance in the news and the prominence given to the items, the mean prominence scores were calculated for the three media across all levels of statistical deviance, social change deviance, and normative deviance. One-way analyses of variance (ANOVAs) were computed in order to test the statistical significance between the prominence scores. Table 12.3 presents the mean scores of these measures.

Deviance was related to prominence in only thirteen of the thirty analyses. The findings indicate that contrary to expectation some of the extremely unusual or deviant news received lower prominence scores and that some very common deviant issues got higher prominence. Moreover, even among the significant relationships reported, limited linear relationships appeared between deviance and prominence. In six of the thirteen cases some of the lowest levels of deviance scored highest on prominence. On the other hand, the highest levels of deviance corresponded with the highest prominent scores in only three of the thirteen cases. In other cases variations in the scores existed, with some common issues getting the maximum prominence, indicating that deviance had only a moderate effect on the size and placement of the news items.

The level and relevance of deviance differed between each medium in both cities. In many cases the prominence scores were lower for higher levels of deviance and vice versa. Each of the deviance levels reflected significant information on how news is treated in each of the Indian media and the levels of deviance that have the maximum impact on prominence. Effects of deviance were found most often in the *Eenadu*. The differences between the

TABLE 12.3 Mean Verbal and Visual Prominence Scores for Intensity of Deviance

| | New Delhi | | | | | Hyderabad | | | | |
| | Newspaper | | Television | | Radio | Newspaper | | Television | | Radio |
Intensity of Deviance	Verbal only (n = 1291)	Verbal plus Visual (n = 361)	Verbal (n = 113)	Visual (n = 113)	Verbal (n = 155)	Verbal only (n = 1911)	Verbal plus Visual (n = 502)	Verbal (n = 140)	Visual (n = 140)	Verbal (n = 132)
Statistical Deviance										
(1) common	220.5	206.9	104.1[c]	116.9[b]	47.0	151.3	97.2	207.9	25.9	20.5
(2) somewhat unusual	224.4	237.6	149.1	201.5	109.5	153.0	107.6	112.1	162.4	40.4
(3) quite unusual	265.6	207.9	279.1	251.6	108.7	157.0	86.6	107.1	113.5	73.7
(4) extremely unusual	247.4	246.7	300.5	96.3	69.0	182.2	52.4	66.5	40.5	–
Social Change Deviance										
(1) not threatening to status quo	224.4	210.4	127.7[a]	124.9[b]	93.3	150.1[b]	96.5	129.3	119.6	55.9
(2) minimal threat	246.4	258.5	157.0	222.4	120.6	118.5	103.1	86.6	132.2	80.2
(3) moderate threat	251.4	223.3	213.4	262.5	76.6	207.3	117.4	113.4	86.0	71.8
(4) major threat	279.7	378.0	242.9	228.0	–	188.0	110.0	101.0	105.1	–

Normative Deviance

(1) does not violate any norms	243.5	213.7	163.6	159.3	94.6	147.4[a]	99.1	108.8	112.5	64.1
(2) minimal violation	155.5	208.2	393.0	196.0	181.8	179.2	94.1	112.5	152.3	74.2
(3) moderate violation	212.6	223.7	189.5	221.0	110.9	214.4	101.1	180.7	97.2	57.3
(4) major violation	222.6	333.2	205.0	234.5	70.7	254.8	107.0	100.0	103.6	–

[a]$p < .05$; [b]$p < .01$; [c]$p < .001$.

prominence scores were significant for both the verbal and verbal plus visual content on all three dimensions of deviance, but the direction of influence was not in correspondence with the hypothesis of the study. Reverse trends were observed in the visuals at all three levels and in the verbal content at the level of statistical deviance. As for the *Hindustan Times*, significant influences were obtained for the visuals, with reversed or mixed trends for the three dimensions of deviance.

As for the broadcast media, only in the case of the verbal measure for television in New Delhi were findings in the expected direction: two of the three analyses (for statistical and social change deviance) yielded significant ANOVAs.

Does deviance have only a moderate effect on the size and placement of the news items? The observations and trends reflect the larger sociopolitical media environment (controlled or uncontrolled), with the media tending to underplay extremely deviant news at the normative levels that would affect law and order and the political situation in the country. Communal riots, religious upheavals, terrorist attacks, and other sensitive issues—although highly deviant—are generally underplayed and are given less prominence to avoid sparking trouble. This was true especially when some major criminal cases and most deviant items were given less coverage.

While major criminal cases received the prominent coverage they deserved, happenings such as petty crime, thefts, and accidents were featured in columns or slots reserved for such news in the media and were coded as being either "common" or "somewhat common" based on their frequency of occurrence. While technically qualifying as sensational and deviant, the media treatment of crime news tries not to glorify the crime or the criminal based on codes of ethics of responsible journalism. Consequently, social change deviance is also underplayed, probably in deference to values that celebrate the status quo. Such news may get relegated to the inside pages of the newspapers in a seemingly deliberate effort to play down events. While deviance was underplayed in verbal content, greater prominence was given to deviant verbal plus visuals in Indian newspapers. Verbal plus visual items, especially of scantily clad Western and Indian women, came under intense criticism from older women and men in the higher and lower socioeconomic status (SES) segments of the focus groups. They were of the opinion that these pictures undermined Indian cultural values and had a negative impact on youth.

Although statistical and social change deviance influenced prominence in both the newspapers as well as New Delhi television, there is no particular influence in the radio newscasts, which continue to report common day-to-day affairs of the state. Interestingly, a pattern fully the opposite of the hypotheses of the study was found in the verbal and visual content in the New Delhi television and the Hyderabad newspaper. Significantly, in all the three media extremely deviant news was often underplayed and does not get prominence as predicted by the hypothesis.

Social Significance in the News

Similar to deviance, the basic assumption was that socially significant news would get greater prominence in the media. Here, too, ANOVAs tested for the differences among the mean prominence scores of the four dimensions of social significance: political, economic, cultural, and public. Table 12.4 presents results of the analyses.

Of the forty analyses, twenty-two yielded significant effects, as seen in Table 12.4. Similar to the case of deviance, the effects for social significance—with some variations—seemed to emerge for newspapers and television but not for radio. According to expectations, the directions of the relationships are perfectly linear in five of the cases and were directly opposite in two cases. In some other cases there is nearly a perfect linear relationship. In addition, the lowest levels of significance correspond with lowest prominence scores in ten of the cases, and the highest levels of significance correspond with the highest prominence scores in twelve of the cases.

In *Eenadu*, prominence was affected linearly by verbal cultural and public significance, while in the *Hindustan Times* the only significant linear component was for economic significance. As for the verbal plus visual measure, in both newspapers political significance was linearly related to prominence, but in the opposite direction from the study hypothesis.

Similar to the case of deviance, the New Delhi television newscasts revealed a perfect or nearly perfect linear relationship regarding political and public significance for both the verbal and visual measures. Given the limited number of stories on television with an emphasis on Indian and international politics, it was not surprising that political and public significance influenced television news in New Delhi. In addition, interest in politics is high among Indians; a majority of the participants in the focus groups expressed their interest and regularly scanned the papers to update political developments. As for television in Hyderabad, items of major cultural significance were very prominent for both verbal and visual content.

However, given the variations, no set patterns were observed in the ways in which the media reporters address the issues of social significance that vary in terms of verbal and visual content. This again reflects the differences in the nature of the media, the editorial policies at the national or regional levels, and presumably the expectations of the readers. Overall, there is a general consensus on public significance, which was an important agenda, to underplay the major deviant issues and give prominence to socially significant issues. Journalists in the focus groups tended to endorse this during the discussions.

While it is presumed that news coverage should be purposeful in order to inform, educate, and even entertain the public (as endorsed by news values including deviance and social significance), it appears that not all news reports are affected by these factors in determining news prominence. This was further examined in the regression analyses.

TABLE 12.4 Mean Verbal and Visual Prominence Scores for Intensity of Social Significance

Intensity of Social Significance	New Delhi					Hyderabad				
	Newspaper		Television		Radio	Newspaper		Television		Radio
	Verbal only (n = 1291)	Verbal plus Visual (n = 361)	Verbal (n = 113)	Visual (n = 113)	Verbal (n = 155)	Verbal only (n = 1911)	Verbal plus Visual (n = 502)	Verbal (n = 140)	Visual (n = 140)	Verbal (n = 132)
Political Significance										
(1) not significant	240.4[a]	205.8	126.8[b]	125.3[c]	72.6	157.2[a]	80.0[b]	164.0	152.4[a]	57.1
(2) minimal	193.0	220.0	123.6	117.5	77.3	141.1	127.0	82.7	84.8	71.3
(3) moderate	217.3	321.6	145.5	206.8	116.8	145.2	161.2	80.6	75.8	64.9
(4) major	288.2	225.4	251.8	288.9	135.6	237.3	130.1	134.8	147.0	–
Economic Significance										
(1) not significant	217.0[c]	207.7[a]	179.2	176.6	99.6	151.5[b]	95.6	113.0	109.8	59.0
(2) minimal	238.3	251.7	187.1	171.3	71.5	127.3	133.1	133.0	144.8	64.1
(3) moderate	263.8	308.1	115.1	141.2	75.8	143.3	108.4	94.3	100.0	77.3
(4) major	517.0	179.0	184.7	203.8	27.0	267.7	153.0	176.5	176.5	–
Cultural significance										
(1) not significant	196.0[c]	187.1	171.4	176.5	90.5	147.5[a]	98.4	110.2[c]	110.2[c]	71.3
(2) minimal	244.2	271.3	58.0	58.0	63.6	154.9	97.8	116.5	116.5	58.1

(3) moderate	377.3	237.9	159.2	142.4	153.6	223.8	95.8	77.4	77.4	67.4
(4) major	347.6	279.0	298.8	298.8	43.7	189.8	174.8	870.0	870.0	—
Public significance										
(1) not significant	155.8[c]	166.0	98.7	131.7[a]	—	95.7[c]	112.0	114.7	88.8	71.3
(2) minimal	204.7	228.5	157.4	193.2	80.1	122.0	95.2	93.2	90.2	54.9
(3) moderate	318.6	247.6	171.0	177.7	88.9	181.8	104.2	125.3	131.5	72.4
(4) major	418.6	266.4	250.2	263.6	119.3	273.7	103.2	118.4	118.4	—

[a] $p < .05$; [b] $p < .01$; [c] $p < .001$.

Deviance and Social Significance as Predictors of News Prominence

In order to test the hypothesis and theoretical implications of deviance and social significance as predictors of news prominence, eight stepwise regression analyses were calculated. For each of the two cities, analyses were done for the prominence of only verbal newspaper content and for the prominence of combined visual and verbal content, as well as for television and radio prominence. All eight analyses explained statistically significant amounts of variance, ranging from 4% to 22% (see Table 12.5).

Regarding verbal newspaper content, 6% and 4% of variance were explained in the *Hindustan Times* and *Eenadu*, respectively. The predictive power of the models was much higher when the visual and verbal measures were combined, accounting for 12% in New Delhi and 17% in Hyderabad. In the *Hindustan Times*, normative deviance as well as economic and cultural significance determine the prominence of verbal content, but when verbal content was combined with visuals, then statistical and cultural significance were the determining elements. In *Eenadu*, statistical deviance, political significance, and public significance were the determining elements of the prominence of verbal content, while statistical deviance and public significance were responsible in the case of the combined measure of verbal and visual content. Also notable is the fact that some of the standardized beta weights are positive and some are negative—a reflection of the direction of the relationship indicated in Tables 12.3 and 12.4, indicating that the higher the relevant deviance or social significance factor, the lower the prominence, or vice versa.

While the pattern of explained variance found for newspapers in both cities was similar, in the case of television it was different: 22% explained variance in New Delhi, and only 5% in Hyderabad. In New Delhi, statistical deviance and political significance determined the prominence of visual and verbal content, while in Hyderabad the only determining element was statistical deviance in a negative relationship with the prominence of verbal content.

As for radio, in New Delhi only 4% of the variance in the prominence of radio content was explained solely by political significance, and in Hyderabad 6% was explained solely by statistical deviance. The prominence of radio content in New Delhi was explained by political significance only and in Hyderabad solely by statistical deviance.

The different trends in the analysis can be explained by three major factors. The visual content in television seems to be extremely deviant, while verbal content is controlled. This was true in both cities, as exemplified by two extremely unusual events. In New Delhi the major concern was regarding a weird "monkey man" presumed to be on a killing spree targeting children, and Hyderabad was full of news about the illegal adoption of babies by some government-approved agencies. Given the unstable political climate with continued mudslinging between (deviant) political parties at the national and regional levels, it is not surprising that statistical deviance and political significance are determinants of prominence, also explaining the trend in New Delhi

TABLE 12.5 Stepwise Regression Analyses of Intensity of Deviance and Social Significance on News Prominence

	New Delhi								Hyderabad							
	Newspaper Prominence Verbal only Total $R^2=.06^b$ ($n=1279$)		Newspaper Prominence Visual and Verbal Total $R^2=.12^a$ ($n=129$)		Television Prominence Total $R^2=.22^c$ ($n=109$)		Radio Prominence Total $R^2=.04^a$ ($n=154$)		Newspaper Prominence Verbal only Total $R^2=.04^c$ ($n=1768$)		Newspaper Prominence Visual and Verbal Total $R^2=.17^c$ ($n=154$)		Television Prominence Total $R^2=.05^b$ ($n=139$)		Radio Prominence Total $R^2=.06^b$ ($n=131$)	
Independent variables	r	Std. Beta	r	Std. Beta	r	Std. Beta	r	Std. Beta	r	Std. Beta	r	Std. Beta	r	Std. Beta	r	Std. Beta
Deviance																
– Statistical, verbal content	-.01	ns	-.02	ns	.40[c]	.30[b]	.00	ns	-.08[b]	-.10[c]	-.30[c]	ns	-.22[b]	-.22[b]	.25[b]	.25[b]
– Statistical, visual content	–	–	-.21[b]	-.27[a]	.26[b]	ns	–	–	–	–	-.39[c]	-.37[c]	.02	ns	–	–
– Social change, verbal content	.01	ns	-.10	ns	.30[b]	ns	.03	ns	.02	ns	-.17[a]	ns	-.08	ns	.15	ns
– Social change, visual content	–	–	-.20[a]	ns	.33[b]	ns	–	–	–	–	-.16[a]	ns	-.08	ns	–	–
– Normative, verbal content	-.07[a]	-.07[b]	-.09	ns	.12	ns	-.01	ns	.01	ns	-.10	ns	.03	ns	.01	ns
– Normative, visual content	–	–	-.18[a]	ns	.16	ns	–	–	–	–	-.11	ns	-.03	ns	–	–

[a] $p < .05$; [b] $p < .01$; [c] $p < .001$; ns = not part of final stepwise regression equation.

(*Continued*)

TABLE 12.5 Stepwise Regression Analyses of Intensity of Deviance and Social Significance on News Prominence (*Continued*)

	New Delhi								Hyderabad							
	Newspaper Prominence (Verbal only) Total $R^2 = .06^b$ ($n = 1279$)		Newspaper Prominence (Visual and Verbal) Total $R^2 = .12^a$ ($n = 129$)		Television Prominence Total $R^2 = .22^c$ ($n = 109$)		Radio Prominence Total $R^2 = .04^a$ ($n = 154$)		Newspaper Prominence (Verbal only) Total $R^2 = .04^c$ ($n = 1768$)		Newspaper Prominence (Visual and Verbal) Total $R^2 = .17^c$ ($n = 154$)		Television Prominence Total $R^2 = .05^b$ ($n = 139$)		Radio Prominence Total $R^2 = .06^b$ ($n = 131$)	
Independent variables	r	Std. Beta	r	Std. Beta	r	Std. Beta	r	Std. Beta	r	Std. Beta	r	Std. Beta	r	Std. Beta	r	Std. Beta
Social Significance																
– Political, verbal content	$-.07^a$	ns	$-.23^b$	ns	$.32^b$	ns	$.19^a$	$.19^a$	$-.04$	$-.05^a$	$-.16$	ns	$-.03$	ns	$.04$	ns
– Political, visual content	–	–	$-.13$	ns	$.39^c$	$.26^b$	–	–	–	–	$-.19^a$	ns	$.01$	ns	–	–
– Economic, verbal content	$.13^c$	$.15^c$	$.19$	ns	$-.03$	ns	$-.06$	ns	$.02$	ns	$-.12$	ns	$.01$	ns	$.12$	ns
– Economic, visual content	–	–	$-.02$	ns	$.02$	ns	–	–	–	–	$-.11$	$.16^c$	$.04$	ns	–	–
– Cultural, verbal content	$.17^c$	$.18^c$	$.28^b$	$.28^b$	$.11$	ns	$.05$	ns	$.03$	ns	$.11$	ns	$.11$	ns	$-.02$	ns
– Cultural, visual content	–	–	$.27^b$	ns	$.10$	ns	–	–	–	–	$-.09^c$	ns	$.11$	ns	–	–
– Public, verbal content	$.07^a$	ns	$.10$	ns	$.23^a$	ns	$.10$	ns	$.15^c$	$.18^c$	$.21^b$	$-.18$	$.07$	ns	$.09$	ns
– Public, visual content	–	–	$-.01$	ns	$.21^a$	ns	–	–	–	–	$-.21^b$	ns	$.13$	ns	–	–

$^a p < .05$; $^b p < .01$; $^c p < .001$; ns = not part of final stepwise regression equation.

television and to some extent in radio. New Delhi is the center of major political activity, so unexpected and unusual political changes continue to surprise people. Lastly, there are no definite trends to quantify prominence in the Indian media in terms of deviance and social significance. The treatment of news differs depending on the political and social structure, government policies and controls, and the diversity of the country. The media tend to be torn between the true values of news reporting in coverage of deviant or socially significant news. Investigative journalism focusing on exposing various political scandals, although reflecting shades of a free press, has never really been as consistent in India as in some other parts of the world.

Cultural issues, especially those of religion and castes, have always been treated with a great deal of caution in a country hypersensitive to these issues. But conflicts between communities have started getting greater coverage in parallel (private) channels that are exercising less restraint. The dilemmas of socially responsible journalism continue to haunt Indian media persons on account of changing deviance thresholds, inter- and intramedia competition, and access to news of India in western media. The participants of the focus groups further confirmed these issues.

People Defining News

In both cities the focus groups provided interesting data on the presentation and consumption of news in the three media. Separate sessions for each of the groups—journalists, public relations practitioners, and media audiences of high and low SES—were organized in New Delhi and Hyderabad.

The journalist groups, composed of twenty-three people from national and regional media, were represented by senior editors, reporters, and columnists from the print medium and news producers and correspondents from radio and television. Men largely represented this group, since only two women responded to the invitation. The public relations groups, composed of twenty-six men and women, were the most dynamic and opinionated. They represented public- and private-sector enterprises as well as being government and office bearers of the Public Relations Society of India. The twenty-eight members of the high SES audience group included a cross-section of professionals, corporate executives, businesspeople, and housewives. The variety of responses from men and women in this group reflected their social economic levels, professional interests, and gender differences in personal news preferences. Lastly, the twenty-seven people in the low SES groups were composed of university employees, factory workers, lower-division clerks, "office boys," and cleaners. These discussions were stimulating because people openly expressed their anger toward the media and the government for not taking up issues for the welfare of the people. They had definite expectations from the media.

The first question asked why people read newspapers, listen to radio news, or watch the news on television. The clear-cut answer was "to make sense of the world around you." Information and knowledge are outcomes of peoples' needs. This aspect emerges in statements defined by high and low SES audiences and public relations persons such as, "News is something that educates you, entertains you and adds to your knowledge of society." One woman in the high SES group said, "News is definitely something new and that we, as people, have to be informed about." Or as one gentleman said, "News increases the knowledge of the people. It is good for development and progress of the people and the country." Another factor that emerged from the exercise was choice. People displayed an understanding of the news consumption schema—that is, although media outputs are mass delivered, news is individually consumed depending on the personal interests of individuals. Age, gender, profession, income, and social class are some of the variables that shape these interests.

Interest in politics is high among Indians, which was also revealed across both levels of audiences during focus group discussions. Journalists believe that there is a tremendous hunger for political information and that people depend on the media for it, wanting to know what the government and politicians are doing for the people. Similar interests were demonstrated by the public relations and media personnel and the audiences. Given the unstable political climate of the country, with coalitions of ideologically dissimilar partners in the government of India and different states, updates on political uncertainties are high on the news agenda.

Most respondents, especially media professionals and public relations practitioners, saw the importance (significance) of an issue, occurrence, or event as the guiding criteria in its treatment and presentation as news in the media. The belief that human nature draws people to negative news was held by a vast majority of respondents (close to 80%, other than journalists). But empirical support for news selection or projection on this basis is weak because journalists are constrained by law, ethics, and peoples' sensitivities. Projection of such news—even based on actual facts—may lead to fear, violence, and other undesirable consequences. The impact of news (positive or negative) was of consequence for the public relations practitioners.

During the course of the focus group discussions, the elements of deviance and social significance became guiding principles of ways in which news items are presented and consumed.

Deviance

Interest and recall of deviant news across most of the Indian focus group participants in both cities relate to their concern as citizens about the sociopolitical unrest and growing rate of crime and corruption in the country. Deviant issues (odd, unexpected, and sensational news) attract attention. Media consumers said that "news is all about the murders, government policies, and what's happening in the country and abroad," and "news is what we hear about the governments' and peoples' corruptions and about entertainment." Among the concepts related to deviance, there are shades of statistical, social change, and normative deviance. Interest in deviant news emerges both in reporting the news by the journalists and from the interests of the audience. Women expressed greater concern and interest in deviant news compared to men. They explained that they had a greater responsibility in alerting and protecting their children and themselves.

Respondents concurred that there is no escape from deviant news. One journalist's opinion was that "negative is what makes news and only striking negative news makes news. The media highlight such news with screaming headlines, bold typefaces, prominent positioning, and striking pictures to draw attention of readers in the print medium and some bold and imaginative reporting of news as it happened or is happening on television."

Journalists had definite views about the definitions of news and their concerns about deviance and social significance and what makes news. Providing information to the public is their principal concern. News itself, they argued, must be based on occurrences, which need to be judged on elements of novelty (unusual, unexpected—perhaps deviance) and certainly for the significance that these elements have an impact on the life of the people. Journalists argued that although disasters make negative news, it is important to cover such issues (and sometimes exaggerate them) so that help from all sources reaches the needy. Crime, they said, is covered so that people are alerted and because it must be curtailed. A certain level of co-orientation was visible from the interest of journalists and media consumers concerning the reportage and consumption of deviant news. Coverage of crime, scoops, scandals, and disasters were explained by journalists as extraordinary (deviant) happenings with elements that have an impact on the life of the people (social significance) sometimes owing to lapses by authorities.

While consumers appeared a little wary and desensitized to what is described as sensational news (e.g., political scandals and corruption) as well as crime news (e.g., rape, murder, atrocities against women), the readership of these items is high, as it enables them to stay alert and to avoid becoming victims. All participants, but especially women, admitted they paid attention to crime news, which had reached alarming proportions especially in New Delhi (deviance and social significance). In New Delhi, paying attention to patterns of crime in news reports is a protection–survival response, and no one can question the social significance of survival. Moreover, such stories catch reader interest, besides boosting circulation and advertising revenues. One high SES participant in Hyderabad justified, "Because of the price war [among newspapers], negative news has a certain attraction; one should look at it from the newspaper's angle also, as they have to sell their paper."

Therefore, issues of deviance are also socially significant. Though most of the focus group discussions revolved around interest in issues of deviance and criticism of the government and the media, the definition of news was very positive regarding the social significance of gaining and sharing knowledge, bringing a sense of happiness and entertainment, as well as ways in improving lives.

Social Significance

News of development and social significance seems to be an alternative or an escape for Indian media practitioners as well as consumers who are almost desensitized to the growing deviant information of political scandals, crime, and corruption, along with natural disasters like floods and earthquakes.

Many participants challenged the phrase "No news is good news." One senior public relations practitioner from Hyderabad remarked, "People would like to know about the issues relating to their benefit, which is lacking in most of the newspapers today." Other people said they look for news to improve and enlighten their lives and to improve their status. They attempted to define news as "something that directly or indirectly affects us as well as something that helps us live a better life and share new things with children" or as "information that has a social purpose and it should have an impact, viewpoint, or a solution to a problem."

For audience members in the higher SES segment, socially significant issues are restricted to what they consider significant, which is not necessarily the entire spectrum of social concerns. At the social level, lifestyle issues (e.g., health and nutrition, grooming, fashion, art, social and cultural events, celebrity news, higher education for their children) become their concerns. At the professional level, they are more concerned with news related to, for example, the economy, investments, trade and business policies, and developments in science and technology. Lower SES participants expect the media to provide information that can better their lives and, above all, that can help them lead a secure and safe life.

Journalists appear to largely understand the needs and expectations of their audiences. Audience, editorial, and business needs place constraints on what journalists actually want to report. As one senior journalist admitted, "Real politics is about people, which remains unwritten and unsaid." Although a majority of the journalists were interested in political developments, some appeared to show a deep interest in topics such as socioeconomic developments, science, education, culture, literature, and art. However, they confirmed that the treatment of these stories (at the time of the study) was superficial and incomplete.

Media consumers felt that it was the responsibility of the media to report news of social significance. In this context, one remarked, "news is something that will directly affect the environment [and] improve health, politics, social life, etc." They did notice and appreciate the shift in the trend of newspaper reporting from being too political to covering a variety of topics. As one high SES respondent observed, "There is a definite change in the newspapers now as they have development, social news, news for women, and are taking up issues related to health, environment, social life, good housekeeping, and entertainment."

A broad assessment of responses leads to the belief that, although deviance influences newsworthiness of an event, social significance appears to be of greater concern among news providers and consumers. In some sense, it also reflects a greater leveling effect: Journalism is more sensitive to audience needs, which is a paradigm shift from the top-down "provider knows best" approach. The risk, however, is that within the media, change is more marketing and circulation driven than editorially driven, if responses of a large number of journalists are to be considered as indicators.

Information, Knowledge, and Social Purpose

There is a definite shift in the focus of the expectations from the media and how news presenters and consumers define news. Consumers define news as something that helps them know about things and improve their lives.

Among the high SES participants, social purpose is more pronounced in information seeking for acquiring knowledge. They also stress the sociocommunicative function of news because they define news as connecting people, friends, business associates, and professionals. It also helps them to be well informed about things around the world in order to discuss them with family, friends, and relatives. Gender-specific, information-seeking behaviors have also been noticed from women, who demand news of and follow not only family-related issues (e.g., housekeeping, health, children, education, nutrition), but also information useful for expanding their worldview (e.g., for improvement of employment opportunities, skills improvement, beauty care).

The importance of news and media exposure among the low SES audience groups was also noteworthy. They have very specific views about politics, international news, crimes, and day-to-day happenings around them. Those who have no or little media access actively acquire information from their more informed peers and have very definite ideas about their expectations from the media. They are not interested in political corruption and "international entertainment," as quoted by one of the respondents, but want information to better their lives and to improve their skills and financial status. While other groups have taken a pragmatic view about negative news, lower SES participants came forward to freely criticize the government for lapses in controlling and responding to situations appropriately and to scrutinize elitist trends in the media they say erode Indian cultural values.

Most Important Events of Life

At the end of each discussion, participants were asked to recall the three most important events in their lives. Here it was noted that negative news does not have the highest recall; instead, anything unusual, socially or culturally significant is etched in memory even without any personal involvement. In a wide spectrum of past and present events recalled, the similarity of responses among the participants gave an insight into the similarity of thinking among members of one country and one culture. The general tendency was to recall national events compared to international ones, which were limited to a few.

Overall, a majority of events recalled was negative. Interestingly, some of the important (deviant and socially significant) events participants across all the focus groups recalled included the assassinations of former prime ministers Indira Gandhi and her son Rajiv Gandhi, the wars with Pakistan, and the two earthquakes, which caused extensive human suffering. Along with these the continued tension in Kashmir, political and defense scandals, and match fixing in cricket were also recalled.

Many positive events were recalled, as participants recognized and applauded the progress made by members of the country, whether in science, technology, space, or sports. Older journalists could recall the most significant milestones—the nation's independence in 1947, India's war victories against Pakistan, the Pokhran underground explosions that paved the way for India's entry into the exclusive nuclear club, and achievements in sports. India winning the World Cup in cricket in 1983 was recalled by at least one or two members in every focus group. Based on participants' explanations, such achievements were remembered because "it made us proud"; "it revived the national spirit within us" or boosted their morale; and "we are as good or better than most people from other countries." The first man landing on the moon and U.S. president Bill Clinton's visit to India were the only two international events mentioned.

Some important observations can be derived from the events recalled. Unlike what would be presumed in individualistic societies of the West, where the most-recalled events would have personal significance, the Asian societies with more collectivist thinking reveal the cultural factors of showing concern, not for themselves alone but for people around them—the members of their community and in their country. Issues significant for the country were also significant personally to the people of the country. Even when deviant issues such as earthquakes, assassinations, and disasters were recalled, they were high in terms of social significance. The reasons stated were "to feel the suffering of fellow Indians," and "the assassinations of the political leaders led to major losses and led to political instability in the country."

Comparing People's News Preferences with What's in the Newspaper

The gatekeeping exercise revealed the extent to which the perceptions of the audiences, public relations personnel, and journalists differed in terms of the prominence they gave to the issues and the prominence given by the media. Spearman's rank-order correlation coefficients were used to determine the relationship between the actual rankings of the prominence of the newspaper items and the subjective newsworthiness rankings of the four groups of people. Table 12.6 presents the correlation coefficients for the two cities. These findings suggest that for people in India, only a few similarities were found among journalists, media practitioners, and high SES groups with even fewer similarities in the low SES groups. Also, there was only minimal or no

TABLE 12.6 Spearman Rank-Order Correlation Coefficients between Newspaper Item Prominence and Focus Group Rankings

	Journalists	Public Relations	High SES Audience	Low SES Audience	Newspaper
Journalists	–	.63[c]	.51[b]	.11	.17
Public Relations	.18	–	.47[b]	.07	.38[a]
High SES Audience	.45[a]	.44[a]	–	–.01	.19
Low SES Audience	–.13	.31	.37[a]	–	–.04
Newspaper	.23	.38[a]	.33	.07	–

Note: New Delhi coefficients are in the upper triangle; Hyderabad coefficient in the lower. SES = socioeconomic status.
[a]$p < .05$; [b]$p < .01$; [c]$p < .001$.

similarity at all between the newsworthiness judgments of the various group members and the actual prominence given to the stories by the newspapers.

The strongest relationship was between the journalists and public relations practitioners (.63) in New Delhi, which could be because of their similarities in professions. A similar relationship is found between the journalists and the high SES group (.51) and also between the public relations practitioners and the high SES groups (.47). These moderate similarities can be attributed to similarity in the socioeconomic status of both groups. In Hyderabad, there was a similarity in the rankings, though at an average level (.45 and .44) among the journalists, public relations personnel, and the high SES group. The other relationship that is significant at low levels was observed between high and low SES audience groups (.37), which is attributable to the fact that many participants expressed total ignorance of many issues in the headlines that meant nothing to their lives.

The data relating to the relationship between the actual newspaper prominence and the perceptions of the various group members yielded only two significant, albeit quite low, correlations, clearly indicating lack of agreement between what appeared in the newspapers and what is of interest to the audiences, journalists, and public relations practitioners. Thus, overall there seem to be marked differences between the newsworthiness of the items in the newspapers and the rankings of the focus group participants. This implicitly reflects the differences in consumer preferences and those projected in the media.

Discussion

The topical structure across the three media in India is generally similar, with certain variations occurring in the extent of coverage. These variations emerge from the national and regional perspectives in the coverage of news and the nature of the medium itself. The topical structures are similar between the newspapers of the two cities and between the radio and television newscasts of the same city, but vast differences exist in the coverage of topics between the national and regional television and radio broadcasts.

Indian newspapers give balanced coverage to different types of national, international, and regional news covering a wide variety of topics. Overall, the five commonly covered topics were internal politics, sports, internal order, business/commerce/industry, and human interest stories in the *Hindustan Times*. In *Eenadu*, sports news was replaced by cultural events. In Hyderabad, almost half of the television newscasts were devoted to internal politics followed by internal order and less than 5% coverage to all other issues, whereas the New Delhi newscasts give international politics maximum coverage, followed by internal politics, internal order, and sports. Radio newscasts, apart from giving prominence to internal politics and internal order, provide a fair amount of coverage on cultural events, economy/health/welfare services, and social relations. Items about politics—more specifically, political leaders—continue to be deviant in India, given its fractured polity and shifting nature of coalitions, along with communal riots, separatists' movements, and border issues. These create an inescapable part of media coverage and continued interest by readers, viewers, and listeners. However, the inclusion and coverage of other social and cultural events reflect the changing trends of news reporting in striking a balance between the negative and the positive, with sports and social issues occupying greater space alongside the dominant internal politics and internal order.

News tends to be location specific, with regional news from Hyderabad covering regional issues and the New Delhi media giving emphasis to issues of national and international importance. This is also true between the English and the regional language media, where differences reflect variation in the selection and presentation of news based on other news values of proximity, valence, and social significance.

Drawing attention to the analyses of variance, deviance and social significance become guiding principles in influencing the size and placement of news items. The influence of the individual dimensions of deviance and significance are statistically significant in about half the analyses. However, these scores differ among each medium and also in the verbal and visual contexts. The direction of these effects varied, and some also have reverse trends, evident when some of the extremely deviant issues got minimal or moderate coverage. A similar trend is observed in visuals, where the most common news items got maximum coverage and vice versa. Most often, statistical and social change deviance affect the prominence of deviant issues. The lower prominence given to deviant news is explained in relation to the volatile political, social, and economic environment of the country, which necessitates that the media play down the most deviant issues that may lead to problems of internal order, especially those that relate to sensitive issues like national and regional politics, communal politics, and religious uprisings.

Indian news was more influenced by social significance than deviance, which was apparent in a majority of the analyses. Political, cultural, economic, and public significance affected the average prominence scores, particularly for newspapers. The regional news of Hyderabad gives prominence to issues of cultural significance. The trends follow a linear direction in the verbal and visual content unlike those of deviance. These findings support the hypothesis that prominence is influenced by issues of social significance.

The multivariate analyses also reflect differences in predicting prominence. The prominence of verbal newspaper content was influenced by normative deviance, as well as by economic and cultural factors. The lower the normative deviance score, the greater the prominence of verbal newspaper content. Statistical deviance was the largest predictor of the prominence of verbal television content, followed by political significance for visuals. This also explains the smaller number of news items in newscasts and the larger number of deviant items in television compared to newspapers. Newspapers give more extensive coverage to news items, with routine news occupying greater space, which may affect the prominence scores when drawing comparisons across the media.

Interest in the news and in the core concepts of this study were well addressed by the producers and consumers of news in the focus group discussions. Differences in opinion are found in terms of consumer interest and gratifications obtained from the media. Deviant news caught attention by its treatment: bold banner headlines, extensive coverage, and sensational pictures. Deviant issues are of concern and are regularly sought after in the media, while issues of social significance would help improve the quality of life. Definitions of news range from general to specific personal and professional interests. Public relations personnel and high SES audience members define news from the deviant and social significance perspectives, while the lower SES audience members read and hear news in order to be better informed about deviant issues and to find ways to protect and improve their lives. In terms of interest and recall of important news items, there appears to be a remarkable co-orientation in the way journalists, public relations practitioners, and media audiences consider subjects to be newsworthy.

In conclusion, the dimensions of deviance and social significance affect the prominence of news in the media and in the definitions of news by the newsmakers and the consumers. Indian journalism and Indian media can be said to share news values with their audiences. However, the predictive power of both of these concepts is higher in television than in newspapers and radio, given the basic characteristics of each medium and the sociopolitical and cultural climate of the country. However, in an overall analysis, the Indian study supports the view that deviance (though underplayed) and, more importantly, social significance are important predictors for the selection and presentation of news, as consumers seek out the unusual and the useful to improve their professional and personal lives.

13
What's News in Israel?

AKIBA A. COHEN AND NOA LOFFLER-ELEFANT

The Media System in Israel

Israel is a small but multicultural society. At the end of 2000 when the data for the study were collected, the population of Israel was nearly 6.3 million inhabitants, of which 81% were Jewish, 17% Muslim, and 2% Christian (Central Bureau of Statistics 2002). At its founding in 1948, the Jewish population of the state was approximately 600,000, and its increase to over 5 million was the result of large waves of immigration from many countries and ethnic backgrounds as well as from natural growth. The non-Jewish population increase was due to natural growth at a rate greater than among Jews.

Given the fact that Israel is considered the homeland of the Jewish people, its media system is predominantly in Hebrew. However, numerous media outlets both in the print and broadcast media have existed—with some modifications over the years—in other languages, including Arabic, Russian, and English, as well as several other European languages.

While Israel has not yet adopted a formal constitution, press freedom has been sanctioned and strengthened over the years by numerous Supreme Court rulings. The current situation is in sharp contrast to the draconian regulations imposed by the British Mandatory authorities that were in effect prior to the establishment of the state. While there is still a formal requirement to obtain a government permit in order to publish a newspaper, such requests have virtually never been denied. Also, the formal authority to shut down a newspaper has been exercised only on rare occasions regarding both Hebrew and Arabic newspapers and has always been in connection with claims of national security. In addition, while formal military censorship still exists, its rules and practices have been relaxed over the years and currently impose minimal restrictions.

A public television channel was first introduced in 1968. The Broadcasting Authority, funded by a license fee, was shaped on the European model of public broadcasting, particularly the British Broadcasting Corporation (BBC). The Israel Broadcasting Authority (IBA) had a monopoly on television broadcasting until 1990. It has always been highly politicized due to governmental appointments of its regulatory bodies. Cable television was introduced in 1990, was operated as franchises, and was funded by subscriptions. Another government-appointed council regulates the cable and direct satellite broadcasting (DBS) systems, which include Israeli as well as numerous foreign channels. Both the cable and the DBS systems, established in 2001, are funded by subscriptions. Neither the domestically produced channels of the cable system nor the direct satellite television system are allowed to present commercials. A second broadcast television channel, also available on cable and DBS and funded by advertising, began broadcasting in 1992. This channel is regulated by a separate body, the Second Authority for Television and Regional Radio (Caspi and Limor 1999; Schejter 1999). Since the current research began, a second commercial television channel (Channel 10) was introduced, as well as a Russian-language channel, both of which are available on cable and DBS.

Despite the expansion of television broadcasts in Israel, radio has been and remains a dominant medium. There are five nationwide public radio stations, all under the supervision of the IBA and funded by its license fee as well as by advertising on some of the channels. Another very popular radio station is Galei Zahal, operated by the Israeli army. Local commercial radio stations began broadcasting during the 1990s under the jurisdiction of the Second Authority for Television and Regional Radio. These local stations are quite popular, especially in peripheral areas of the country. In addition, numerous private radio stations broadcast to different segments of the Israeli society, such as ultra-Orthodox Jews, Arabs, and Russian immigrants.

There are three major and three minor nationwide Hebrew daily newspapers in Israel. There are also several newspapers in other languages, including two English newspapers, the *Jerusalem Post* and the English version of the elite Hebrew daily, *Haaretz*, published by the International Herald Tribune; one Arabic daily and three Arabic weeklies; and three Russian dailies and numerous Russian weeklies. In addition, many of the several hundred local weeklies are published by the national conglomerates that publish the three main Hebrew dailies. These weeklies, usually published on Fridays, range from several pages, mainly in the *kibbutzim* and *moshavim* (collective settlements), to large editions including several supplements, published in the major towns and cities.

Israelis are heavy news consumers. About 90% read one newspaper daily, and on weekends many read two or more. Television newscasts often lead the television ratings, especially on days when major events take place in the country. This could probably be explained by the complex security situation in Israel, along with its numerous domestic crises and conflicts. Finally, all radio channels broadcast news bulletins on the hour and brief headlines on the half-hour. Radio sets are turned on everywhere, including on public transportation, and people listen to several bulletins each day.

It is important to note the unique character of local media in Israel. Most of the media outlets in Israel are national in scope, mainly due to Israel's small size and sought-after social solidarity. Thus, for example, if a soldier from a small town was killed in military action, his photograph would appear in all the national newspapers and on the national television newscasts. The local media play a relatively minor role compared to national media. The local media usually serve as a supplement to, rather than a substitute for, the consumption of national media. In many respects the local media can be considered as a form of community media.

Local newspapers started appearing during the 1980s (Caspi 1986), with a promising prospect of fulfilling the faults of national newspapers. This was mainly welcomed in peripheral areas of Israel. According to more current research, the periphery in Israel suffers from underrepresentation in the national media; when it is covered, it is usually biased in emphasizing "disorder" (Avraham 2000, 2002).

The Study Sample

The two Israeli cities chosen for the study were Tel Aviv and Be'er Sheba. Although not the capital, Tel Aviv is the main city when it comes to economic and cultural life, and is the center of a heavily populated metropolitan area. Most of the country's media outlets, especially newspapers, are based in Tel Aviv and its surroundings. While the national television and radio stations are based in Jerusalem—the capital city as well as the largest in terms of population—each channel has a major studio in Tel Aviv as well. There is no local television station in Tel Aviv, but a local newscast is broadcast daily on one of the cable channels. There are several local commercial radio stations based in Tel Aviv, as well as two local weekly newspapers. Both newspapers are published by two of the leading national dailies.

The Tel Aviv newspaper selected for the study was *Ha'ir* [The City], published each Friday and provided as a supplement to readers of the *Haaretz* daily and also sold separately in the entire metropolitan area. The radio program selected for analysis was *Mish'al Al Ha'Boker* [Mish'al (the name of the host) This Morning], aired on the most popular local radio station in Tel Aviv, Nonstop Radio. Since there is no bona fide newscast on the local station, the program selected is one that deals with current affairs. It is also syndicated to several of the other radio stations across the country. According to data provided by the regulator of commercial television and radio, the reach of the station was approximately 12% at the time of the study. Finally, the television news program selected, *Tahana Merkazit* [Central Station], is a 30-minute daily newscast broadcasted on one of the cable channels.

Be'er Sheba is a peripheral city in the southern part of the country and serves as the main center of its southern region of the country, which is sparsely populated mainly by small towns and villages. There are four local newspapers in Be'er Sheba, two of which are published as supplements of two of the national newspapers. The newspaper selected for the study, *Kol Hanegev* [The Whole Negev] is published weekly as a supplement of *Yedioth Aharonoth* [The Latest News], the most popular national paper for residents of Be'er Sheba and the Negev region. The radio station selected for the study was Radio Darom [Southern Radio], based in Be'er Sheba. According to data published by the regulator of commercial television and radio, this station is the second most popular local station in the country, with a reach of nearly 16% at the time of the study. The radio program analyzed in the study, *Hayom Ba'Darom* [Today in the South} deals with regional current affairs. The television newscast selected, *Sihat Ha'ir* [Talk of the Town], was at the time of the study broadcast two times each week on one of the cable channels of the Be'er Sheba franchise.

Because the two newspapers selected for the study are weeklies (as are all other local newspapers in Israel), these media outlets were sampled on seven consecutive Fridays rather than in a composite week, as was the case in most of the other countries of the study. The Tel Aviv sample consisted of 852 items: 666 in the newspaper, 107 on radio, and seventy-nine items on television. In Be'er Sheba there are 1,223 items: 936 in the newspaper, 155 on radio, and 132 on television.

Topics in the News

As Table 13.1 shows, the newspapers both in Tel Aviv and Be'er Sheba dealt quite extensively with entertainment (17.5% in Tel Aviv and 13% in Be'er Sheba). Interestingly, television and radio news did not cover entertainment at all. A possible explanation for this finding is that local newspapers in both cities are weeklies, unlike television and radio, which have daily news programs in Tel Aviv. As weeklies, local newspapers supplement the national newspapers and cover different, softer topics. Other topics heavily covered by the newspapers were cultural events (mainly in Tel Aviv, with 17.2%) and sports (14.7% in Tel Aviv and 17.4% in Be'er Sheba). Again, this kind of information (at the local level) cannot be found in the national newspapers.

Another interesting finding is the extensive coverage of "internal order" issues in the Be'er Sheba newspapers compared with the Tel Aviv newspaper (11% versus 3.2%). Be'er Sheba radio devoted 16.5% to this topic, with only 8.8% on the Tel Aviv radio station. Only on television is this topic more prevalent in Tel Aviv than in Be'er Sheba (13.5% versus 8.7%), probably because issues concerning internal order in Tel Aviv are more likely to be reported in the national media, whereas events occurring in Be'er Sheba are only reported in the national media in exceptional cases.

The radio newscast in Tel Aviv is different from the other electronic media newscasts in the study. As noted, the program analyzed is syndicated to several local stations nationwide. This fact is important in order to understand the limited variance of its topics, with 39% dealing with internal politics and another 32.5% with international politics (usually also related to Israel). Other

TABLE 13.1 Distribution of General Topics of News Items by City and Medium

Topics	Newspaper		Television		Radio	
	Tel Aviv	Be'er Sheba	Tel Aviv	Be'er Sheba	Tel Aviv	Be'er Sheba
Entertainment	17.5	13.0	.0	.0	.0	.0
Cultural Events	17.2	8.8	9.9	15.1	.0	.8
Sports	14.7	17.4	7.8	30.7	3.5	14.5
Internal Politics	9.7	7.9	11.3	7.8	39.0	20.2
Human Interest Stories	6.4	6.1	9.2	6.4	.0	1.2
Communication	5.9	2.9	.0	1.4	.0	.0
Business/Commerce/ Industry	3.4	3.0	.7	.9	.0	2.0
Housing	3.3	3.0	4.3	.9	.0	.8
Internal Order	3.2	11.0	13.5	8.7	8.8	16.5
Education	2.9	6.6	2.1	6.0	.0	4.8
Social Relations	2.8	4.2	4.3	6.4	6.6	9.7
Fashion/Beauty	2.1	1.2	4.3	2.3	.0	.0
Economy	1.8	3.9	2.8	2.8	.4	6.5
Transportation	1.8	1.2	3.5	.0	.0	.8
International Politics	1.4	.4	5.0	2.8	32.5	8.9
Health/Welfare/ Social Services	1.3	5.1	2.1	1.4	.0	.8
Environment	1.2	.3	8.5	.5	.0	.0
Labor Relations and Trade Unions	1.1	1.4	.0	1.8	.4	2.4
Disasters/Accidents/ Epidemics	.7	.7	5.0	.5	.4	1.6
Military and Defense	.5	1.3	1.4	.5	8.3	4.4
Science/Technology	.4	.1	.0	.5	.0	.0
Ceremonies	.3	.4	2.8	2.3	.0	4.0
Weather	.2	.0	.7	.0	.0	.0
Energy	.1	.1	.7	.0	.0	.0
Other	.1	.0	.0	.0	.0	.0
Population	.0	.1	.0	.5	.0	.0
Total[a]	100.0	100.0	100.0	100.0	100.0	100.0
	(n = 920)	(n = 1438)	(n = 141)	(n = 218)	(n = 228)	(n = 248)

Note: Distributions given in percent.

[a]Total percentage may not actually be 100.0 due to rounding error.

topics referred to internal order (8.8%), military and defense (8.3%), social relations (6.6%), and sports (3.5%). Most of the other topics were not covered at all. Once again, the range of topics covered is an indicator of the different nature of this program in comparison with other local media, which have more national characteristics and less emphasis on local matters.

In the Tel Aviv television newscast the coverage of environmental issues is greater than in any other medium (8.5% compared to 0 to 1.2% elsewhere). In this sense the local television newscast has managed to find a unique niche among the other media. Sports is the main topic for television news in Be'er Sheba (30.7%), which is much higher than in the other media in that city.

Spearman rank-order correlation coefficients were used to examine the relationship between the rankings of the various topics in the different media (Table 13.2). There is a high correlation between newspaper topics in Be'er Sheba and Tel Aviv (.88). Relatively high correlations also appeared between television and radio newscasts in both cities (.65 and .80, respectively). These findings are interesting as one may assume that the differences between the two cities would lead to differences in the focus of local news. As noted, Be'er Sheba is a small, peripheral city underrepresented in the national media, while Tel Aviv is the central city in Israel. It could be expected that media in Be'er Sheba would focus on internal, municipal issues not found in the national media, while in Tel Aviv no such need would exist. Surprisingly, the high correlations between the two cities do not support this notion. On the contrary, they indicate that the media in the two cities focus on the same areas. It should be noted in this context, however, that in order to learn more about the differences between the two cities, the nature of the coverage also should be examined.

TABLE 13.2 Spearman Rank-Order Correlation Coefficients between Rankings of News Topics in Various Media

	Tel Aviv Newspaper	Be'er Sheba Newspaper	Tel Aviv Television	Be'er Sheba Television	Tel Aviv Radio	Be'er Sheba Radio
Tel Aviv Newspaper		.88[c]	.46[a]	.58[b]	.14	.35
Be'er Sheba Newspaper			.45[a]	.68[c]	.31	.56[b]
Tel Aviv Television				.65[c]	.43[a]	.55[b]
Be'er Sheba Television					.48[a]	.72[c]
Tel Aviv Radio						.80[c]
Be'er Sheba Radio						

[a] $p < .05$; [b] $p < .01$; [c] $p < .001$.

Analyzing the correlations between the different media in both cities adds interesting findings, also related to the differences between the cities. In Tel Aviv, a large and diverse city, relatively low correlations are found between the media. This is not surprising considering the former discussion on prevalent topics. Tel Aviv newspapers put much emphasis on culture and the coverage of cultural events. A city like Tel Aviv is an endless source for cultural activities and developments. The local television newscast's focus on environmental issues is a unique niche of coverage. As noted already, the radio newscast is more nationally oriented (since it is broadcast by several local radio stations). Thus, the different focus of each medium in Tel Aviv leads to relatively low intercorrelations among them (newspapers and television .46, newspapers and radio .14, radio and television .43). The fact that the lowest correlation in Tel Aviv is found between the newspaper and the radio station could be explained by the extremely different nature of the two media. The first is a local weekly focusing on culture, while the second is a syndicated daily program focusing on current affairs. Be'er Sheba, as already mentioned, is a small peripheral city with the same vibrancy of Tel Aviv. An assumption could be made, therefore, that very few hard news events happen there each week. When such news does occur, all media cover it. The findings support this assumption with the relatively higher correlations between the media in Be'er Sheba (newspapers and television .68, newspapers and radio .56, and television and radio .72).

Deviance in the News

The basic hypothesis of the study was that the more deviant a news item is the more prominently it will be covered in the media. In other words, deviant items will take up more space or time and will be more prominently placed compared to other items. Table 13.3 presents the mean prominence scores for each of the three components of deviance: statistical deviance, social change deviance, and normative deviance.

Significant differences between the levels of deviance are obtained in ten of the thirty analyses conducted, of which only one involved visual content (social change deviance of visual content on television in Be'er Sheba). In twenty-one of thirty analyses, no cases were coded as "extremely unusual." In nine of these, none were coded as "quite unusual." A possible explanation for this is that with the exception of radio in Tel Aviv, all of the media in the Israeli case are local media, covering mainly news at the municipal or regional level where no highly unusual events normally occur. In six of ten significant analyses, prominence rises in a linear fashion as deviance rises. In another analysis the lowest prominence score is found for the least deviant items and highest scores for the most deviant ones.

Social change deviance is the best indicator of prominence, with higher social change deviance receiving higher prominence scores in four of the analyses. In Tel Aviv, the effects of deviance are most often found regarding the newspaper, where prominence is significantly different for all components of

TABLE 13.3 Mean Verbal and Visual Prominence Scores for Intensity of Deviance

Intensity of Deviance	Tel Aviv					Be'er Sheba				
	Newspaper		Television		Radio	Newspaper		Television		Radio
	Verbal only (n = 661)	Verbal plus Visual (n = 339)	Verbal (n = 79)	Visual (n = 79)	Verbal (n = 106)	Verbal only (n = 926)	Verbal plus Visual (n = 330)	Verbal (n = 132)	Visual (n = 132)	Verbal (n = 155)
Statistical Deviance										
(1) common	377.2[c]	778.1	199.3	204.4	.0	233.3[c]	812.7	142.8	148.7	56.2[c]
(2) somewhat unusual	744.9	863.2	289.2	419.4	190.6	444.1	1120.2	255.3	.0	117.2
(3) quite unusual	1896.6	.0	248.2	176.4	283.0	808.3	119.7	.0	.0	231.2
(4) extremely unusual	680.0	.0	.0	206.4	390.7	132.0	.0	.0	.0	.0
Social Change Deviance										
(1) not threatening to status quo	413.8[b]	775.3	207.1	206.4	208.0	386.3	836.3	137.9[c]	143.9[b]	81.1[c]
(2) minimal threat	766.6	879.9	.0	.0	303.1	463.3	470.0	851.4	783.0	158.9
(3) moderate threat	1246.8	.0	152.4	.0	306.2	719.3	.0	.0	.0	362.2
(4) major threat	.0	.0	.0	.0	.0	.0	.0	.0	.0	.0

Normative Deviance

(1) does not violate any norms	389.9[b]	783.0	163.6[c]	196.0	254.4	370.4[a]	828.5	127.3[b]	139.2	66.7[b]
(2) minimal violation	643.2	694.2	387.4	251.8	245.2	312.1	1510.2	236.5	238.5	97.0
(3) moderate violation	612.5	.0	221.9	248.4	291.4	624.8	470.0	470.3	294.0	190.2
(4) major violation	1119.5	.0	232.3	633.6	245.0	.0	.0	157.3	.0	.0

[a]$p < .05$; [b]$p < .01$; [c]$p < .001$.

deviance, indicating a linear increment for social change deviance. For normative deviance, there is no linear relationship, but the lowest category of deviance has the lowest score and the highest category the highest score. Most of the analyses for the other media in Tel Aviv are not statistically significant; however, the analysis of normative deviance for the verbal content of the television newscast is significant but does not present a linear relationship between deviance and prominence. In Be'er Sheba, the data for the radio newscast presents a linear relationship for all components of deviance. Also, the television newscast in Be'er Sheba presents a significant linear relationship for social change deviance and the prominence of both verbal and visual content.

Social Significance in the News

The research hypothesis here was that the more a news item is socially significant, the more prominent it will be in the media. The findings are presented in Table 13.4. Of the forty ANOVAs conducted, twelve resulted in significant differences, of which only two are significant for television in the two cities. In Tel Aviv, effects of social significance are most often found regarding the newspaper, where prominence is significantly different for three components of deviance. In two of the components, economic significance and cultural significance, there are linear relationships with prominence. The fourth component, public significance, is not statistically significant. Most of the analyses for the other media in Tel Aviv are not statistically significant except for political significance of verbal content in the television newscast.

In Be'er Sheba, the radio newscast presents linear relationships for two components of social significance, political and public significance. The third component, economic significance, is not statistically significant but shows a linear relationship with the prominence scores. Also in Be'er Sheba, there are significant linear relationships regarding the television newscast for three of the four components of social significance: political, economic, and public significance.

Looking broadly at the data, some interesting similarities can be seen between the statistically significant differences for deviance and for social significance. Social significance was theoretically composed of four components—political, economic, cultural, and public. In three of the four components, statistically significant differences are found regarding the Tel Aviv newspaper. In three other components, significant differences arose regarding the Be'er Sheba radio station. The Tel Aviv television newscast presents two significant differences, and the television newscast in Be'er Sheba presents three significant differences for verbal content and one for visual content.

Comparing these data to those on deviance in Table 13.3, there is much similarity in the findings. There too, many of the analyses are significant for the Tel Aviv newspaper and for the radio station in Be'er Sheba. As for radio in

TABLE 13.4 Mean Verbal and Visual Prominence Scores for Intensity of Social Significance

Intensity of Social Significance	Tel Aviv					Be'er Sheba				
	Newspaper		Television		Radio	Newspaper		Television		Radio
	Verbal only ($n = 662$)	Verbal plus Visual ($n = 340$)	Verbal ($n = 79$)	Visual ($n = 79$)	Verbal ($n = 107$)	Verbal only ($n = 927$)	Verbal plus Visual ($n = 327$)	Verbal ($n = 132$)	Visual ($n = 132$)	Verbal ($n = 155$)
Political Significance										
(1) not significant	416.4[a]	779.1	165.2[b]	203.4	163.3	351.6[b]	846.8	128.2[b]	147.0	66.2[b]
(2) minimal	395.3	703.2	253.8	228.6	.0	346.3	1047.2	236.5	270.0	81.3
(3) moderate	849.6	980.0	320.5	419.4	246.0	713.6	.0	305.3	259.2	163.9
(4) major	775.5	.0	.0	.0	397.6	138.0	.0	.0	.0	.0
Economic Significance										
(1) not significant	416.0[b]	777.9	196.3	206.4	272.0	381.3	852.7	131.4[c]	148.7	104.8
(2) minimal	586.3	.0	392.4	.0	61.7	476.9	384.5	633.4	.0	106.3
(3) moderate	1814.8	.0	400.5	.0	265.0	381.1	.0	919.8	.0	153.9
(4) major	.0	.0	.0	.0	.0	132.0	.0	.0	.0	.0

[a] $p < .05$; [b] $p < .01$; [c] $p < .001$.

(*Continued*)

TABLE 13.4 Mean Verbal and Visual Prominence Scores for Intensity of Social Significance (*Continued*)

Intensity of Social Significance	Tel Aviv					Be'er Sheba				
	Newspaper		Television		Radio	Newspaper		Television		Radio
	Verbal only (n = 662)	Verbal plus Visual (n = 340)	Verbal (n = 79)	Visual (n = 79)	Verbal (n = 107)	Verbal only (n = 927)	Verbal plus Visual (n = 327)	Verbal (n = 132)	Visual (n = 132)	Verbal (n = 155)
Cultural Significance										
(1) not significant	363.2[c]	772.8	213.2	208.9	283.3	304.8[b]	847.0	173.1	168.8	124.1[a]
(2) minimal	577.5	821.6	189.6	195.4	87.1	476.2	944.8	92.8	95.2	62.9
(3) moderate	1326.1	1509.0	174.2	188.0	215.0	701.2	.0	.0	.0	196.7
(4) major	.0	.0	206.4	.0	.0	.0	.0	.0	.0	.0
Public Significance										
(1) not significant	405.7	780.0	180.1	193.4[b]	299.3	322.2[a]	851.1	126.2[c]	145.0[a]	66.9[b]
(2) minimal	541.3	534.9	256.2	220.8	117.8	460.9	649.0	180.9	545.4	143.3
(3) moderate	742.2	.0	239.3	437.4	258.6	530.8	.0	456.9	241.2	163.0
(4) major	960.0	.0	264.6	.0	477.0	814.9	.0	.0	.0	.0

[a] $p < .05$; [b] $p < .01$; [c] $p < .001$.

Tel Aviv and the newspaper in Be'er Sheba, there are no significant results in either case. The significant difference for television visual content appeared in both cities only for public significance. This finding seems to suggest that deviance and social significance are somewhat related. In other words, when a medium indicates the presence of deviance it also suggests the presence of social significance. Thus, in line with the research hypothesis, it is suggested that deviance and social significance are two components of a broader concept of newsworthiness.

Deviance and Social Significance as Predictors of News Prominence

In order to simultaneously test the influence of all the dimensions of deviance and social significance on news prominence, eight stepwise regression analyses are calculated. For each of the two cities, separate analyses are calculated for the prominence of verbal newspaper content, for the newspaper visuals and verbals combined and for television and radio prominence. As can be seen in Table 13.5, other than two nonsignificant analyses (radio prominence in Tel Aviv and verbal newspaper prominence in Be'er Sheba), all regression analyses produced statistically significant results. In three cases, however, the amount of explained variance is rather small—5% or less.

Deviance and significance explained more variance in television than in any other medium—45% in Be'er Sheba and 20% in Tel Aviv. In Be'er Sheba, four verbal components weighed significantly in predicting prominence: social change deviance, normative deviance, economic significance, and public significance. In Tel Aviv, two components accounted for the explained variance: normative deviance and political significance. No visual components were predicted in either case.

Of the two radio stations, only in Be'er Sheba is there a significant amount of variance explained (17%). The components making significant contributions to the prominence of verbal news items were social change deviance and normative deviance. In Tel Aviv, 9% of variance in the prominence of verbal content is explained by statistical deviance and cultural significance. Of the explained variance in prominence, 4% is due to combining verbal and visual content. The same analysis for Be'er Sheba newspapers produces 5% of explained variance.

Comparing the different dimensions of deviance and social significance leads to some interesting findings. As could be suggested from the ANOVAs reported, none of the variables was a significant predictor of visual content prominence. The ANOVAs show that almost no significant differences exist between the different categories of deviance and social significance for visual content. As for statistical and normative deviance, each played a significant role in three analyses. Statistical deviance is an effective component of the explained variance for the newspapers in both cities: in Tel Aviv for both verbal as well as for the combined visual and verbal prominence; and in Be'er Sheba for the prominence of combined visual and verbal content.

TABLE 13.5 Stepwise Regression Analyses of Intensity of Deviance and Social Significance on News Prominence

	Tel Aviv								Be'er Sheba							
	Newspaper Prominence Verbal only (Total R^2=.09c, n=661)		Newspaper Prominence Visual and Verbal (Total R^2=.04b, n=330)		Television Prominence (Total R^2=.20c, n=79)		Radio Prominence (Total R^2=ns, n=106)		Newspaper Prominence Verbal only (Total R^2=ns, n=924)		Newspaper Prominence Visual and Verbal (Total R^2=.05c, n=319)		Television Prominence (Total R^2=.45c, n=132)		Radio Prominence (Total R^2=.17c, n=155)	
Independent Variables	r	Std. Beta	r	Std. Beta	r	Std. Beta	r	Std. Beta	r	Std. Beta	r	Std. Beta	r	Std. Beta	r	Std. Beta
Deviance																
- Statistical, verbal content	.25c	.22c	.19b	.19b	.10	ns	.17	ns	.06	ns	.23c	.23c	.13	ns	.32c	ns
- Statistical, visual content	—	—	.03	ns	.02	ns	—	—	—	—	.06	ns	ns	ns	—	—
- Social change, verbal content	.13b	ns	.09	ns	-.04	ns	.13	ns	-.01	ns	.03	ns	.45c	.31c	.38c	.31c
- Social change, visual content	—	—	.01	ns	ns	ns	—	—	—	—	-.02	ns	.28b	ns	—	—
- Normative, verbal content	.14c	ns	.07	ns	.20	.26a	.06	ns	-.02	ns	.06	ns	.21a	.23b	.28c	.17a
- Normative, visual content	—	—	-.02	ns	.26a	ns	—	—	—	—	.07	ns	.17	ns	—	—

Social Significance

– Political, verbal content	.07	ns	.01	ns	.36[b]	.36[b]	.17	ns	-.02	ns	.11[a]	ns	.26[b]	ns	.26[c]	ns
– Political, visual content	–	–	.00	ns	.14	ns	–	–	–	–	.05	ns	.07	ns	–	–
– Economic, verbal content	.12[b]	ns	.12[a]	ns	.25[a]	ns	-.06	ns	-.02	ns	.03	ns	.51[c]	.47[c]	.10	ns
– Economic, visual content	–	–	ns	ns	ns	ns	–	–	–	–	-.03	ns	ns	ns	–	–
– Cultural, verbal content	.21[c]	.17[c]	.09	ns	-.07	ns	-.09	ns	.03	ns	.10	ns	-.19[a]	ns	-.01	ns
– Cultural, visual content	–	–	.06	ns	.03	ns	–	–	–	–	.01	ns	-.17	ns	–	–
– Public, verbal content	.10[b]	ns	.05	ns	.17	ns	.02	ns	-.03	ns	.12[a]	ns	.36[c]	.19[b]	.24[b]	ns
– Public, visual content	–	–	-.03	ns	.31[b]	ns	–	–	–	–	-.01	ns	.12	ns	–	–

[a]p < .05; [b]p < .01; [c]p < .001; ns = not part of final stepwise regression equation.

Normative deviance effectively explains television prominence in both cities and radio prominence only in Be'er Sheba. Social change deviance is not effective in Tel Aviv but is a very effective predictor of prominence for television and radio in Be'er Sheba in both cases (.31).

Interestingly, each component of social significance is effective for only one medium in each of the two cities. Political significance is the main predictor of television prominence in Tel Aviv (.36) together with normative deviance. Economic significance is very effective for television in Be'er Sheba (.47). Public significance is a more moderate predictor for the same medium (.19). The fact that cultural significance is an effective predictor (.17) for the newspaper in Tel Aviv could be expected, considering the extensive coverage of topics related to culture and cultural events by the Tel Aviv newspaper.

As noted, the largest percentage of explained variance is found for the prominence of television content in Be'er Sheba. This can be explained by four components, with the main predictors being social change deviance (.31) and economic significance (.47). Normative deviance and public significance were moderate predictors. For the other two media, the explained variance of prominence is due to one or two components at most.

People Defining News

Organizing the focus groups in Israel was not an easy task, especially with the journalists. In Tel Aviv four groups were conducted and in Be'er Sheba five groups, because of the difficulty in assembling print and broadcast journalists together in one session. Overall the groups ranged in size from seven to ten participants, with the two groups of Be'er Sheba journalists consisting of five per group. The audience focus groups, both low and high socioeconomic status (SES), were conducted in the home of one of the participants, who was responsible for inviting the other group members. The meetings with the journalists and the public relations practitioners were held in the offices of the participants who agreed to host the groups. The Be'er Sheba sessions took place between May and September 2000, and the Tel Aviv groups were conducted between July and November 2001 (three of these groups took place following the September 11, 2001, events in the United States). One thing typifying the Israeli focus groups was the relatively high degree of debating among the group members, which is characteristic of how Israelis converse.

The first stage of the focus groups dealt with the question of the three most important news items in the participants' lives. The participants generally recalled wars, terrorist acts, peace negotiations, and other political events. The assassination of former Israeli Prime Minister Yitzhak Rabin was mentioned by many of the people in all groups, except in the low SES audience group in Be'er Sheba. It is interesting to note that after the respondents revealed the events they had indicated in writing, most of those who did not mention Rabin felt the need to apologize or explain why.

As noted, three of the focus groups took place after the 9/11 tragedy, which led many participants to mention it, yet most of the events mentioned in all the groups were Israeli events rather than global ones. Only a few events outside Israel were mentioned, such as the fall of the Berlin Wall, the demise of the Soviet Union, and the deciphering of the human genome.

Quite a number of participants mentioned the first ever visit in 1977 by Egyptian President Anwar Sadat to Israel and described the scene as he arrived at the Israeli airport. In this connection some said the historical development rather than the media created the event. Indeed, many participants mentioned events that led to social or political change. Thus, for example a female public relations person in Be'er Sheba tried to characterize the events she mentioned: "My common denominator is the personal feeling that everything is going to change, for better or for worse ... a feeling of empowerment or loss of it." A Tel Aviv public relations person who mentioned the human genome put it this way: "This event symbolized the end of an era and the beginning of a totally new one." Another person referred to the metaphor of the "end of an era" with regard to Israel's withdrawal from Lebanon. In a Tel Aviv high SES audience group one of the members said, "These are planned events. ... What was and what will be are not the same." Finally, another respondent surmised, "In my relatively short life, these are the most significant events that took place, that have historical significance, and will continue to be meaningful."

The next theme of the focus group discussions surrounded the question of what kinds of information are relevant to the participants. Specifically, reference was made to positive versus negative news, as well as news topics that receive too much or too little coverage.

The Tel Aviv public relations people mainly present an instrumental attitude toward the media, describing newspaper reading as a process of collecting media opportunities. They viewed the distinction between positive and negative news as invalid. According to them, the main criterion was the relevance of the news item to their professional and personal lives. They considered the abundance of bad news as a mere characteristic of the media and as being more appealing to audiences rather than good news.

The Be'er Sheba public relations people also identified negative news as engaging the audience. A debate ensued in the group as to whether news is bad because reality is bad or because of the tendency of journalists to provide the public with what it wants. One of the female participants remarked, "There's something about people that attracts them to blood." This group claimed there is exaggerated coverage of violence and politics and too little coverage of such events related to scientific and medical achievements and the environment. Another female group member felt that the media have become hysterical and have thereby given legitimacy to hysterical behavior among members of the public in just about every realm of life.

The Tel Aviv journalists, similar to the public relations people, argued that what catches their attention is information relevant to their beats. Several of the group members explained that information that gets their attention is generally new and deviant. In an attempt to identify what kind of information gets his attention, one participant said, "The sole condition that interests me is whether the item will tell something new or something that is extraordinary, uncommon. ... The word 'new' is the key word. Everything that happens that was previously unknown to me is interesting." Another participant agreed and added, "Anything can get your attention, but only on condition that there's an especially good or bad story to it, something extreme." The group heatedly debated the issue of covering good news. Some of the participants claimed that reporting about a surplus in the municipal budget or about a successful police operation does not constitute news because city workers and police officers are supposed to do their jobs for the benefit of the public and therefore should not receive special attention. Other participants argued that as audience members they would want to receive such information and that unfortunately it is generally lacking. A similar argument took place among the Be'er Sheba journalists, some of whom claimed that the role of the media is to report on things that are not in order and that this is the expectation of the audience. There was consensus among the Tel Aviv journalists that the audience prefers bad and sensational news but that at the same time the audience is critical of the way the media fulfill their role.

According to the Be'er Sheba journalists, the issue was not positive versus negative news but rather that the audience is interested in news "that would excite them." The litmus test for interesting news seemed to be at the emotional level. In Be'er Sheba the proximity issue also came up. The editor of a local television news program explained, "What's close is what catches people." Another participant added, "If a Be'er Sheba resident wins several million in the Lotto, I'm sure that this item would devour any rape, murder, or corruption story. This is a story that the man in the street can identify with." Several journalists in Be'er Sheba complained about the fact that the national press does not cover events taking place in the periphery of the country. Some complained that too much attention was given to crime and too little to education, the environment, and entertainment.

The low SES audience group participants in Be'er Sheba made the same claim that there was too much coverage of crime and politics and too little of education, culture, and entertainment. The group members generally rejected the distinction between good and bad news, claiming that it is important for them to be exposed to information that is useful to them. Positive information, they said, could improve their quality of life, while negative information can help them become aware of things they need to avoid. A similar notion was also suggested by the high SES audience group in Be'er Sheba. Most of the participants agreed that, although they enjoy reading good news stories, they are also attracted to bad news stories out of a sense of identification and desire to learn what to beware of. As one of the participants put it, "Every bad news item that stems from something bad, something negative, something that didn't go as it should, could actually be a sort of beginning of a process of improving the situation so that later on we can hear good news."

As in the debate among the journalists, some of the participants in the low SES audience group complained that the media emphasize bad news, but some of the other participants claimed that this is the role of the media and that the difficult reality is not the invention of the media. One of the participants put it this way: "I think that the foundations of a healthy society are that we take the good for granted and the bad as something deviant that we need to cope with."

One member of the high SES audience group said she was attracted to information that represents something new, whether positive or negative. Others in the group argued in this vein that good news items are not very common and that is why they become salient when they are present. As for the general population—above and beyond the focus group members themselves—the participants felt that people prefer negative and shocking news.

The group members complained that the media deal too much with certain areas, such as politics, without reporting anything new. They demanded that the news be more up to date and not repetitive. This group also complained that the media do not deal sufficiently with education and sports or the events that take place in the southern part of the country and the periphery in general.

Some of the participants in the high SES audience group in Tel Aviv clearly characterized the kinds of information that attract attention according to the hypotheses of the study. Definitions included new or sudden information, extreme events (even at the personal level), deviation from the norm, and the oft-cited example "man bites dog." One group member stressed that other dimensions catch his attention and that he is attracted to information he finds important socially, politically, and economically; this notion came at the heel of a complaint by a female participant regarding an item about a Nigerian beauty queen brought to the attention of the group by the person who felt it was an important social issue.

Toward the end of the focus group discussions, the moderator asked participants directly how they would define *news*. The instinctive response in most of the groups was "something new."

In the Be'er Sheba public relations group, people said that of all new events occurring, items that would become news are those with the potential of changing attitudes or perceptions, arousing responses, and being relevant to the general public. One of the participants, the spokesperson for the local university, gave the following example: "At our commencement there was a blind student and an 80-year-old student. Of the thousands of students graduating, those two became news items." The group members agreed that these two graduates change the normal way of thinking. The print journalists agreed: "A worthy news item is one that people will talk about." The broadcast journalists drew a distinction between what is important and what is interesting.

Important things, they argued, are those events that have an impact on what happens to us in our daily lives—interesting things are "sexy" or items that bring high ratings. Finally, the high SES audience participants suggested that for events to become news they must deviate from social norms.

In Tel Aviv, the public relations participants emphasized that commercial considerations bring the media to make news of what appears to be interesting and important for as broad a public as possible. This group also spoke of the deviance that makes news. "It doesn't have to interest many people, but because it's deviant it will be included; it doesn't have to be new but interesting and eye-catching." The public relations people agreed that their perceptions of the nature of news were different from that of the general public in that most news items are the outcome of self-serving interests. The Tel Aviv journalists claimed that the characteristics of news change from one medium to another.

In both cities, the audience groups described news as information that would not reach the public without the media. The main characteristics of news are the interest of the public in the items and their uncommon nature. The audience group also made reference to the commercial considerations in determining what makes news. Finally, the Tel Aviv high SES audience group talked about "quality" news as representing real change, presented in low-key and neutral language and without aiming toward the audience's emotions but rather at their cognitions.

Comparing People's News Preferences with What's in the Newspaper

The respondents in each of the focus groups went through three rounds of the gatekeeping task in which they arranged ten news items (based on their headlines) according to their perception of how they should be lined up in the newspaper. Once they were done with sorting the items, the moderators asked the group members which criteria they used in their rankings.

One criterion indicated by people in each of the groups in both cities was the public perspective of the item—that is, the extent to which the information was significant for the general population, information that could be useful or threatening for people, and information dealing with private or public life. Some of the participants placed strong emphasis on items dealing with education, while others said they placed items that had the potential of affecting the most people higher in their line-up. Public relations practitioners, journalists, and the Be'er Sheba audiences felt it had to do with the geographic proximity of the events: the closer they are, the more highly they would be placed in the line-up.

The final analysis involves correlations between the rankings of items, both between the different focus groups and between each group and the actual ranking of the items in the city newspaper from which the items were taken. Table 13.6 presents the Spearman rank-order correlation coefficients for the two cities. For technical reasons, the data for the gatekeeping task done by the

TABLE 13.6 Spearman Rank-Order Correlation Coefficients between Newspaper Item Prominence and Focus Group Rankings

	Journalists	Public Relations	High SES Audience	Low SES Audience	Newspaper
Journalists	–	.78[c]	.81[c]	.90[c]	.57[b]
Public Relations	−.44[a]	–	.78[c]	.84[c]	.56[b]
High SES Audience	NA	NA	–	.80[c]	.49[b]
Low SES Audience	−.65[c]	.75[c]	NA	–	.67[c]
Newspaper	.25	−.16	NA	−.24	–

Note: Tel Aviv coefficients are in the upper triangle; Be'er Sheba coefficients in the lower. SES = Socioeconomic status; NA = not available.
[a] $p < .05$; [b] $p < .01$; [c] $p < .001$.

high SES audience group in Be'er Sheba cannot be present, because it was not recorded properly.

The rank-order correlations between the four focus groups in Tel Aviv are all positive and high, ranging from a minimum of .78 to a maximum of .90. This indicates high agreement between the different groups in terms of which items should be more prominent and which should be less prominent. In Be'er Sheba, however, correlations are lower, with negative coefficients between the rankings of the journalists with the rankings of the low SES audience group and public relations practitioners. These findings indicate that the public relations people perceived the desired line-up differently from the journalists.

As for the actual placement of the sets of ten items, here too the findings in the two cities are dramatically different. In Tel Aviv, the correlations are all positive and ranged between .49 with the high SES audience participants and .67 with the low SES participants. In Be'er Sheba, on the other hand, the correlations for the public relations practitioners and the low SES participants are −.16 and −.24, respectively, indicating little if any relationship between the rankings. The correlation for the journalists is low (.25), indicating only very slight agreement in the rankings of the journalists and the real newspaper items.

This seems to indicate that in Be'er Sheba, the peripheral town in which people complained about the way their city was represented in the media, a lack of satisfaction indeed exists with the way the newspaper actually presents its stories. Interestingly, the public relations practitioners "joined forces" with the audience members in expressing their dissatisfaction. On the other hand the journalists, who are involved in the process of setting the agenda of the newspaper, while not rejecting the newspaper presentation, did not seem to be very satisfied with it either.

Discussion

Israel is a highly politicized and conflict-ridden society. Israel has also gone through many distressing events as a society with short- as well as long-term impacts on its people. Along with the traumatic experiences that have plagued Israel over the years, and despite the numerous social divisions with which the society has been attempting to cope—Jews versus Arabs, religious versus secular, left versus right—its people have always rallied together in times of crisis.

These points seem to be well reflected in the kinds of news items people mentioned in the focus groups as the most important events in their lives. Many of these were on a major national scale, such as the waging of war and the signing of peace treaties, along with the political strife leading to the assassination of Prime Minister Rabin (an event recalled by many). This social context also seems to explain why much of the discussion that ensued in the focus groups directly and indirectly dealt with the notions of deviance and social significance, even if not always mentioning the terms specifically by name.

Also, as noted at the outset, the major media outlets in the country are national newspapers, television channels, and radio stations. These media typically devote considerable time and space to national and international issues, mainly those involving Israel, its Arab neighbors, and relations with the world powers, mainly the United States. In the present study, however, the media selected for analysis are of the local genre; hence, many of these major stories having impact on the life of the entire country are typically absent from their coverage.

Moreover, when dealing with the periphery of the country, consisting of many small towns and villages, the national media only report on exceptional events taking place in those locations or deemphasize the coverage of such events in comparison to the reportage given to local events in the major cities. As "compensation" for this gap in coverage, the local media focus on softer news events, including an abundance of cultural events and sports. In even greater contrast to the lack of reporting of such events in the local media is the focus group respondents' mention of national events, possibly indicating their relative inadequacy.

These facts might also relate to the relatively low correlations obtained in the gatekeeping tasks in both cities but especially in Tel Aviv. Specifically, this may be due to the Israelis' high interest in national and international politics that are quite absent from the local media, especially the local newspapers. (The Tel Aviv radio program is an exception, as stated already, because it is actually a nationally syndicated program dealing with broader national affairs.)

Israelis are also quite socially minded and highly critical of various authorities. They often complain about the policies of the central government, municipalities, and other social agencies. They demonstrate often, and they threaten to take legal action—and quite frequently actually do—usually by asking for injunctions against these agencies. All of this seems to tie in with the emphasis placed by the focus group participants on the need for social change and the function of news in aiding to promote such processes when asked to define news.

Finally, the findings in the Israeli study seem to coincide with the love–hate relationship many Israelis have with the media. As noted, they are very heavy media consumers, especially of news, but they often express outrage at how the media deal with various phenomena in their country. This is probably more typical of people living in the periphery of the country—in places like Be'er Sheba—as they only find reference to their towns and villages in the dominant national press when negative events occur in their communities. It could therefore be expected that the local media, especially in Be'er Sheba, would play a significant role in supplementing the national media. These expectations seem not to have been realized, however, leading the people to be critical of their local media.

Several explanations might be offered for this presumed failure. One is that most of the local media outlets are part of national newspaper conglomerates or are under the scrutiny of the centralized national operators and regulators of the broadcast media. Another explanation has to do with the characteristics of the staff of local media, which are usually composed of young journalists at the beginning of their professional lives trying to build up their portfolios before moving on to the national media. Their admitted ultimate goal in most cases seems to be serving national needs rather than local needs. Thus, the local news media scene in Israel, while quite prolific—especially the print media—is not appreciated enough by its potential consumers even though they buy the local newspaper each weekend.

14
What's News in Jordan?

MOHAMMED ISSA TAHA ALI

The Media System in Jordan

Jordan is a small Arab country with limited natural resources. Despite its small size, it has played a pivotal role in the struggle for power in the Middle East. Jordan's significance results partly from its political and military links to the West. Its strategic location, at the crossroads of what most Christians, Jews, and Muslims call the Holy Land, also gives Jordan a relatively strong role. Unlike many of its eastern neighbors, Jordan has no oil of its own. Its economic strength depends on foreign aid, transfers of workers abroad, and moderate exports of cement, phosphates, fertilizers, medical products, and agricultural produce. The annual per capita income is $1,745.

Jordan's population was approximately 5.3 million in 2002, growing rapidly at 2.8% per year. Life expectancy is 65 years for men and 70 years for women. The population distribution pattern across geographic regions is heavily influenced by the quality of land and water availability. Nearly 90% of the population resides in northwest Jordan. Administratively, Jordan is divided into twelve governates, with Amman (which is also the name of the capital city) being the largest of the twelve.

There is a dearth of literature concerning the media in Jordan. In 2000, Jordan had six daily newspapers, four of which appear in Arabic: *Al-Rai* [The Opinion], *Al-Dustur* [The Constitution], *Al-Arab Al-Yawm* [Arabs Today], and *Al-Aswaq* [The Markets]. An additional two dailies, the *Jordan Times* and the *Star*, are published in English.

Jordan has traditionally been a country where the media are under tight state control. While the press is mostly privately owned, it is subject to censorship. The Ministry of Culture and Information is responsible for most of the press censorship on a daily basis, but editors generally exercise self-censorship to minimize conflicts with the authorities. Nonetheless, the Arabic language newspapers have been suspended at various times for publishing articles considered objectionable. In this context, the press was seen as an extended public relations arm of the government. The government also operates an official news agency, Petra, in addition to several international news services that maintain bureaus

239

in Amman, including Agence France-Presse, the Associated Press, Reuters, Dutch News Agency, The Russian Information Telegraph Agency of Russia-Telegrafnoe Agentstvo Sovietskoye Soiuza (ITAR-TASS), and many others.

However, since 1998 most of these restrictions have been relaxed. In fact, there has been no strict daily control over any publications and cases are usually raised only when complaints about an article reach the department.

In addition to the daily newspapers, there are also nineteen weekly newspapers including those owned by opposition parties critical of the government. Among the weeklies that claim the largest circulation are *Shihan* [Mountain], *Al-Bilad* [The Country], *Al-Sabil* [The Road], *Al-Ordon* [Jordan], and *Al-Hilal* [The Crescent].

Jordan radio and television are controlled by the government. Jordan Television is a state-run corporation. It operates Channel One, Channel Two (as a foreign-language channel), and the Jordan Satellite Channel, which transmits the programming of Channel One. Jordan Radio operates services in Arabic, English, and French. Radio and television are on the air twenty hours per day in Arabic and fifteen hours per day in English. The news in foreign languages is mainly directed to Jordanians, since many people speak English. In 2000, there were an estimated 550,000 privately owned radio receivers. Both radio and television present advertisements and are funded by the government. Naturally, since both accept advertising, a large amount of funding comes from commercial resources. The Jordanian government now allows private television and radio stations, ending a long era of a state-operated monopoly. It is also now possible for domestic and foreign investors to establish private television and radio stations, subject to strict regulation. The government is also encouraging Arab commercial satellite broadcasters to relocate to its media-free zone, though investors remain concerned about safeguards against censorship.

All the Jordanian media—whether dealing with politics, the economy, social issues, or sports—did not have an identity of their own. Such a media identity requires several characteristics, including distinct media institutions, professionalization, specialization, knowledge of the issues, and a well-trained network of journalists aware of and dedicated to the issues. These components were lacking in the Jordanian media.

In 1989 the late King Hussein began a drive toward democratization in Jordan. As a result, the number of daily and weekly newspapers increased from four to around twenty, but their quality lagged behind. At the time, an owner or publisher of a newspaper could be almost anyone possessing an academic degree and capital of about $5,000. Under laws introduced last year the qualifications were raised to eight years of professional experience, at least five of which had to be in Jordan, and the required capital for a daily newspaper increased to $350,000. The main reasons for introducing these changes were the unprofessional and unethical standards of journalists in the country.

The Study Sample

The two cities selected for the Jordanian study were the capital, Amman, and the smaller city of Irbid, located fifty miles north of Amman. Amman has a population of 1.9 million inhabitants, comprising some 38% of the population of the country; Irbid's population is 900,000, comprising some 18% of the total population. Irbid is one of the few Jordanian cities that could meet the parameters for the content analysis. The daily newspaper selected in Amman was *Al-Rai*, which enjoys the largest circulation. No daily newspaper is published in Irbid; hence, a weekly, *Shihan*—the popular outlet in the city—was selected.

The dates for the composite week for the newspaper sample were according to the general plan of the study, beginning Monday, November 13, 2000, and ending Sunday, December 31, 2000. For the small city, because the newspaper *Shihan* is a weekly newspaper, the sample began Sunday, November 12, 2000, and continued for seven consecutive weeks.

Because newspapers and broadcast media have different news cycles, the collection of television and radio news content took place on the day previous to the selected newspaper sampling day. Thus, because our newspaper sample started on a Monday, the television and radio sample started on a Sunday, with every following broadcast sample day being the day previous to the respective newspaper sample day.

For the television sample in Amman, Jordan's Channel One was selected. This channel provides the leading domestic television newscast and has the highest reputation and largest viewership in the country. The dates selected were in accordance with the general plan of the study, beginning Sunday, November 12, 2000, and ending Saturday, December 30, 2000. The main news bulletin was aired at 8:00 p.m. In Irbid, since no television station providing a local television newscast exists, a short news bulletin on regional issues airing at 6:00 p.m. on Channel One was used instead.

The radio stations providing the leading newscasts were selected. For Amman, the main radio newscast broadcast on Radio Jordan was selected. This daily program was on the air at 6:00 p.m. Given the limited choice of radio stations in Jordan and the fact that no radio station provides a local newscast, in Irbid the evening newscast was used, which aired at 9:00 p.m. and typically included items of regional and local news.

The six Jordanian news outlets yielded a total of 3,352 news items during the study period. The two newspapers accounted for the vast of majority of news items—2,761 (1,620 in *Al-Rai* and 1,141 in *Shihan*)—followed by radio items—297 (166 for Amman and 131 for Irbid). Television news items accounted for 294 (227 in Amman and sixty-seven in Irbid).

Topics in the News

As can be seen from Table 14.1, international politics and internal politics dominate the topics of news in both television and radio in Jordan. This is the subject that the media usually cover by reporting on visits and statements from heads of state or individual politicians. Being a small country located in a politically disturbed region, this finding is not surprising. The tragic death of King Hussein, the accession of his son to the throne, and the deterioration of the peace process in the Middle East were all reflected in this trend. Apart from internal and international politics, it could be noticed that there are differences among the three media in addressing similar topics even within the same geographic area. This is clear in the following points.

First, for the Amman newspapers the three most prevalent topics after international politics were sports (15.4%), the economy (9.9%), and cultural events (8.8%). The main topics after international and internal politics for television were sports (10%), business/commerce/industry (5.3%), and economy (5%); on the radio the three main topics were military and defense (6.2%), education (3.3%), and weather (2.9%).

Second, the three most prevalent topics after international politics in Irbid's newspaper were cultural events (20.6%), internal politics (12.5%), and internal order (9.8%). The main topics on television news were education and internal politics (both with 16.7%), human interest stories (9.5%), and weather (7.1%); on the radio, after international and internal politics were the economy (4.1%) and military and defense and the weather (both with 2.1%).

The relative importance of the topics varied considerably among the media. For example, while sports activities were heavily reported in the newspapers, they were seemingly less important on the television and were completely ignored on the radio. The same trend was found with regard to entertainment. Also, weather was covered relatively more by radio and television news but was ignored to some extent by the newspapers.

It could be concluded that while there are no meaningful demographic, economic, and social differences between Amman and Irbid, there were various differences among newspaper, television, and radio news reporting concerning the news topics.

Comparing the number of different topics appearing in the three media, large differences were found among them. Based on the detailed topic categories, the newspaper items made reference to 166 different topics, compared to fifty-four topics covered on television items and only thirty on radio items. This is not only because newspapers can carry more information than the other two media but also due to the nature of government-controlled television and radio news, both geared to reflect government rather than private activities.

Table 14.2 shows that the rank-order correlation of the topical structures between the newspapers in both cities is high (.87), thus indicating much similarity between the topics. As for radio, the topical structures are also

TABLE 14.1 Distribution of General Topics of News Items by City and Medium

Topics	Newspaper		Television		Radio	
	Amman	Irbid	Amman	Irbid	Amman	Irbid
International Politics	18.7	15.7	42.3	25.0	58.9	80.7
Sports	15.4	3.8	10.0	.0	.0	.0
Economy	9.9	3.2	5.0	.0	.5	4.1
Cultural Events	8.8	20.6	.0	1.2	.5	.0
Internal Politics	8.2	12.5	15.3	16.7	21.5	5.5
Business/Commerce/ Industry	7.0	5.0	5.3	2.4	1.0	.0
Health/Welfare/ Social Services	5.4	5.3	4.6	4.8	1.0	.0
Social Relations	4.4	3.2	1.4	4.8	1.4	.7
Education	3.9	1.8	.4	16.7	3.3	1.4
Internal Order	3.0	9.8	2.5	2.4	1.4	1.4
Entertainment	2.6	7.2	.7	.0	.0	.0
Military and Defense	2.2	1.4	.7	.0	6.2	2.1
Science/Technology	2.0	.4	.4	1.2	.0	.0
Transportation	1.4	.8	.4	.0	.0	.0
Ceremonies	1.2	1.3	3.6	1.2	.0	.7
Human Interest Stories	1.2	2.9	1.4	9.5	1.4	.0
Communication	.9	1.8	.7	2.4	.0	.0
Disasters/Accidents/ Epidemics	.7	.4	.4	.0	.0	.0
Environment	.7	.9	.0	.0	.0	.7
Labor Relations and Trade Unions	.7	1.1	.0	.0	.0	.7
Weather	.7	.0	4.3	7.1	2.9	2.1
Population	.5	.7	.7	.0	.0	.0
Housing	.3	.0	.0	.0	.0	.0
Fashion/Beauty	.2	.1	.0	.0	.0	.0
Energy	.1	.2	.0	4.8	.0	.0
Other	.0	.0	.0	.0	.0	.0
Total[a]	100.0	100.0	100.0	100.0	100.0	100.0
	($n = 1764$)	($n = 1210$)	($n = 281$)	($n = 84$)	($n = 209$)	($n = 145$)

Note: Distributions given in percent.

[a]Total percentage may not actually be 100.0 due to rounding error.

TABLE 14.2 Spearman Rank-Order Correlation Coefficients between Rankings of News Topics in Various Media

	Amman Newspaper	Irbid Newspaper	Amman Television	Irbid Television	Amman Radio	Irbid Radio
Amman Newspaper		.87[c]	.70[c]	.36	.58[b]	.38
Irbid Newspaper			.59[b]	.39[a]	.53[b]	.29
Amman Television				.46[a]	.57[b]	.43[a]
Irbid Television					.70[c]	.35
Amman Radio						.68[c]
Irbid Radio						

[a]$p < .05$; [b]$p < .01$; [c]$p < .001$.

significantly related but to a lesser degree than for newspapers (.68). The topical structures for television were only somewhat related, albeit significantly (.46). This means that strong relationships exist in topics of newspapers, while moderate relationships exist in topics of television. Finally, the values of correlation coefficients for topics in radio seem to be smaller than the other two types of media, although generally statistically significant.

Looking at intercorrelations among the three media in each of the two cities, topical structure for the media in Amman seems more uniform than in Irbid. In Amman the average rank-order correlation among the three media is much higher (.62) with all three correlations being statistically significant, whereas in Irbid the average correlation coefficient is only .34, with two of the three correlations not even reaching statistical significance.

Thus, it could be concluded that the correlations among the topics in different media are weak and sometimes nonsignificant, especially in the small city. In the case of Jordan, this is quite reasonable given the structure of ownership of the three types of media: variation among newspapers on the one hand and state-owned television and radio on the other.

Deviance in the News

One of the hypotheses of the study posits that deviant news is more prominently placed than other news. This was tested by means of one-way analyses of variance (ANOVAs) of the mean prominence scores for each level of statistical deviance, social change deviance, and normative deviance across the various media in the two cities. The results of the ANOVAs are presented in Table 14.3.

As can be seen, only six of the thirty ANOVAs were significant, including four for normative deviance, two for statistical deviance, and none for social

TABLE 14.3 Mean Verbal and Visual Prominence Scores for Intensity of Deviance

	Amman					Irbid				
	Newspaper		Television		Radio	Newspaper		Television		Radio
Intensity of Deviance	Verbal only ($n = 1598$)	Verbal plus Visual ($n = 343$)	Verbal ($n = 227$)	Visual ($n = 118$)	Verbal ($n = 166$)	Verbal only ($n = 1082$)	Verbal plus Visual ($n = 389$)	Verbal ($n = 67$)	Visual ($n = 43$)	Verbal ($n = 131$)
Statistical Deviance										
(1) common	227.1[b]	519.3	123.8	136.7	77.7	234.3[a]	473.5	93.1	136.8	71.9
(2) somewhat unusual	307.5	166.7	320.4	396.2	98.7	227.2	220.4	206.5	286.1	79.1
(3) quite unusual	292.0	–	142.1	240.8	76.6	257.9	194.0	212.1	294.5	83.0
(4) extremely unusual	295.7	–	118.7	238.7	76.9	396.1	220.3	171.8	212.2	81.4
Social Change Deviance										
(1) not threatening to status quo	254.1	513.8	184.2	244.4	89.0	231.7	463.4	195.3	268.0	85.5
(2) minimal threat	222.6	–	112.8	360.0	47.6	291.1	213.9	152.0	244.0	79.5
(3) moderate threat	237.0	–	161.9	182.0	74.5	278.2	158.3	179.3	243.3	71.8
(4) major threat	–	–	139.0	313.8	79.1	647.3	–	72.5	72.5	61.0
Normative Deviance										
(1) does not violate any norms	244.9[a]	512.3	195.2	253.8	97.2	230.7[b]	466.7	359.5[a]	461.6[a]	88.0
(2) minimal violation	263.5	164.0	161.1	221.7	63.8	218.3	281.0	163.1	233.5	71.5
(3) moderate violation	363.7	102.0	151.0	277.1	76.0	367.0	272.1	140.5	183.1	74.8
(4) major violation	357.3	–	58.8	200.0	56.8	174.2	–	66.2	81.8	49.9

[a] $p < .05$; [b] $p < .01$; [c] $p < .001$.

change deviance. In five of the six cases the ANOVAs for the verbal dimension were significant compared with only one in the visual dimension. Also, four of the six significant ANOVAs were regarding the newspapers in the two cities.

It was generally expected that in the cases of significant ANOVAs there would be greater prominence the more deviant the items. There was a slight tendency for this linear relationship for both statistical and normative deviance in the case of the Amman newspaper as well as for statistical deviance in the Irbid newspaper (but no such pattern for normative deviance in Irbid).

However, the most interesting finding is that the relationship was reversed in the case of normative deviance on Irbid television. In other words, the higher the normative deviance, the less prominent the items were. It should be noted that a similar trend, albeit not statistically significant, was found for television in Amman (especially with regard to the verbal dimension). What this may suggest is that the state-controlled television makes a point of deemphasizing items the greater the amount of normative deviance contained in them. This gives even further support to the impact of the differential patterns of media ownership in Jordan.

Social Significance in the News

As with deviance, the basic assumption regarding social significance was that socially significant news items are more prominently presented than news items with little or no social significance. In order to test this hypothesis, the mean prominence scores for all the media were analyzed by means of forty ANOVAs, calculated for each of the components of social significance—political, economic, cultural, and public.

Across the board, there seems to be relatively little variation among the four levels of each of the social significance variables in the two cities. And yet as can be seen in Table 14.4, five of the forty ANOVAs produce significant differences. It is interesting to note that all of the significant differences were for newspapers: two in Amman and three in Irbid. Also, three of the five significant ANOVAs related to public social significance—in Amman regarding verbal contents and in Irbid regarding both verbal and visual contents. The main reason for these effects is probably the large number of public activities reported in both large and small cities on a daily basis. This may suggest that the relevance of news items for the political process and decision making seems to be an important criterion of newsworthiness in the Jordan newspapers.

Also, three of the statistically significant differences were found for public significance and one each for political and cultural significance. No differences were found relating to economic significance.

A close examination of the mean scores of verbal versus visual prominence in the case of the newspaper and television indicates that in many cases the prominence was higher for the visual dimension, especially among the television items. Also, as in the case of the deviance measures, in quite a few cases

TABLE 14.4 Mean Verbal and Visual Prominence Scores for Intensity of Social Significance

Intensity of Social Significance	Amman					Irbid				
	Newspaper		Television		Radio	Newspaper		Television		Radio
	Verbal only ($n=1598$)	Verbal plus Visual ($n=343$)	Verbal ($n=227$)	Visual ($n=118$)	Verbal ($n=166$)	Verbal only ($n=1082$)	Verbal plus Visual ($n=389$)	Verbal ($n=67$)	Visual ($n=43$)	Verbal ($n=131$)
Political Significance										
(1) not significant	242.0[c]	510.1	190.7	236.7	92.3	228.0	468.2	128.7	173.0	88.6
(2) minimal	225.1	–	160.9	171.9	100.1	287.9	275.6	207.0	305.0	84.4
(3) moderate	428.4	–	156.8	305.8	69.3	263.5	330.7	209.7	286.2	66.6
(4) major	874.8	–	154.7	386.4	71.3	251.1	86.0	380.9	470.3	78.4
Economic Significance										
(1) not significant	262.3	509.9	189.4	285.0	90.8	238.3	463.3	219.3	291.9	93.1
(2) minimal	158.6	–	90.9	108.3	52.7	239.8	212.5	159.9	224.5	60.8
(3) moderate	189.7	–	33.9	50.0	67.9	358.8	180.0	116.6	173.8	57.3
(4) major	216.6	549.0	136.5	141.7	43.5	45.0	–	40.0	40.0	55.7
Cultural Significance										
(1) not significant	252.3	510.1	176.4	253.4	84.8	262.9[c]	459.0	180.2	237.0	85.3
(2) minimal	288.8	–	76.7	120.0	68.6	254.1	228.3	264.2	322.8	52.2

[a]$p<.05$; [b]$p<.01$; [c]$p<.001$.

(Continued)

TABLE 14.4 Mean Verbal and Visual Prominence Scores for Intensity of Social Significance (*Continued*)

Intensity of Social Significance	Amman					Irbid				
	Newspaper		Television		Radio	Newspaper		Television		Radio
	Verbal only ($n = 1598$)	Verbal plus Visual ($n = 343$)	Verbal ($n = 227$)	Visual ($n = 118$)	Verbal ($n = 166$)	Verbal only ($n = 1082$)	Verbal plus Visual ($n = 389$)	Verbal ($n = 67$)	Visual ($n = 43$)	Verbal ($n = 131$)
(3) moderate	186.0	–	100.0	120.0	–	119.2	239.5	159.2	237.1	0.0
(4) major	328.6	–	115.0	142.5	75.0	181.5	–	177.8	263.3	72.5
Public Significance										
(1) not significant	224.9[c]	526.1	218.8	328.2	96.2	183.1[c]	477.7[a]	290.4	367.2	101.5
(2) minimal	236.3	240.6	158.6	179.5	69.4	243.6	249.3	151.0	207.3	85.2
(3) moderate	321.6	222.5	147.7	265.8	80.9	408.4	205.7	200.3	269.4	70.3
(4) major	1001.0	–	82.0	175.0	57.0	714.5	170.0	108.5	167.0	34.2

[a]$p<.05$; [b]$p<.01$; [c]$p<.001$.

there appeared to be a negative, but nonsignificant, relationship between the level of the dimensions of social significance and the prominence scores.

In sum, the notion that socially significant news items would be more prominently displayed received some support regarding newspapers but not for television and radio. Again, this could be explained by the fact that both Jordanian radio and television stations are owned and controlled by a government that prefers relatively little variability in the political, economic, cultural, and public significance of news.

Deviance and Social Significance as Predictors of News Prominence

To test the effects of all the dimensions of deviance and social significance on news prominence, eight stepwise regression analyses were carried out—four for each city. The dependent variables were prominence of verbal newspaper content, combined verbal and visual newspaper content, prominence of television content, and radio content. The findings are presented in Table 14.5.

The data for Amman indicate significant effects for the two newspaper measures of prominence: verbal only as well as verbal and visual. For the verbal measure of prominence, the explained variance was only 5%, with two components with a value of .14 (political and public significance) and two additional components (normative deviance and cultural significance) with a value of only .06. For the combined verbal and visual measure of prominence, however, three components—normative deviance and political and public significance—contributed to the 16% explained variance, with values ranging from .20 to .22.

As for the city of Irbid, the only meaningfully significant regression is for the prominence of television, for which 18% of the variance was explained. The component contributing to this was verbal normative deviance, which had a value of −.43. This coincides with the negative relationship indicated in Table 14.3. While the regression analyses for verbal newspaper and the combined verbal and visual newspaper variables in Irbid produced 4% and 2% explained variance, respectively, these findings are hardly meaningful.

In conclusion, the regression analyses provide evidence for the effects of deviance and social significance on prominence for newspapers in Amman and for television in Irbid, with no effect for radio news in either city.

People Defining News

Four Jordanian focus groups were conducted in Amman during April 2000 followed by four other focus groups in Irbid in May 2000. In each city, the groups consist of journalists, public relations practitioners, and the media consumers (low and high socioeconomic status, or SES). The discussions in the focus groups centered on the most important events in peoples' lives, the salience and social significance of news, and news valence. In the latter part of the discussions the respondents undertook the gatekeeping task.

TABLE 14.5 Stepwise Regression Analyses of Intensity of Deviance and Social Significance on News Prominence

	Amman								Irbid							
	Newspaper Prominence Verbal only		Newspaper Prominence Visual and Verbal		Television Prominence		Radio Prominence		Newspaper Prominence Verbal only		Newspaper Prominence Visual and Verbal		Television Prominence		Radio Prominence	
	Total $R^2=.05^c$ (n = 1594)		Total $R^2=.16^c$ (n = 320)		Total $R^2=$ ns (n = 118)		Total $R^2=$ ns (n = 166)		Total $R^2=.04^c$ (n = 1079)		Total $R^2=.02^a$ (n = 331)		Total $R^2=.18^b$ (n = 43)		Total $R^2=$ ns (n = 131)	
Independent Variables	r	Std. Beta	r	Std. Beta	r	Std. Beta	r	Std. Beta	r	Std. Beta	r	Std. Beta	r	Std. Beta	r	Std. Beta
Deviance																
– Statistical, verbal content	$.07^b$	ns	$.26^c$	ns	.04	ns	–.07	ns	.06	ns	.07	ns	.01	ns	.03	ns
– Statistical, visual content	—	—	ns	ns	.04	ns	—	—	—	—	–.01	ns	.01	ns	—	—
– Social change, verbal content	–.01	ns	ns	ns	.02	ns	–.06	ns	$.06^a$	ns	.02	ns	–.12	ns	–.08	ns
– Social change, visual content	—	—	ns	ns	.02	ns	—	—	—	—	–.04	ns	–.12	ns	—	—
– Normative, verbal content	$.07^b$	$.06^a$	$.23^c$	$.22^c$	.00	ns	–.13	ns	.06	ns	$.13^a$	$.13^a$	$-.43^b$	$-.43^b$	–.09	ns
– Normative, visual content	—	—	ns	ns	.00	ns	—	—	—	—	.05	ns	$-.43^b$	ns	—	—

Social Significance

– Political, verbal content	.15^c	.14^c	.25^c	.22^c	.05	ns	–.11	ns	.04	ns	–.00	ns	.34^a	ns	–.05	ns
– Political, visual content	—	—	ns	ns	.05	ns	—	—	—	—	.02	ns	.34*	ns	—	—
– Economic, verbal content	–.05	ns	–.08	ns	–.08	ns	–.13	ns	.04	ns	.03	ns	–.19	ns	–.14	ns
– Economic, visual content	—	—	ns	ns	–.08	ns	—	—	—	—	ns	ns	–.19	ns	—	—
– Cultural, verbal content	.02	.06^a	–.08	ns	–.03	ns	–.03	ns	–.12^c	–.07^a	–.02	ns	.03	ns	–.07	ns
– Cultural, visual content	—	—	ns	ns	–.03	ns	—	—	—	—	–.04	ns	.03	ns	—	—
– Public, verbal content	.15^c	.14^c	.28^c	.20^c	–.05	ns	–.09	ns	.19^c	.17^c	.06	ns	–.08	ns	–.17	ns
– Public, visual content	—	—	ns	ns	–.05	ns	—	—	—	—	.01	ns	–.08	ns	—	—

$^a p < .05$; $^b p < .01$; $^c p < .001$; ns = not part of final stepwise regression equation.

In both cities, each focus group consisted of ten persons with two exceptions: a group of nine journalists in Amman and seven journalists in Irbid. Generally speaking, almost 90% of participants were young, between 19 and 22 years of age. In Irbid, no females participated, but the Amman groups had five females—three low SES audience members and two public relations practitioners.

Most Important News Events in Life

Participants in each of the focus groups were asked to recall and note in writing the three most important news events in their lives. In all, fifty-six news stories were mentioned, ranging from well-known international events to minor regional or local occurrences. Nearly 57% of the events were negative in nature. The largest group of negative events addressed death of prominent persons or celebrities, wars, killing of innocent people, and catastrophes. Positive news, however, addressed subjects like political and cultural events, sports, economic reforms, and scientific achievements.

Several key factors emerged that could explain why these particular events were recalled. First and foremost was the fear and threat associated with remembering certain news events (e.g., the death of King Hussein of Jordan, the outbreak of the Gulf War, the assassination of Israeli prime minister Yitzhak Rabin, and the death of Princess Diana in the United Kingdom). Second, positive news brought happiness to the lives of people so that they could still remember those good times. The main stories of positive news included the start of the Palestinian uprising, the Israeli withdrawal from South Lebanon, holding the Arab Summit in Amman, and signing of the Peace Treaty between Israel and Jordan.

Deviance

When discussing the nature of news, almost all of the focus group participants referred to certain aspects of deviance having an effect on people. For example, one journalist in Amman stated, "No matter what the subject of news is, it is really the impact of news on different people that matters. Some people may feel delighted about scoring a goal in a football match, while others may switch off the television or radio set." However, there were some difficulties in defining deviance in the news, especially in differentiating between deviance of Jordanian news and deviance of news in general. The participants linked their comments with the impact that news has on people. Sad news, for example, was not considered deviant, while happy or good news was considered as such. Once again, news that does not involve humans or human life was not considered deviant. As one low SES audience participant put it, "Why should I care about the tragic death of a panda in a zoo?"

When asked why good news is deviant, several respondents attributed this to the history of the individual. "It is something that is in the nature of everybody," one Irbid low SES audience participant replied, and everyone in the group agreed. Another low SES audience participant in Irbid correlated this with the topic of a news item: "Financial news such as stock market reports and the exchange rates for the Jordanian currency have no deviance at all, in my opinion. The Jordanian Dinar exchange rate has been fixed since 1989. It is something that does not affect me, so I take it for granted," he said. Another type of remark referred to the "biological structure" of people who accept or reject deviance in news. This discussion about the biological structure of humans raised more questions than answers. One male public relations participant in Amman said, "The more you feel that news items will affect your life, the more deviant it is."

Social Significance

Discussing social significance among Jordanian focus group participants resulted in several observations. First, most participants relate social significance to the social status of people. In other words, it would seem that if news recipients belong to the low SES audience group, they will feel that a specific news item is more socially significant than members of the high SES audience group. Second, when asked to talk about themselves and not about other people, social significance seems to draw less attention of the participants. As one journalist in Amman put it, "Social significance of news is not that important in Jordan since most people, especially news consumers, belong to the middle class." All journalists in Amman seem to agree with this notion. Third, the prominence of news plays a decisive role in determining the social significance of news. The frequency of social news supports this idea, as most focus group participants in Irbid agreed that "social events in this small city are rare except on big issues." Again, another journalist in Irbid stated, "I don't feel that journalists have social responsibility to follow and rank the social significance of news. Maybe this is the task of PR [public relations] people, especially those working in large private establishments." "On the contrary, both journalists and PR people have the responsibility to check the social significance of news before sending it to press," commented a public relations practitioner in Irbid. The same person added, "PR people in the government have more responsibility to look at the social significance of news before the information is published. Another public relations person in Irbid added that it is quite common that the Director General has the final say about the exact text of news items, not the PR person."

This discussion seems to lead to the conclusion that the social significance of news is more obvious in large cities than in small cities. Also, when social differences between people are small, social significance of news is less important.

News Valence

The concept of news valence was clear to nearly all the focus group participants. Several observations can be made. First, all participants, especially journalists, agreed that news consumers are more attracted to negative (or bad) news than to positive (or good) news. This is not surprising since two bad news items dominated the media in Jordan during the research period: the tragic death of King Hussein of Jordan and the continuous hardship of the Palestinians.

Second, almost all journalists were convinced that the prominence of news depends on whether the news items are negative or positive. One journalist said, "A newspaper with no bad news would have a smaller circulation."

Third, nearly all the focus group participants agreed that negative news for some people is considered positive for others. This is clear when discussing the uprising of the Palestinians (*Intifada*), a major event that dominated Jordan media.

Fourth, while discussing the reasons behind the people's interest in negative news, participants agreed that sooner or later some negative news would affect their social and economic life. One public relations person in Amman stated that "most news items in the Jordanian media are negative: war, fighting, assassinations, occupations, demolition of houses, acts of religious fanatics, and many others. We hardly hear positive news except those concerning the weather and new appointments of this or that person. All these bad news stories affect our living conditions and the lives of our relatives."

When confronted with the question of the reason behind preferring negative news, some participants said that it might be because they have been used to it since childhood. As one high SES audience person in Amman put it, "Every day we hear bad news, so that this news became part of our nature. This is why we love to hear bad news, especially if it took place far away from our country. Maybe other nations want to hear bad news about Arabs too, you never know."

Sociocommunicative Function of News

News is a hot issue and a necessity for everyday life, not only for the Jordanian focus group participants but also for the Jordanian society, which is why many participants emphasized the necessity of following the news in order to be well informed with what is going on. Otherwise, it might be difficult to communicate with colleagues and friends and sometimes even to be socially isolated. As one high SES in Amman said, "When my colleagues start to discuss an event I immediately enter into the discussion. Not only me but also everyone needs to know what is going on. There are at least three major events that take place every day. In talking to others, you usually refer to what you heard or read in the media. I only feel isolated and dislike entering into discussion when the topic is about horoscopes or sports." Nearly all the focus group participants in Irbid discussed the effect on social relations of being aware of news. One public relations participant in Irbid said, "When men meet, they talk about three things: bad living conditions, politics, and religion. All three topics are covered by each media type, with less emphasis on the latter topic.

So that if you are uninformed with what's going on, you will find it hard to communicate with others. However, when women meet they talk about cooking, personal care, and religion. These topics are also covered by the media, so it's necessary for every woman to read or hear about these things in order to communicate well with other women."

In conclusion, the sociocommunicative function of news appeared to be among the key drivers of news interest among almost all focus group participants in both cities.

Comparing People's News Preferences with What's in the Newspaper

The study hypothesized that a positive relationship exists between the rankings of newsworthiness among journalists, public relations practitioners, and audience members. This hypothesis was tested by means of the gatekeeping task described earlier in the book. The hypothesis was tested by applying Spearman's rank-order correlation between the actual rankings of the prominence of the newspaper items and the subjective rankings of the participants in each of the focus groups.

As can be seen in Table 14.6, all the correlation coefficients for the capital city of Amman were significant at the .05 level. The strongest relationships were between high and low SES audience groups (.77), followed by journalists and high SES audience (.76), and between journalists and public relations practitioners (.75).

In the smaller town of Irbid, four of six correlations were significant at the .05 level. The highest correlations were between low SES audience members and public relations practitioners (.83) and between low SES audience members and journalists (.77). Thus, these findings tend to suggest that Jordanians—both media professionals and laypeople—have similar views concerning newsworthiness.

The data relating to the relationships between the actual media coverage (in the newspapers) and the various groups of people yielded five significant correlations among the eight correlations calculated. Yet, while there was clearly a

TABLE 14.6 Spearman Rank-Order Correlation Coefficients between Newspaper Item Prominence and Focus Group Rankings

	Journalists	Public Relations	High SES Audience	Low SES Audience	Newspaper
Journalists	—	.75[c]	.76[c]	.72[c]	.48[b]
Public Relations	.72[c]	—	.48[b]	.63[c]	.38[a]
High SES Audience	.22	.40[a]	—	.77[c]	.69[c]
Low SES Audience	.77[c]	.83[c]	.35	—	.51[b]
Newspaper	−.13	.09	.73[c]	.05	—

Note: Amman coefficients are in the upper triangle; Irbid coefficients in the lower. SES = socioeconomic status.
[a] $p < .05$; [b] $p < .01$; [c] $p < .001$.

relationship between the definition of newsworthiness by the group members and the actual prominence of the newspaper items, these correlations were generally lower than between the various focus groups.

In Amman, all correlation coefficients were statistically significant. However, some of these correlations are below 50%, such as those for journalists (48%) and for public relations people (38%). The high correlations for news consumers imply that the general public has similar views concerning prominence of news, while lower correlations for news producers imply less agreement on news prominence.

In Irbid, on the other hand, only one correlation is significant, .73 for the views of high SES audience members. The other three correlations were both low and insignificant, implying a weak relationship between the views of journalists, public relations people, and low SES audience members concerning the prominence of news items.

In sum, there is strong empirical evidence that the implicit definitions of newsworthiness of various news stories provided by the focus group participants are more similar with each other than with the actual prominence given to these stories in the local newspapers. There is also clear evidence supporting the hypothesis of the study concerning the strong relationships between the actual rankings of the prominence of the newspaper items and the subjective rankings of the three groups of people, although this evidence is clearer in the big city.

Discussion

Looking at the main topics of the news in the Jordanian media, it is evident that international politics and internal politics dominate the news on both television and radio. In Amman the three most prevalent topics in each media were (1) newspaper: sports, the economy, and cultural events; (2) television: sports, business/commerce/industry, and the economy; and (3) radio: military and defense, education, and weather. In Irbid, the three most prevalent topics in each media were (1) newspaper: cultural events, internal order, and entertainment; (2) television: education, human interest stories, and weather; and (3) radio: the economy, military and defense, and weather. These data help draw the conclusion that the relative importance of the topics also varied considerably among the media.

Examining the relationships among the distributions of the topics in the different media reveals weak and sometimes nonsignificant correlations, especially in Irbid, the small city. In the case of Jordan, this is quite reasonable given the structure of ownership of the three types of media: variation among newspapers on the one hand and state-owned and controlled television and radio on the other.

The analysis of deviance in the news indicates a slight tendency for a linear relationship for both statistical and normative deviance in the case of the Amman newspaper as well as for statistical deviance in the Irbid newspaper.

However, the most interesting finding is that the relationship was reversed in the case of normative deviance on Irbid television. This may suggest that the state-controlled television makes a point of deemphasizing items the greater the amount of normative deviance contained in them, giving even further support to the impact of the differential patterns of media ownership in Jordan.

Concerning social significance in the news, relatively little variation seems to take place among the four levels of each of the social significance variables in the two cities. The notion that socially significant news items would be more prominently displayed received some support regarding newspapers but not for television and radio.

The test of the effects of all the dimensions of deviance and social significance on news prominence provided evidence for the effects of deviance and social significance on prominence for newspapers in Amman and Irbid, with no effect for radio news in either city.

The focus group discussions about how people define news resulted in several observations:

1. Nearly 57% of the events were negative in nature.
2. A fear and threat was associated with remembering certain negative events, while positive news brought happiness to the lives of people so that they could still remember those good times.
3. Almost all the focus group participants referred to certain aspects of deviance affecting people.
4. Most participants relate social significance to the social status of people.
5. The prominence of news plays a decisive role in determining the social significance of news, which seems to lead to the conclusion that the social significance of news is more obvious in large cities than in small cities. When social differences between people are small, social significance of news is less important.
6. News consumers are more attracted to negative (or bad) news than to positive (or good) news.
7. The prominence of news depends on whether the news items are negative or positive.
8. The sociocommunicative function of news appeared to be among the key drivers of news interest among almost all focus group participants in both cities.
9. Strong empirical evidence exists that the implicit definitions of newsworthiness of various news stories provided by the focus group participants are more similar with each other than with the actual prominence given to these stories in the local newspapers. There is also clear evidence supporting the hypothesis of the study concerning the strong relationships between the actual rankings of the prominence of the newspaper items and the subjective rankings of the three groups of people.

15

What's News in Russia?

NATALIA BOLOTINA

The Media System in Russia

In order to understand and appreciate the current Russian media scene, it is important to briefly discuss the system that prevailed in the former Soviet Union. News in Russia has gone a long way from an item in a handwritten newspaper to today's commercial product, distributed in a mass and exciting form that expresses views, beliefs, and interests of various social groups. According to Vladimir Dal, one of the most prominent Russian-language researchers, "News is a quality, a feature of all that is new, a piece of information about a happening, an adventure, the first information about something" (Dal 1989, 549).

Until the mid-nineteenth century when the telegraph was invented, the newspaper was people's only source of news. People's connection to the outside world was primarily through the newspaper, which provided information about other people's lives and about the world. Newspapers conveyed the thoughts and emotions of those at the events being described.

In Soviet times, the perception and definition of news rested on the idea that information serves people, groups of people, parties, or classes and is an important part of propaganda and the process of getting people to act. The media, including books, magazines, newspapers, radio, and television, were referred to as powerful devices of the class struggle for those in power as well as the most important "weapons" of ideological impact on an individual. Interestingly enough, most Soviet researchers tended to consider news as a very broad notion including information, rumor, as well as domestic and personal life stories. According to this approach, information was a news item that had social and political value.

The Soviet media system, which served as a model for the media systems of the entire socialist camp, was based on the following four principles: (1) the marking of a very clear and strict division of the mass media into the printed press and broadcast outlets; (2) the subordination of the media outlets to central (party and state) ideological control, which was an embodiment of a unique and complex combination of centralized, local, and internal censorship despite the considerable editorial independence concerning neutral political issues; (3) the existence of newspapers and magazines as the centerpiece in the system of "ideology work," caused by fragmentation of the audience for print media, and aimed at different social levels and groups; and (4) the existence of a centralized (nonmarket) media economy where the role of commercial advertising was minimal (or almost entirely absent), while publishing, based on central planning, provided the state with considerable profits (Kenez 1985; Remington 1988).

News in the Soviet press was opposed to news in the bourgeoisie press, which was considered sensational in its emphasis on celebrities and scandals, economics and politics, sports, and private events. In the view of several Soviet scholars, the owners of the Western media outlets made journalists comment on events in a manner they considered correct. On the other hand, the Soviet media were obliged to avoid meaningless information. The fact that an event was unusual, interesting, and even sensational did not mean that it was worthy of being published in a Soviet newspaper. Authors always had to remember who was going to read the news and thus implicitly respond to the following questions: Why should it be published? Will it teach the reader anything? Will it set an example or inspire anyone? An article with no social or educational value was considered shallow and would not be published in a Soviet newspaper. Journalists had to adhere to Vladimir Lenin's belief that the state is strong only when the masses are aware of everything, can make judgments about everything, and can commit in full consciousness.

While supplying Soviet people with information on the most important happenings in the world, the Soviet media aimed at the economic, political, and cultural development of the people. In the first instance, news had to have social and political values. The topics covered by the Soviet mass media emerged from its social role. It should be noted that despite its declared openness, some themes could not be presented to the Soviet audience. Glavnaya Literaturnaya Redaktsiya (GLAVLIT), the central literary authority, the strict censorship body, was in control of all outgoing information about calamities, accidents, crime rates, and so forth. This was considered the kind of news the Soviet people should not consume.

It is critical to mention the structural changes in Russian mass media as a result of the demise of the Soviet Union in 1989, which resulted in a free press becoming part of the new ideologically oriented media of the post-Perestroika era. This entailed a totally new set of market-audience relations established in the field of mass media, resulting in new features of the current contents and audience perceptions of the Russian media.

In general, some current worldwide media trends are becoming relevant in Russia as well, such as the increase in entertainment and a declining interest in news in general and foreign news in particular. This has been made possible in Russia thanks to the abolition of censorship and opening up of new topics of interest. Thus, television broadcasting has moved toward more entertainment at the expense of news, but newscasts now present more broad coverage of global events, adopting a kaleidoscope style. Radio has followed a similar pattern by becoming more oriented to the interests of specific groups and by reducing newscasts and increasing music. As for newspapers, they now focus more on local content. This has led, in part, to the domestication of foreign news—that is, the tendency of journalists to present foreign news in a manner relevant to the domestic audience.

Other important trends are the blending of commercial interest and news as well as the integration of journalism with advertising, which lead to new kinds of manipulation previously unheard of involving the realization of goals and values of certain political and commercial groups.

All of this has changed the patterns of consumption of the media in Russia. Today, about 40% of Russians watch the daily newscasts of the central television channel, while the audience for the national (federal) newspapers accounts for only 20%. Due to the prevailing economic instability in Russia, local television has become the most important source of information for about 40% of the people, while only 19% of the audience consider local newspapers to be the most important source of information (Zassoursky et al. 2002).

The number of media outlets in Russia is constantly growing despite financial and political instability in the country. According to the National Statistical Review, the *Russian Press in the Russian Federation as of 2000*, the total number of Russian newspapers amounted to 5,758, with a total circulation of 7.1 million, and 2,781 magazines, with a total circulation of 496,593. According to the Russian National Association of Broadcasters, in 1999 there were 100 state-owned television companies (eighty-eight of which were regional) and 150 nonstate-owned companies. Nine television channels (only two of which are state-owned) can be received by more than half of the entire Russian population. According to some estimates, the number of broadcasters at the regional level is approximately 1,000 (Union of Journalists of Russia 1997, 1998).

Russian Radio is also expanding its boundaries. One of the examples is the Moscow media market. In Moscow in 1990 there were just three nonstate-owned radio stations in Moscow. By 1991 there were ten radio stations, while by early 1997 the Federal Television and Radio Broadcast Board issued 500 radio broadcast licenses.

The Study Sample

The study was based on data gathered by analyzing the contents of selected media and from focus groups conducted in two Russian cities: Moscow, the capital of Russia, home of approximately 10 million people, and Tula, one of the major regional industrial centres with a population of approximately 544,000 people.

In Moscow, the newspaper selected for analysis was *Izvestia* [The News], one of the oldest and most respected newspapers in the country. The television station selected was Obschestvennoe Russiyskoe Televidenie (ORT), one of about ten major Russian television channels. This highly rated national channel, partially run by the state and partially run by private business concerns, is viewed by a diverse audience. The radio station selected was Radio Rossiya [Radio Russia]. It is a highly regarded national channel, is run by the state, and presents quality broadcasts. The newscast analyzed for ORT in Moscow was *Novosty* [The News], which airs at 7 p.m. and runs for thirty minutes. The radio news program was the 7 p.m. news bulletin that runs for approximately twenty minutes.

In Tula the newspaper chosen was the *Tula Izvestia* [The Tula News] one of the major newspapers in the Tula region. It first appeared in 1991 but at the time of this study was not published on Saturdays and Sundays. Its circulation at the time of the study was somewhat above 54,000.[11] The television channel selected for the study was State Television, one of three channels in Tula whose audience accounts for approximately 2 million viewers. The radio station selected for the study, Moja Tula [My Tula], is owned jointly by the state and by the Tula media group. The television newscast analyzed was the *Reporter*, which is broadcast daily at 6 p.m. and runs for approximately ten minutes.[12] The radio news program studied is also named the *Reporter*. It should be noted that the length of both the television and radio news programs vary from day to day depending on the events of the day.

Topics in the News

The analysis of the Russian media was based on 1,160 newspaper items in the Moscow newspaper and 536 in Tula; 197 items on television in Moscow and fifty-eight items in Tula; and 203 radio items in Moscow and 104 in Tula. The large difference in the number of items between the two cities is due to several factors. First, the Moscow newspaper has many more pages than the Tula newspaper. Second, the television and radio news programs in Tula are typically dedicated to local news only, whereas the Moscow programs cover world news, national news, and Moscow news. Finally, the Moscow media are better funded; hence, they have better access to various information sources and accordingly provide more news.

It should also be noted that the Moscow edition of *Izvestia* is not published on Sundays, and in Tula the edition is not published on Saturdays and Sundays; hence, the composite week was modified by adding a Monday in Moscow and a Monday and Tuesday in Tula.

As can be seen in Table 15.1, the main topics reported in the Moscow newspaper were internal politics (14.2%), cultural events (12.7%), and sports (11.4%). In Tula the leading topic was internal order (13.6%), followed by internal politics (12.3%). In the television newscasts the lead topic in Moscow, as in the case of newspapers, was overwhelmingly internal politics (25.4%), followed by international politics (18.8%). Thus, nearly half of the newscast was devoted to politics. The television topics in Tula receiving the most coverage

TABLE 15.1 Distribution of General Topics of News Items by City and Medium

Topics	Newspaper		Television		Radio	
	Moscow	Tula	Moscow	Tula	Moscow	Tula
Internal Politics	14.2	12.3	25.4	8.6	16.3	21.2
Cultural Events	12.7	6.3	4.1	3.4	2.0	5.8
Sports	11.4	3.2	1.0	10.3	24.1	23.1
International Politics	9.1	4.3	18.8	.0	14.3	.0
Communication	7.4	4.1	1.5	.0	.5	.0
Business, Commerce, and Industry	7.0	3.5	1.0	6.9	.5	3.8
Economy	6.7	9.7	6.6	6.9	4.9	7.7
Human Interest Stories	5.9	9.7	2.0	1.7	2.0	4.8
Internal Order	4.9	13.6	13.2	15.5	7.4	6.7
Entertainment	2.5	2.4	.0	.0	.0	.0
Social Relations	2.3	1.3	.5	.0	2.5	1.0
Ceremonies	2.2	5.0	4.6	6.9	1.0	3.8
Science/Technology	2.0	.7	2.0	3.4	3.4	1.9
Disasters/Accidents/ Epidemics	1.6	2.2	3.6	3.4	7.9	2.9
Transportation	1.4	3.0	1.0	.0	.5	.0
Housing	1.1	2.8	1.0	3.4	1.0	1.9
Weather	1.1	4.1	1.5	3.4	4.4	6.7
Energy	1.0	.4	2.5	.0	1.5	1.0
Health, Welfare, and Social Services	1.0	3.9	.5	6.9	1.0	2.9
Other	1.0	.0	.0	.0	.0	.0
Fashion/Beauty	.8	.0	.0	.0	.0	.0
Environment	.7	.6	.0	.0	1.0	1.0
Military and Defense	.7	2.6	7.6	10.3	2.5	1.9
Labor Relations and Trade Unions	.5	.6	.5	3.4	1.0	.0
Education	.4	2.8	.0	3.4	.0	1.9
Population	.2	.7	1.0	1.7	.5	.0
Total[a]	100.0	100.0	100.0	100.0	100.0	100.0
	(n = 1160)	(n = 536)	(n = 197)	(n = 58)	(n = 203)	(n = 104)

Note: Distributions given in percent.

[a]Total percentage may not actually be 100.0 due to rounding error.

were internal order (15.5%), followed by sports (10.3%). Interestingly, compared with Moscow, no items at all dealt with international politics. Radio news presented a different picture in the two cities, with sports taking first place in Moscow (24.1%) and Tula (23.1%). The next highest category in both cities was internal politics with 16.3% in Moscow and 21.2% in Tula.

Comparing the distribution of the topics in the media, the Spearman rank-order correlation coefficients are presented in Table 15.2. The correlations among the three media in Moscow are positive, with a mean correlation of .59, and in Tula they are only slightly higher, with a mean of .66. These correlations suggest that the media in the two cities place somewhat different emphases in terms of the topics they cover.

It is interesting to note that the highest rank-order correlations in both cities were between the topic distributions on television and on radio in Moscow (.73) and in Tula (a relatively high .79). These relatively higher correlations between television and radio in the two cities can be explained by the fact that there are certain structural relationships between the two broadcast media. In Tula both radio and television are run by the same organization, although they do have separate newsrooms. In Moscow, as mentioned already, the radio channel analyzed is owned by the state, and the television channel studied is partially run by the state. Because of this commonality of state ownership, in both cities the information policy of the media organizations would tend to be similar, as would the sources used by the various channels, and the topics they covered would therefore be similar.

In addition, looking at the correlations between the two cities for each of the three media, it seems that the relationship was strongest between the two newspapers (.70), was slightly weaker between the radio stations (.64), and was lowest between the two television stations (.47). This could probably be

TABLE 15.2 Spearman Rank-Order Correlation Coefficients between Rankings of News Topics in Various Media

	Moscow Newspaper	Tula Newspaper	Moscow Television	Tula Television	Moscow Radio	Tula Radio
Moscow Newspaper		.70[c]	.53[b]	.21	.50[b]	.48[a]
Tula Newspaper			.65[c]	.52[b]	.47[a]	.67[c]
Moscow Television				.47[a]	.73[c]	.50[b]
Tula Television					.49[a]	.79[c]
Moscow Radio						.64[c]
Tula Radio						

[a]$p < .05$; [b]$p < .01$; [c]$p < .001$.

explained by the scarcity of financial resources. Television news, compared to radio news and newspapers, is the most costly. Given the very limited resources, especially in Tula compared to Moscow, the two television stations cannot cover the same range of topics—thus the relatively lower correlation coefficients obtained between the two cities. In newspapers, on the other hand, the cost factor is less manifest given the overall lower expenditure involved, while the cost of radio news production lies inbetween television and newspaper production. Thus, the highest correlation was found for the newspapers while the correlation for radio was inbetween that of television and newspapers.

Deviance in the News

One of the main assumptions of the study was that the more deviant the news items, the more prominently they would be displayed in the media. Table 15.3 shows the findings for the Russian sample. Of the thirty one-way analyses of variance (ANOVAs) computed, only seven were significant, but none displayed a complete linear relationship.

Of the seven significant findings, five dealt with the Moscow newspaper and two with the Moscow radio. None of the ANOVAs were significant for Tula. In nearly all the cases for the Moscow newspaper, the least deviant news items were the least prominent, as expected; however, in only one case (normative deviance) was the highest category of major violations of norms the most prominent for verbal content. In fact, in most of the cases, the middle levels of minimal or moderate violations of norms were the most prominent. A similar pattern was obtained for the significant ANOVAs for the Moscow radio station.

Examining the nonsignificant ANOVAs in Moscow, however, reveals that some of these analyses were in the predicted direction. In Tula, on the other hand, several of the ANOVAs, while not statistically significant, were actually in a linear pattern but in the opposite direction to the expectation—namely, the lower the deviance, the higher the prominence.

Two tentative conclusions are suggested here. First, in Russia highly deviant news items, although sometimes brought to the attention of the public, are not necessarily relevant to people's lives; hence, they receive less prominence than was predicted by the study hypothesis. Second, the statistically insignificant finding in the opposite direction, particularly in Tula, can be attributed to the fact that the most mundane and common news stories—hence the least deviant ones—are the items of most relevance to the readers and therefore are the most prominently displayed.

Social Significance in the News

The assumption concerning social significance was similar to that for deviance—namely, the higher the social significance of the items, the greater the prominence would be. Table 15.4 presents the Russian data. Of the forty

TABLE 15.3 Mean Verbal and Visual Prominence Scores for Intensity of Deviance

	Moscow					Tula				
	Newspaper		Television		Radio	Newspaper		Television		Radio
Intensity of Deviance	Verbal only (n = 810)	Verbal plus Visual (n = 311)	Verbal (n = 105)	Visual (n = 107)	Verbal (n = 147)	Verbal only (n = 361)	Verbal plus Visual (n = 107)	Verbal (n = 28)	Visual (n = 20)	Verbal (n = 74)
Statistical Deviance										
(1) common	303.7	539.6[b]	30.0	155.3	31.3[a]	197.8	365.0	17.1	213.9	22.8[a]
(2) somewhat unusual	374.8	705.6	119.8	115.0	52.5	219.7	298.0	168.5	237.9	77.8
(3) quite unusual	418.6	1042.6	138.9	129.5	66.2	213.8	266.4	345.0	–	122.5
(4) extremely unusual	541.6	906.0	155.7	101.7	41.8	60.7	455.4	217.5	135.0	–
Social Change Deviance										
(1) not threatening to status quo	308.1[c]	611.6[c]	88.0	135.8	33.3[b]	206.8	322.8	179.1	201.8	68.7
(2) minimal threat	398.1	1214.7	129.8	114.2	68.7	280.1	278.9	160.3	99.1	132.4
(3) moderate threat	590.0	906.6	133.2	165.5	51.3	190.9	148.6	136.8	13.1	96.3
(4) major threat	445.1	–	153.2	140.0	65.1	105.5	–	–	6.0	130.5

Normative Deviance

(1) does not violate any norms	337.7[b]	668.8[c]	134.0	133.6	45.7	217.0	322.0	163.0	198.0	77.7
(2) minimal violation	507.3	675.1	135.0	125.6	69.4	294.5	52.7	136.8	75.3	34.0
(3) moderate violation	393.1	2622.5	120.5	134.2	63.7	170.6	274.4	258.3	6.0	208.0
(4) major violation	518.1	614.0	144.8	140.0	35.2	94.2	—	6.0	21.9	68.3

[a] $p < .05$; [b] $p < .01$; [c] $p < .001$.

TABLE 15.4 Mean Verbal and Visual Prominence Scores for Intensity of Social Significance

Intensity of Social Significance	Moscow					Tula				
	Newspaper		Television		Radio	Newspaper		Television		Radio
	Verbal only (n = 809)	Verbal plus Visual (n = 308)	Verbal (n = 105)	Visual (n = 107)	Verbal (n = 147)	Verbal only (n = 360)	Verbal plus Visual (n = 107)	Verbal (n = 28)	Visual (n = 28)	Verbal (n = 74)
Political Significance										
(1) not significant	326.8[c]	650.8	99.8	123.4	33.5[c]	160.2[b]	287.4	182.2	182.5	56.1[c]
(2) minimal	428.2	924.4	109.8	157.5	48.1	315.8	406.8	176.0	145.9	112.3
(3) moderate	379.9	628.4	152.5	113.0	85.0	288.2	266.0	64.8	2.2	130.2
(4) major	711.4	858.4	166.9	95.5	71.6	327.9	–	–	–	189.0
Economic Significance										
(1) not significant	315.4[c]	660.3	118.6	126.9	36.3[c]	173.7[b]	287.7[a]	164.0	189.5	50.0[c]
(2) minimal	445.7	934.5	176.7	226.7	60.6	177.0	510.1	261.4	3.6	73.9
(3) moderate	428.9	789.6	172.5	–	95.7	300.0	81.0	141.9	2.2	117.7
(4) major	647.6	–	117.9	–	78.7	335.5	871.3	3.7	6.0	174.8
Cultural Significance										
(1) not significant	353.2	674.6	137.5	133.7	47.0	189.3	278.5	131.6	152.0	85.1
(2) minimal	380.1	749.2	94.2	102.5	47.6	268.5	483.7	221.8	246.2	88.6

(3) moderate	737.6	514.3	107.3	146.4	68.9	206.4	763.5	157.2	4.0	64.8
(4) major	706.0	449.3	150.7	123.3	44.5	417.9	—	135.0	135.0	68.8
Public Significance										
(1) not significant	633.4[a]	257.2[c]	86.7	141.1	30.8[b]	172.9[a]	263.7[a]	92.7[a]	221.3[a]	49.1
(2) minimal	999.4	429.3	160.9	123.3	65.9	130.3	344.3	244.9	186.4	65.1
(3) moderate	633.2	578.6	113.1	132.4	67.4	228.8	472.2	179.9	22.8	74.7
(4) major	—	339.7	134.1	112.2	59.1	272.9	656.8	7.5	5.1	116.4

[a] $p < .05$; [b] $p < .01$; [c] $p < .001$.

ANOVAs computed, sixteen were significant. Of these, seven were in Moscow and nine in Tula. As with the case of deviance, in none of these cases was there a clear linear relationship.

In Moscow, three significant ANOVAs were found for the verbal content of the newspaper and three for the radio news. However, in only political significance and economic significance were the lowest and highest values associated with the lowest and highest prominence of verbal content, respectively. In fact, the middle levels of social significance were generally the highest in prominence. Thus, in Moscow there was little support of the assumption of the study.

In Tula the situation was slightly better in terms of the expected outcome. In five of the nine significant ANOVAs there was a linear relationship, indicating that the higher the social significance, the higher the prominence. Of these, four of the five related to the Tula newspaper. In addition, in two other cases the lowest and highest levels of social significance were the lowest and highest in prominence, respectively. It should be noted, however, that in the case of the public significance of television visuals, there was also linear but in the opposite direction—that is, the more social significance, the less the prominence. Thus, in Tula the expected relationship was more clearly pronounced than in Moscow.

Deviance and Social Significance as Predictors of News Prominence

Based on the analyses of the relationships between deviance and social significance (Tables 14.3 and 14.4), a further analysis was conducted to determine the amount of variance in prominence scores based on all the components of deviance and social significance. This was accomplished by means of multiple regressions, conducted for each medium in both cities. In the case of newspapers, separate analyses were done for the prominence of verbal content as well as for combined verbal and visual content. The findings are presented in Table 15.5.

Seven of the eight analyses produced statistically significant results, with the amount of explained variance in prominence ranging from 5% for the verbal content of newspapers in Moscow to 45% for radio content in Tula. The only statistically nonsignificant analysis was for television content in Tula. Of the various components of deviance and social significance, political significance was related to five of the prominence scores for verbal content, and economic significance was related to four of the verbal prominence variables.

The 45% explained variance in the prominence of verbal content on the Tula radio station was due to economic significance (.41), political significance (.30), and statistical deviance (.20). As for the prominence of the combined visual and verbal newspaper content in the Tula newspaper, the 30% variance explained was due to the following five components (in declining order of the standardized beta weights): public significance, social deviance (in a negative direction), cultural significance, economic significance, and political significance.

TABLE 15.5 Stepwise Regression Analyses of Intensity of Deviance and Social Significance on News Prominence

Independent Variables	Moscow								Tula							
	Newspaper Prominence Verbal only Total R² = 0.05[c] (n = 807)		Newspaper Prominence Visual and Verbal Total R² = 0.10[c] (n = 298)		Television Prominence Total R² = 0.09[a] (n = 105)		Radio Prominence Total R² = 0.19[c] (n = 1147)		Newspaper Prominence Verbal only Total R² = 0.07[c] (n = 360)		Newspaper Prominence Visual and Verbal Total R² = 0.30[c] (n = 103)		Television Prominence Total R² = ns (n = 20)		Radio Prominence Total R² = 0.45[c] (n = 73)	
	r	Std. Beta	r	Std. Beta	r	Std. Beta	r	Std. Beta	r	Std. Beta	r	Std. Beta	r	Std. Beta	r	Std. Beta
Deviance																
– Statistical, verbal content	.08[a]	ns	.21[c]	ns	.12	ns	.13	ns	–.03	ns	–.06	ns	.23	ns	.30[a]	.20[a]
– Statistical, visual content	–	–	.20[c]	.16[b]	–.13	ns	–	–	–	–	–.05	ns	–.06	ns	–	–
– Social change, verbal content	.16[c]	.09[a]	.27[c]	.24[c]	.10	ns	.22[b]	ns	–.05	ns	–.10	ns	.05	ns	.22	ns
– Social change, visual content	–	–	.23[c]	ns	.01	ns	–	–	–	–	–.11	–.30[b]	–.20	ns	–	–
– Normative, verbal content	.09[a]	ns	.22[c]	ns	.02	ns	.06	ns	–.08	ns	–.13	.ns	–.01	ns	.08	ns
– Normative, visual content	–	–	.16[b]	ns	–.01	ns	–	–	–	–	–.06	.ns	–.11	ns	–	–
Social Significance																
– Political, verbal content	.12[c]	ns	.22[c]	ns	.21[a]	.23[a]	.37[c]	.27[b]	.19[c]	.16[b]	.20[a]	.20[a]	.08	ns	.53[c]	.30[b]
– Political, visual content	–	–	.06	ns	.04	ns	–	–	–	–	.10	ns	.27	ns	–	–

[a]p < .05; [b]p < .01; [c]p < .001; ns = not part of final stepwise regression equation.

(Continued)

TABLE 15.5 Stepwise Regression Analyses of Intensity of Deviance and Social Significance on News Prominence (*Continued*)

	Moscow								Tula							
	Newspaper Prominence Verbal only $R^2 = 0.05^c$ (n = 807)		Newspaper Prominence Visual and Verbal $R^2 = 0.10^c$ (n = 298)		Television Prominence $R^2 = 0.09^a$ (n = 105)		Radio Prominence $R^2 = 0.19^c$ (n = 1147)		Newspaper Prominence Verbal only $R^2 = 0.07^c$ (n = 360)		Newspaper Prominence Visual and Verbal $R^2 = 0.30^c$ (n = 103)		Television Prominence $R^2 = ns$ (n = 20)		Radio Prominence $R^2 = 0.45^c$ (n = 73)	
Independent Variables	r	Std. Beta	r	Std. Beta	r	Std. Beta	r	Std. Beta	r	Std. Beta	r	Std. Beta	r	Std. Beta	r	Std. Beta
– Economic, verbal content	.14[c]	.09[a]	.19[b]	ns	.09	ns	.36[c]	.26[b]	.17[c]	.15[c]	.25[a]	ns	.15	ns	.57[a]	.41[c]
– Economic, visual content	–	–	.11	ns	.18	.20[a]	–	–	–	–	.26[b]	.26[b]	ns	ns	–.11	–
– Cultural, verbal content	.09[a]	.12[b]	.04	ns	–.02	ns	.07	ns	.08	.14[a]	.28[b]	.28[b]	.26	ns	–	–
– Cultural, visual content	–	–	.04	ns	–.01	ns	–	–	–	–	.26[b]	ns	.03	ns	–	–
– Public, verbal content	.17[c]	.09	.24[c]	ns	–.05	ns	.27[b]	ns	.12[a]	ns	.20[a]	ns	.22	ns	.28[a]	ns
– Public, visual content	–	–	.13[a]	ns	–.07	ns	–	–	–	–	.27[b]	.33[b]	–.20	ns	–	–

[a]$p < .05$; [b]$p < .01$; [c]$p < .001$; ns = not part of final stepwise regression equation.

The noteworthy case in Moscow was radio news, where 19% of the variance in prominence was explained by two components of social significance, political and economic. Finally, the 10% variance in the prominence of verbal and visual newspaper content was due to social change and statistical deviance.

The overall findings, then, are that in Tula, deviance and social significance explained more variance than in Moscow. This could be due to the fact that in Tula the media play a more meaningful role in peoples' lives, given their more limited concerns about local matters to which the media can successfully cater. It is also possible that, since there are numerous media outlets in Moscow compared to Tula, the prominence in Tula is better explained by Moscovites being able to obtain information from other sources, not being limited—as are the residents of Tula—to the specific outlets analyzed in the study. Finally, more explained variance for radio in both cities is due to the nature of the news broadcast on the quality radio stations in Russia. Given the limited scope of news on such stations, most if not all of what is reported is focused on the most important information of the day and not on trivial matters.

People Defining News

Four focus groups were conducted in Moscow in summer 2001, and four groups were held in Tula in autumn 2001. The focus groups consisted of eight to ten participants per group. Each group had both women and men of different age groups, except for the low socioeconomic status (SES) audience group in Tula, which consisted of women only. The journalists' focus group in Moscow was composed solely of reporters from *Izvestia*. In Tula, the journalists' group was composed of television and radio reporters, because the newspaper reporters refused to participate. All of the public relations practitioners in the Moscow group came from a single commercial public relations firm that serves various commercial and public organizations. In Tula, on the other hand, the participants in the public relations group came from the public relations departments of various public and private organizations.

It should be noted at the outset that, as the focus groups got under way, it became clear that the journalists and public relations practitioners shared many notions since public relations persons often perceive themselves as producers of news. Also, the audience members, whether of high or low SES, had much in common. Hence, to a large extent the forthcoming analysis of the focus groups discussions was done by partially combining in each city the opinions and comments of the two professional groups and the two audience groups.

Important Events in One's Life

The focus group participants were asked about the most important news items they could recall. The participants in all the audience groups expressed similar views on what news is and mentioned almost the same themes. These included the following:

- Weather forecasts
- Information related to their professions
- Cultural events (e.g., festivals and theatrical premiers)
- Major political and economic events that took place both at the international and domestic levels, such as the disturbances in Göteborg, Sweden, on the occasion of the antiglobalization conference
- Meetings of the Russian president with foreign leaders
- Various security issues such as the United States dropping out of the Anti-Missile Defense Treaty, problems of war and peace, conservation of nuclear waste, and accidents connected with the biological weapons
- Educational issues and those connected with young people, such as an experiment in the Tula regional schools and drug addiction among young people

In other words, attention of the focus group participants was attracted by themes that in one way or another had an effect on their everyday life, by events of global social and political significance, and by topics that undermine social values and the status quo, influencing their normal course of life. As some of the respondents indicated, they mentioned topics that everybody cares about. It is interesting to point out that when mentioning these themes, especially regarding television, participants often used eloquent images that stirred strong emotions like joy and excitement, fear, disgust or rage and used expressions like "it worried me," "so scary," and "shocking." According to the participants, these strong images usually make them think more of their future, have concern for the country, fear for their life, be concerned over the health of loved ones, and be concerned about the uncertainty they may experience. These observations about the emotional impact of images apply to both audiences and professionals.

According to all the focus group participants, personal interests determine information interests. Some of them believed that those who reflect upon global processes pay special attention to foreign news, which helps the reader follow the main currents in the world economy and politics.

Issues on internal politics and economics proved to be of great significance to the participants, especially issues having to do with the republics of the former Soviet Union, which are now independent states. This seemed to be quite a sensitive topic for most of the respondents, including the journalists and the public relations practitioners; the collapse of the Soviet Union caused a major crash in beliefs, ideals, and stereotypes, affecting the personal lives of the people. When the USSR broke up, some people had to leave their homes and move to new places; some lost relatives, and others could not visit their families and friends for both political and financial reasons.

One of the hottest topics in Russia at the time of the study—hence mentioned by the focus group participants—was the Chechen war. Interestingly enough, quite a few of the respondents did not wish to hear anything on this matter and tended to avoid it. According to them, the general audience got used to the information about Chechnya, and to the images of the people getting shot and tanks being blown up and, thus, generally speaking did not pay much attention to it.

Participants with children tended to search for information about young people. Interestingly enough, at times professional and personal interests overlapped. Teachers and doctors in the focus groups who had children always followed news connected with education, homeless children, drug addiction, and children's diseases. Another example was a high SES audience participant from Moscow who for professional reasons follows all kinds of statistics—including statistics of homeless children—since as a mother she considered it to be rather important.

Because of an "iron curtain" that existed for many decades, Russians are highly interested in all sorts of news related to these topics. For the younger generation, news on traveling mostly relates to the question, "Where should I go on vacations?" In contrast, for elderly people—few of whom could afford an overseas trip—travel news is their way to learn about faraway lands and other nations' lifestyles, habits, and traditions.

Social Significance and Good or Bad News

Most of the topics mentioned by the focus group participants had to do with social significance. This is probably due to the cultural tradition according to which the media must deal with socially important issues and neglect trivial events. "Man bites dog" stories would never have been published in a Soviet newspaper because they are sensational, meaningless, and therefore useless. News must educate, add to people's knowledge, and be socially valuable. At some point, when *Perestroyka* had just begun, the Russian media were flooded with "shallow" sensational stories. This was a kind of novelty for the audience. Nowadays, when the country is going through an extremely difficult period both economically and politically, people naturally pay more attention to news that is relevant to themselves and to their families' everyday activities.

Some focus group participants thought that if something happened in Russia, everyone would discuss it. Generally speaking, bad news was said to draw more attention since it is provocative and unusual, while others insisted that good news was in the center of public interest. Their reasoning was that "if something good happens there is hope that things will improve." Those who admitted they prefer positive news explained that people cannot do without bad news since it makes them think, draw conclusions, and act. Negative information usually raises problems; in contrast, positive information helps readers to relax. Since there is so much evil in the world, positive news helps people overcome difficulties and get on with their lives. At the same time,

negative information affects people a great deal, and they can't help responding to it. It definitely adds to people's experience and helps to avoid troubles.
An interesting point was made by one of the focus group participants: She
believed that useful information is perceived as negative by emotional people;
people always pay attention to negative stories including fighting calamities,
homeless people, orphans, and accidents.

Most of the respondents thought that all information carries both positive
and negative aspects and that people really need both elements in order to
draw conclusions and to develop opinions on various matters. At times,
respondents found it rather difficult to distinguish between positive and negative news except, of course, in obvious cases such as the Kursk submarine tragedy and the Chernobyl disaster.

In performing the gatekeeping exercise, the focus group participants from
both Tula and Moscow indicated that they selected and highly ranked news
items that appealed to them both professionally and on a personal level. In
addition, the main criterion for some focus group participants was whether a
story appealed to their emotions. While talking about how they ranked positive and negative news items, participants in all of the focus groups used
words that described strong emotions a propos to the events: "I was proud of
my country," "crucial discovery," "absolutely disgusting." Also, the public relations practitioners in Moscow were sure that people outside Moscow
and those who reside in the city have different information priorities; that is,
people in the outlying regions are more concerned about their daily lives and
are not interested in global affairs.

As noted previously, the focus group members also underlined the dual
nature of information. An example mentioned by one person was the reopen-
ing of a theater in Voronezh following some renovations. This event was obviously seen as positive, but the item also had negative connotations since the
theater in question had been in very poor condition and was falling apart.
Another example concerns the decision of the United States to drop out of the
Anti-Ballistic Missile Defense Treaty, viewed by some as something neutral
and by others as negative.

All news items that are new and unusual can be of interest to the public,
but negative news seems to attract even more attention. Most of the participants believed that, although people like positive stories, they pay more attention to negative ones. In their work, journalists take into account this seeming
paradox. Most of the participants, except for the journalists, agreed that there
was much more negative information than positive information in the
Russian media and stressed that some newspapers in Russia specialize in negative information. In fact, a major part of domestic news coverage in Russia is
of a negative nature, and people tend to remember negative information
better than positive information.

In discussing their rankings of news stories, all of the participants of the
audience focus groups—but none of the journalists and public relations

practitioners—said they believed that the mass media relish negative information and turn it into something attractive. A woman from the low SES audience group in Tula said, "Journalists talk too much about negative things, and instead of fighting evil, they cause it." Some focus group participants believed a journalist's job is to provide both negative and positive information. It was also alleged that journalists make money by providing negative information.

Some of the focus group participants were convinced that information undermining moral values should not be broadcast or published because it harms children. In addition, journalism also assumes that children are unable to critically analyze information. Again, a low SES woman in Tula explained, "They take things wrongly and at times negative information turns into some sort of instruction, a plan to follow."

Following are some examples of negative information given by the focus group participants:

- The war in Afghanistan—people were concerned that the war was spreading around the world, and they feared uncertainty and chaos.
- The Chernobyl disaster—people lacked information about the tragedy that affected the entire country and some of the respondents' families. A low SES woman in Tula said, "My son was born then, and now I observe the impact of this tragedy on his health."
- The dissolution of the Soviet Union—uncertainty caused by the changes; having pity on the country that unlike Europe was in deep trouble. A young woman from the low SES audience group in Moscow said, "Everybody was getting together while our country was falling apart; I realized that we were not going to live in peace and friendship anymore."
- The NTV television crisis—a business conflict that turned out to be a political scandal
- The end of the Mir space station—represented the epoch of Russian superiority in space
- Terrorist attacks and explosions in Moscow

Examples of positive news included the following:

- Amazing scientific achievements—A low SES audience participant in Moscow stated that "one can program human qualities and put an end to the most dangerous diseases," and a low SES man in Moscow said that "Canadian scientists succeed in measuring neutrino."
- The first space flight—A woman from the high SES audience group in Moscow said, "We were all beside ourselves with joy and excitement."
- A Chinese Zoo visit to Moscow
- The celebration of Moscow's 850th anniversary

Among the most significant topics highlighted by the news producers, journalists, and public relations practitioners were those related to internal politics, both at the local and federal levels as well as international politics:

- A hunger strike in the Tula region
- Bombings in Georgia
- A scandal in the Federal Parliament (the Duma)
- A rise in oil prices
- A summit in Bonn
- Actions of the State Commission on Pardons
- The military campaign in Afghanistan
- The dire situation of the state communal services and pensions

In Tula public relations practitioners' opinions, crime, violence, and sex are the issues determining television ratings: "Two corpses get about thirty lines on the fifty-eighth page; an article about five corpses found would be published on the forty-sixth page; and 280 corpses would get a column on the front page," related a Tula male public relations practitioner in his 30s. The flow of innocent blood and pieces of human flesh grasp the public's attention. Aside from the aforementioned themes, stories that cover calamities, accidents, and the well being of society; community life; and charity and donations—everything appealing to human emotions like compassion and clemency—are of great interest to the audience.

While most of the participants from the journalist and public relations practitioner groups preferred positive news and believed that it shows life is changing for the better, they nevertheless felt that one cannot get along without negative information. One of the participants noted that journalists do not perceive positive information as something that actually affects the audience. According to one of the audience participants, what matters most to journalists is objectivity and trustworthiness.

Almost all focus group participants noted that information does not necessarily have to be sensational. It should add to people's knowledge and be useful. The packaging of the stories and trustworthiness are perceived as two of the most important features of news. Story packaging plays an essential role in creating information preferences. The way the facts of the story are presented determines people's interest and holds their attention. Truthful information is something the respondents really value, and some of them regret that the contemporary Russian media lack reliable information.

As mentioned earlier, images have a tremendous impact on the audience, and if a story is supported by vivid visuals it always shakes up the audience. Among the most powerful visuals noted by the focus group participants were the September 11, 2001, attacks on the U.S. World Trade Center as broadcast on Russian television, in which viewers witnessed the death of thousands of people on the other side of the world. Another example mentioned by the participants was that of a man shot in front of the camera in Vilnius, Latvia. According to respondents, their news interests depend neither on whether the items are considered hard or soft nor on whether they are positive or negative. News of global—at times even revolutionary—nature is what makes respondents look at things from a different perspective and breaks stereotypes.

Toward the end of the focus group sessions, all participants were asked to provide a composite definition of *news*. The responses given in this section indicated that although the journalists and public relations practitioners believed that audience members had different perception of news, in reality regardless of their professional and social status, people thought of news in a similar manner. Following are some of the characterizations provided by the participants.

Journalists

- "News is something that can make the chief of our department become interested in. News is something that can be sold to authorities or colleagues."
- "Something we never knew before. Any information is potentially news."
- "News is something we witnessed. News must have references to time and place, comments on how important an event is."
- "News should be truthful, interesting to everyone, well packaged, with no advertising, compressed, capacious, useful and competent, fresh, bright and have some kind of piquancy. It ties one up to a media outlet, makes an individual think."

Public Relations Practitioners

- "I work with news every day, but I can't define it."
- "Only people at the Faculty of Journalism know what news is."
- "Something that didn't exist the day before and just happened."
- "A piece of information that is being perceived at a given moment as something extraordinary, reported by a media outlet and referred to as a fact or an event."
- "A statement or an opinion can be news."
- "Something stated in a headline that holds one's attention all day long. Something audiences discuss."
- "News can't be objective as human beings produce it. Objectivity is a quality that is quite unique for our media."
- "It should be short, laconic, sensational, correct, trustworthy, fresh, and answer three questions: When? Where? What?"
- "News is (the) kind of information that gives us hope and helps us to change the future."

Audiences

- "No news is good news."
- "News is a part of our life, an element of communication between human beings."
- "A significant event one has never heard before. It should be unusual, brief, objective and fresh, neutral, that is not being thrust upon a person, with no comments, something that creates a change, affects one's life, stirs emotions, leaves memories, helps to assess things around you, enriches a person's experience and helps to develop an opinion and move to action. It doesn't matter whether it is positive or negative."
- "Everyone perceives news in his own unique way. One's perception of news depends on one's mood. News is a new word that touches upon an individual's interests."

Comparing People's News Preferences with What's in the Newspaper

In the gatekeeping task, participants in the focus groups were asked to arrange ten items taken from the Moscow and Tula newspapers, based on their headlines, in the order in which they believed the items should be presented if they, the participants, were editors of their respective city newspaper. This was repeated two more times for two different newspaper days. Table 15.6 presents Spearman rank-order correlation coefficients between the rankings of the members of the various focus groups, as well as the rankings of the group members with the actual line-ups in the two newspapers.

Although mostly statistically significant, the data indicate moderate correlations between the members in the four Moscow focus groups (with a mean correlation of .56) and even weaker correlations among the groups in Tula (.36). These findings indicate moderate agreement among the four groups in the two cities, especially in Tula.

TABLE 15.6 Spearman Rank-Order Correlation Coefficients between Newspaper Item Prominence and Focus Group Rankings

	Journalists	Public Relations	High SES Audience	Low SES Audience	Newspaper
Journalists	—	.50[b]	.40[a]	.56[b]	.22
Public Relations	.50[b]	—	.66[c]	.63[c]	.05
High SES Audience	.05	.35	—	.58[b]	.08
Low SES Audience	.21	.49[b]	.53[b]	—	.34
Newspaper	.24	.03	−.08	.12	—

Note: Moscow coefficients are in the upper triangle; Tula coefficients in the lower. SES = socioeconomic status.
[a]$p < .05$; [b]$p < .01$; [c]$p < .001$.

Moreover, the rank-order correlations between the subjective evaluations of the focus group members with the actual presentation of the stories in the newspapers were very low. In Moscow, the mean rank order correlation was a low .17, and the range was from .05 to .34. In Tula, the range was from –.08 to .24, with a very low mean rank correlation of .12. In other words, there was virtually no agreement between the focus group members (including the journalists) with the real prominence rankings of the items in the newspapers of the two cities.

The intriguing question is why? Two possible conjectures are offered here. First, due to their past experience with the media, Russians still have a basic mistrust of the media despite the fact that the entire media landscape in Russia has been transformed radically in recent years. It seems that this sense of mistrust can linger for quite some time. A second and somewhat related factor might be that, given the incredibly rapid pace of social and political changes taking place in Russia, what was news a short while ago is no longer as important as events that are happening at the moment. The items presented to the participants in the gatekeeping task were based upon newspaper items reported several months prior to the time the focus groups were conducted. As a result, people may have treated these items as less relevant, leading to little variance among them in terms of their newsworthiness. Accordingly, the little variance in each individual's mind created "random" rankings that upon being averaged for the correlational analysis produced very low correlations with the real rankings of the newspaper items.

Discussion

The Russian content analysis established that the most heavily reported topics in the Moscow newspaper were internal politics, culture, and sports, while in Tula the dominant events were concerned with internal order and internal politics. As far as the television newscasts were concerned, the lead topics in Moscow were internal and international politics, so that nearly half of the newscast was devoted to politics. In Tula, on the other hand, the leading topics were internal order, military, and defense issues as well as sports. As for radio, in both cities sports items and internal order lead the newscasts. Indeed, the analysis of the focus groups also showed that internal order and politics as well as cultural events and sports are uppermost in the minds of the people; hence, a degree of congruence exists between the informational needs people express and what the media offer them. Indeed, the events and issues receiving the most prominence in the focus group discussions dealt with such events as tanks being blown up in Chechnya, corruption among Russian government officials, elections in one of the Russian regions, and the resignation of President Boris Yeltsin. In addition, in the course of the focus group discussions, the audience members in Moscow expressed more interest in world events than in Tula—a fact that also corroborates the findings of the content analysis across the three media.

Another important point is that the focus group discussions support some of the ANOVA results regarding deviance. In particular, intensely deviant news did not necessarily receive the most attention. This was particularly the case with regard to the focus groups in Tula where the participants expressed particular interest in topics that had to do with or influenced their everyday life.

Another interesting point coming out of the Russian data was a discussion in the focus groups regarding a particular feature of present-day Russian media. Both journalists and public relations practitioners in Moscow and Tula expressed the notion that all information can be potential news. They furthermore all seemed to agree that any news can be turned into the hottest topic of the day and that there was actually less news than what appears in the media. This led the focus group participants to the noteworthy conclusion that news in the Russian media is perceived by many as a means of manipulation.

Another important point made by the public relations and journalist focus group participants in the two cities was that people dealing with news on the professional level do not perceive it the same way as audience members. Accordingly, the news priorities of journalists differ somewhat from those of the audience. As one of the Moscow journalists observed, most of the stories that appear in the media are pieces oriented to editors-in-chief and that the news of the day does not necessarily reflect the real interests of the audience. This disparity clearly supports the gap exhibited by the relatively low rank-order correlation coefficients between the focus group members and the actual prominence of the newspaper items in the gatekeeping task.

In sum, while the Russian media have come a long way in recent years from the former Soviet approach, the news as presented in Russia is still perceived with a degree of suspicion and doubt both by the producers of news and by its consumers.

16

What's News in South Africa?

DANIE DU PLESSIS

The Media System in South Africa

The South African media system is as diverse as the population and history of the country. The press and the broadcasting systems developed differently and have different histories. In addition, the political influences of different regimes over the course of 350 years were responsible for strange turns in the development of the media system. As recently as the early 1990s a peaceful but revolutionary transformation of the whole of society took place when an all-inclusive democracy replaced the minority rule by whites in the country.

In 1996 a new constitution was adopted by Parliament, and for the first time in history the supreme power in the country was entrenched in the constitution, with a constitutional court guarding people's basic rights such as freedom of speech and freedom of expression; these values guided the media system in the country. The first *Press Freedom Indexes,* released in 2002 by Reporters without Barriers, ranked South Africa 26th in the world together with Austria and Japan (Government Communication and Information System 2002, 133).

South Africa is an extremely heterogeneous country with a population of approximately 44 million. Of these, 76.7% classified themselves as African, 10.9% as white, 8.9% as colored (of mixed descent), and 2.6% as Indian/Asian (Statistics South Africa 2000). To cater to South Africa's diverse peoples, the constitution provides for eleven official languages: Afrikaans, English, isiNdebele, isiXhosa, isiZulu, Sepedi, Sesotho, Setswana, siSwati, Tshivenda, and Xitsonga. According to the Census 1996 figures, isiZulu is the mother tongue of 22.9% of the population, followed by isiXhosa (17.9%), Afrikaans (14.4%), Sepedi (9.2%), and English (8.6%).

The huge diversity of the population constitutes a major problem for the development of any media system. Just under 50% of the population, for example, has no speaking knowledge of English or Afrikaans (the former two official languages) (Schuring 1993). This implies that the media market is fragmented and therefore less competitive because smaller audiences need to be serviced.

Since the beginning of South Africa's press history, newspapers and press groups have been organized around the defining factor of race. The first newspaper, published in 1800, was directed at readers in the white community, with black interests ignored (Roelofse 1996, 85). The first paper aimed at a black South African audience appeared in 1837 and was printed by the Wesleyan Mission Society.

The history of the black press followed five distinct phases (Roelofse 1996, 82):

1. A missionary period, representing a time of missionary publications (from the 1830s)
2. The independent period (from 1880), when the first independent newspapers by blacks for blacks were established
3. The white-owned period (from 1930), when a lack of capital and equipment caused a white takeover of the black press, although it still aimed at a black audience
4. A period of English press dependence on the black reader
5. Since 1995, a period in which mainstream newspapers (including traditionally white newspapers groups) are being taken over by black business interests

The white press industry developed in two streams with two distinct developments: the development of English press groups and of Afrikaans press groups.

During the apartheid years—especially since the 1960s—the government provided a legal framework that favored a self-regulatory function of the press with the Press Council (followed by the Media Council) that served as instruments for indirect control. Newspapers had to be registered by the state and could be deregistered if they did not conform to policy, which led to the publication of many "alternative" publications. Even during this time the South African newspapers were better able than many African newspapers to withstand political, ideological, and commercial pressures. Since the early 1990s the press was liberalized with the removal of most of the legislation controlling the media.

Today the South African press consists of seventeen daily newspapers (mostly located in different metropolitan areas), seven Sunday papers, and approximately 158 community newspapers.

The largest English daily newspaper, aimed at a black audience, is the *Sowetan,* with a daily circulation of almost 200,000. The largest Afrikaans newspaper is *Die Burger,* published in Cape Town with slightly more than 120,000 printed on a Saturday. The *Sunday Times* has a circulation of more than 500,000 every Sunday.

Broadcasting was traditionally in the hands of the government. The first radio broadcast in South Africa took place in 1923 as a private initiative, and then the South African Broadcasting Corporation (SABC) came into being in 1936 as a public broadcaster, broadcasting in Afrikaans and English. In 1960, the SABC introduced radio stations for indigenous black languages (Wigston 1996, 312). Currently the SABC's national radio network comprises nineteen stations, which combined reach an average daily adult audience of more than 20 million.

In the early 1990s, following the political changes in South Africa, the Independent Broadcasting Authority—established in 1993—permitted independent radio stations. In 2000, the Independent Communications Authority of South Africa (ICASA) replaced the Independent Broadcasting Authority. Several of the SABC commercial radio stations were sold to private investors and community radio stations. Currently more than twelve commercial radio stations broadcast programs (mostly in specific regions) and more than fifty community radio stations operate under license from ICASA.

In 1976 the SABC introduced television as a public broadcaster. Today the SABC's network consists of three channels that broadcast in eleven languages (with English as the dominant language). There are more than 4 million licensed television households in South Africa. About 50% of all programs transmitted by the SABC are produced in South Africa. Some ninety-eight news bulletins are broadcast in all official languages weekly (South Africa 2002, 136). All television stations broadcast nationally, which is why only one television newscast was sampled for the study.

In 1986 South Africa's first private subscription television service, M-Net, was launched. M-Net focuses on sport, movies, and general entertainment; its programming does not include any newscasts.

In 1995 Multichoice Africa was formed as the first African company to offer digital satellite broadcasting. It is currently present in more than fifty countries across Africa and offers more than fifty-four video and forty-eight music channels with more than 1.4 million subscribers.

E-TV, established in 1998 and starting with broadcasts in 1999, is a private commercial television service (free to air) that is dependent on advertising, does not charge subscription fees, and includes news broadcasts.

The Study Sample

The cities chosen for this study were Johannesburg and Bloemfontein. The two cities are perceived to represent two completely different communities in South Africa. The media chosen in the two cities were also deliberately chosen to represent different communities in the two cities. In Johannesburg, a newspaper and radio station were selected to represent a black urban audience, and in Bloemfontein the newspaper and radio station that concentrate on a white Afrikaans-speaking audience were selected.

Johannesburg is in the Gauteng Province and is South Africa's largest city, with more than 3 million residents in the larger metropolitan area, which includes the black township Soweto. Because Gauteng, and specifically

Johannesburg, is the economic power base of South Africa, it is the most diverse province and city in the country, with all population groups represented and complemented with people from all over Africa and the rest of the world. In the Gauteng province, six daily newspapers compete with eight commercial radio stations (in addition to the public broadcasting stations).

The *Sowetan*, with the highest circulation (approximately 200,000) and readership (more than 2 million) of any newspaper in South Africa, is the newspaper selected for the study in Johannesburg. With its beginnings as a struggling newspaper reflecting the lives and viewpoints of black people under the system of apartheid, the *Sowetan* gave voice to the liberation struggle, with a staff including numerous political activists. After 1994 political freedom was achieved, and the *Sowetan* changed with the people it served. As its readers became more educated and affluent, it adapted to its new role in order to still serve the same community, but in different ways. Since 1981 the newspaper has been published in English six days a week, from Monday to Saturday. Although the *Sowetan* is distributed nationally, its focus is on Gauteng, and its target audience is a middle-class progressive black urban readership.

There was a slight deviation from the planned composite week of the study. Because December 31, 2000, was a Sunday—the *Sowetan* is not published on Sundays—and Monday, January 1, 2001, was a public holiday, the edition for January 2, 2001, was selected for the sample.

Owned by SABC, the radio station selected was Metro FM, a commercial entertainment radio station targeting a black metropolitan audience. Although Metro FM broadcasts nationally, its focus is on Johannesburg and Gauteng, where most of its listeners reside. Very short news bulletins (approximately five minutes) are broadcast every hour, and the 7 p.m. newscast was selected as the main newscast for the day.

The collection of data for Metro FM was secured in advance by requesting the radio station to record and keep copies of the 7 p.m. newscasts for the selected dates. The arrangement was twice confirmed. Unfortunately recordings of only four newscasts could be secured afterward. The newscasts for November 12, December 6, and December 22, 2001, were lost and despite all efforts could not be replaced.

In keeping with the study's objective to maximize variance, Bloemfontein was chosen as the second city. With a population of approximately 500,000 people, it is the sixth largest city in South Africa. Bloemfontein, which houses South Africa's Supreme Court of Appeal, is the capital of the Free State Province and has a well-established institutional, educational, and administrative infrastructure. The Free State Province and its people, who speak mainly Sesotho (blacks) and Afrikaans (whites), are widely seen as conservative. The economy there is mainly driven by agriculture and mining (with a growing manufacturing sector), with a low level of urbanization. Only one Afrikaans daily, *Die Volksblad*, is produced in Bloemfontein, and its target is a white audience with distribution in the Free State and parts of the Northern Cape Province but most of which is in Bloemfontein. Because it is the only local daily newspaper, its readers include a high percentage of Sesotho and English speakers.

Here, too, there was a slight deviation from the planned composite week. Because December 31, 2000, was a Sunday—*Die Volksblad* also is not published on a Sunday—and Monday, January 1, 2001, was a public holiday, the first edition for 2001 was published on January 3, 2001, and thus was selected for the sample.

The only commercial (and regional) radio station in Bloemfontein is OFM, a bilingual Afrikaans–English radio station catering to as broad an audience as possible in the Free State Province. It is an entertainment station with brief (five minutes) newscasts every hour. The main newscast is at 6 p.m. every day.

Because there are no regional television broadcasts in South Africa, the SABC3's 8 p.m. main newscast was selected for the study. The 30-minute broadcast contains news items and concludes with a weather forecast.

The five South African outlets provided a total of 1,986 news items during the study period. The two newspapers provided for the most items (1,146 items in *Die Volksblad* and 613 items in *Sowetan*). The SABC3 newscasts yielded 137 items and the radio stations OFM (fifty-nine items) and Metro FM (twenty-nine items) the least.

Topics in the News

As can be seen in Table 16.1, if all of the broad categories of news topics are analyzed for all the media together the primary ones (more than 10%) are sports (19.4%), followed by business/commerce/industry (11.8%), internal order (10.7%), and internal politics (10.3%). However, if single topics, which were condensed into the broader categories, are analyzed, only news items dealing with crime (6.8%), reports on sports competitions and results (7.0%), and reports on individual athletes (6.7%) made up more than 5% of the topics in all news items. In addition, crime was the most common single topic (number of news items) on the front pages or first third of all media newscasts included in the study. This clearly indicates the priorities in news reporting in South Africa, namely sports and crime.

Proximity plays a very important role in the reporting of crime events. In the content analysis of the media it emerged that a majority of the reports on crime (50.6%) dealt with crime events taking place in the same or surrounding communities. In addition, if only the front pages of newspapers or the first third of newscasts were taken into account, this number increased to 57.7% of news items referring to crime events taking place in the same or surrounding communities. On the front pages or in the first third of newscasts, crime stories dealing with events in the same community dramatically increased from 26.2% (of all crime stories) to 42.3%. The media therefore respond

TABLE 16.1 Distribution of General Topics of News Items by City and Medium

Topics	Newspaper		Television		Radio	
	Johannesburg	Bloemfontein	Johannesburg	Bloemfontein	Johannesburg	Bloemfontein
Sports	27.5	15.7	11.8	11.8	3.7	3.5
Internal Order	11.3	9.4	10.9	10.9	22.2	12.9
Human Interest Stories	9.0	10.0	3.8	3.8	.0	3.5
Internal Politics	8.9	8.8	16.6	16.6	3.7	36.5
Cultural Events	6.8	4.8	1.5	1.5	.0	.0
Disasters/Accidents/ Epidemics	6.3	2.5	6.2	6.2	16.7	3.5
Business/Commerce/Industry	5.0	16.3	9.8	9.8	.0	5.9
Economy	4.7	5.3	3.0	3.0	5.6	.0
International Politics	3.5	2.9	10.9	10.9	24.1	4.7
Entertainment	2.9	2.2	.6	.6	.0	.0
Education	2.6	3.7	2.4	2.4	.0	4.7
Transportation	2.0	2.4	4.7	4.7	5.6	7.1
Health/Welfare/Social Services	1.8	3.8	3.3	3.3	1.9	4.7
Housing	1.6	.2	1.5	1.5	3.7	.0
Social Relations	1.3	2.1	2.4	2.4	3.7	7.1
Labor Relations and Trade Unions	1.2	.8	3.3	3.3	1.9	.0

	(n = 767)	(n = 1680)	(n = 338)	(n = 338)	(n = 54)	(n = 85)
Communication	.9	1.8	.0	.0	.0	.0
Other	.9	.4	.0	.0	.0	.0
Environment	.7	1.0	1.8	1.8	1.9	1.2
Fashion/Beauty	.4	.0	.0	.0	.0	.0
Military and Defense	.4	1.0	1.8	1.8	1.9	.0
Science/Technology	.3	1.1	.3	.3	.0	2.4
Energy	.1	.8	.0	.0	.0	.0
Weather	.1	.7	.3	.3	.0	1.2
Population	.0	.0	.9	.9	3.7	1.2
Ceremonies	.0	2.4	2.4	2.4	.0	1.2
Total[a]	100.0	100.0	100.0	100.0	100.0	100.0

Note: Distributions given in percent.
[a]Total percentage may not actually be 100.0 due to rounding error.

intuitively to the need of audiences to be informed about crime events taking place close to them.

Sports made up the majority of reports in the Johannesburg newspaper (27.5%) and is the second most common topic in the Bloemfontein newspaper (15.7%) and television (11.8%), possibly because all of South African society (all population groups) is preoccupied with sports (as they are with crime), although different types of sports appeal to the black and white components of society. In the *Sowetan*, horse racing and soccer dominate, and in *Die Volksblad* rugby, cricket, and track and field athletics are the major sports topics.

In the Bloemfontein newspaper, business/commerce/industry received the most coverage (16.3%), possibly because most of the readers of this newspaper are affluent whites who can afford to invest their money or who actually are involved in industry as entrepreneurs; in contrast, the majority of readers of the Johannesburg newspaper are, in general, less affluent black workers or black middle-class upcoming professionals. This again emphasizes the racial divide in media usage as indicated in the first section of this chapter.

Internal order, which includes crime, is predictably one of the prominent topics both in Johannesburg and Bloemfontein. That the Gauteng province is also the crime capital of South Africa is reflected in the slightly more attention devoted to internal order by the Johannesburg media—especially the Johannesburg radio (22.2%) and the Johannesburg newspaper (11.3%).

Television distinguished itself from the other media with less emphasis on human interest stories and more prominence on topics dealing with internal politics and international politics. Bloemfontein radio paid much attention to internal politics (36.5%) and social relations (7.1%), perhaps because during the time of this research local elections were held across the country, possibly indicating that the Bloemfontein radio was following a specific agenda to influence local elections.

The significant Spearman rank-order correlation coefficients in Table 16.2 indicate that there are differences between the media in the two cities and also between the different media. The highest correlation can be found between the Johannesburg newspaper and the Bloemfontein newspaper (.83). The topical structure of the two newspapers is therefore relatively similar. There is also a degree of correlation (.77) between the Bloemfontein newspaper and television and to a lesser degree (.73) between the Johannesburg newspaper and television.

Johannesburg radio's topical structure differs from all other media, having the lowest correlation with the Bloemfontein newspaper. However, it does correlate slightly positive (.64) with television, which may be because the SABC owns both the radio station and the television station. Some synergy in news gathering and presentation may be responsible for the similarity between the two.

TABLE 16.2 Spearman Rank-Order Correlations between Rankings of News Topics in Various MediaCoefficients among Rankings of News Topics in Various Media

	Johannesburg Newspaper	Bloemfontein Newspaper	Johannesburg Television	Bloemfontein Television	Johannesburg Radio	Bloemfontein Radio
Johannesburg Newspaper		.83[c]	.73[c]	.73[c]	.38	.49[a]
Bloemfontein Newspaper			.77[c]	.77[c]	.20	.63[b]
Johannesburg Television				—	.64[c]	.74[c]
Bloemfontein Television					.64[c]	.74[c]
Johannesburg Radio						.38
Bloemfontein Radio						

[a] $p < .05$; [b] $p < .01$; [c] $p < .001$.

Deviance in the News

The study assumes that deviant news items will be larger or longer and will hence be more prominently placed compared to news items that are not deviant. This basic assumption was tested by calculating prominence scores for news in the different media for each level of statistical deviance, social change deviance, and normative deviance. One-way analyses of variance (ANOVA) were also computed to test the statistical significance of the differences between the prominence scores. As noted, because television is not broadcast in regions, only one national television news broadcast was included in the study.

Table 16.3 presents the mean scores. In only seven of the thirty analyses conducted, significant differences between the levels of deviance were obtained. This does not mean that all the significant differences indicate linear relationships between deviance and prominence. In only five of the seven cases the lowest level of deviance corresponds with the lowest prominence scores. Across the analyses, statistically significant or not, in twelve of the thirty cases

TABLE 16.3 Mean Verbal and Visual Prominence Scores for Intensity of Deviance

| | Johannesburg | | | | | Bloemfontein | | | | |
| | Newspaper | | Television | | Radio | Newspaper | | Television | | Radio |
Intensity of Deviance	Verbal only (n = 587)	Verbal plus Visual (n = 143)	Verbal (n = 137)	Visual (n = 130)	Verbal (n = 29)	Verbal only (n = 925)	Verbal plus Visual (n = 244)	Verbal (n = 137)	Visual (n = 130)	Verbal (n = 59)
Statistical Deviance										
(1) common	277.5	538.4	113.6	120.7[a]	53.0	239.1	271.8[b]	113.6	120.7[a]	55.0[b]
(2) somewhat unusual	193.2	710.2	201.0	222.4	48.5	201.3	434.7	201.0	222.4	49.2
(3) quite unusual	274.2	926.4	230.3	240.4	46.4	212.7	394.7	230.3	240.4	50.4
(4) extremely unusual	97.7	–	307.5	–	150.0	300.3	282.3	307.5	–	207.0
Social Change Deviance										
(1) not threatening to status quo	275.2	558.2	157.6[a]	166.6	60.7	187.4[b]	347.0	157.6[a]	166.6	52.2
(2) minimal threat	180.9	347.0	249.8	256.8	43.8	285.1	407.3	249.8	256.8	56.3
(3) moderate threat	211.0	635.5	261.5	258.3	41.4	253.7	256.0	261.5	258.3	47.7
(4) major threat	503.0	1677.4	175.0	175.0	–	153.3	–	175.0	175.0	–

Normative Deviance

(1) does not violate any norms	277.2	569.7	173.6	178.9	56.4	226.4	333.5	173.6	178.9	41.8
(2) minimal violation	276.3	251.7	230.6	235.7	–	162.4	551.5	230.6	235.7	60.0
(3) moderate violation	134.9	–	228.1	241.8	44.0	171.5	434.4	228.1	241.8	57.0
(4) major violation	137.3	482.6	212.2	218.5	42.4	269.4	–	212.2	218.5	100.3

[a]$p < .05$; [b]$p < .01$; [c]$p < .001$.

the highest levels of deviance correspond with the highest prominence scores. Also, in the case of television the lowest levels of deviance corresponded in all six calculations with the lowest prominence scores.

In looking at the different types of deviance (ten calculations for each of statistical deviance, social change deviance, and normative deviance), statistical deviance stands out as a marker for prediction of prominence. In eight of the ten calculations for mean prominence, the highest level of deviance corresponded with the highest prominence scores. This is in stark contrast with the other types of deviance, where only two of ten cases showed the same relationship. In a similar comparison, the lowest levels of deviance correspond with the lowest prominence scores in six of the ten calculations for both statistical deviance and normative deviance but not for social change deviance.

Social Significance in the News

As with deviance, the study assumes that socially significant news items will be more prominent in the media than news items that are not socially significant. This hypothesis was tested by calculating the mean prominence scores for the news items in the different media. Of the forty analyses, sixteen produced significant effects (Table 16.4). Of these, the lowest level of significance corresponded with the lowest prominence scores in twelve of the cases. The highest level of significance corresponded with the highest prominence scores in five of the cases. Television accounted for twelve of the sixteen cases. In ten of the twelve cases the lowest level of significance corresponded with the lowest prominence scores.

Across all of the analyses, statistically significant or not, in four of the eight calculations for the Johannesburg newspaper and in five of the eight calculations for the Bloemfontein newspaper the highest levels of social significance corresponded with the highest prominence scores. In Johannesburg television, seven of the eight calculations indicated a correspondence between the lowest levels of social significance with the lowest prominence scores.

As for political significance, in six of the ten mean prominence scores for political significance, the highest levels of social significance corresponded with the highest levels of prominence, while the lowest levels of social significance corresponded with the lowest prominence scores in five of the ten cases. As for the economic significance dimension, four of the six calculations yielded results where the lowest levels of social significance correspond with the lowest prominence scores, and only in two of the ten cases did the highest levels of social significance correspond with the highest levels of prominence. In seven of the ten mean prominence scores for public significance, the lowest levels of social significance corresponded with the lowest prominence scores, while in four of the eight cases the highest levels of social significance corresponded with the highest levels of prominence. Cultural significance was a predictor of prominence in three of the ten analyses. In Johannesburg cultural

TABLE 16.4 Mean Verbal and Visual Prominence Scores for Intensity of Social Significance

Intensity of Social Significance	Johannesburg					Bloemfontein				
	Newspaper		Television		Radio	Newspaper		Television		Radio
	Verbal only (n = 592)	Verbal plus Visual (n = 142)	Verbal (n = 137)	Visual (n = 130)	Verbal (n = 29)	Verbal only (n = 924)	Verbal plus Visual (n = 242)	Verbal (n = 137)	Visual (n = 130)	Verbal (n = 59)
Political Significance										
(1) not significant	259.6	551.9	158.4	159.3[a]	43.9	196.9[c]	332.1	158.4	159.3[a]	55.9
(2) minimal	236.5	–	197.1	211.0	60.8	288.9	439.2	197.1	211.0	47.1
(3) moderate	138.6	188.3	223.7	265.5	62.8	208.9	600.0	223.7	265.5	51.4
(4) major	288.1	1025.0	231.3	150.8	45.0	590.9	–	231.3	150.8	30.0
Economic Significance										
(1) not significant	260.2	539.5	132.3[b]	136.9[b]	40.1	207.8[a]	343.5	132.3[b]	136.9[b]	51.2
(2) minimal	212.4	677.7	232.0	261.9	69.0	179.8	303.0	232.0	261.9	56.7
(3) moderate	288.3	703.9	252.1	253.0	62.0	263.3	246.1	252.1	253.0	58.0
(4) major	256.1	243.5	240.6	150.8	–	344.4	–	240.6	150.8	–

[a] $p < .05$; [b] $p < .01$; [c] $p < .001$.

(Continued)

TABLE 16.4 Mean Verbal and Visual Prominence Scores for Intensity of Social Significance (*Continued*)

Intensity of Social Significance	Johannesburg					Bloemfontein				
	Newspaper		Television		Radio	Newspaper		Television		Radio
	Verbal only (n = 592)	Verbal plus Visual (n = 142)	Verbal (n = 137)	Visual (n = 130)	Verbal (n = 29)	Verbal only (n = 924)	Verbal plus Visual (n = 242)	Verbal (n = 137)	Visual (n = 130)	Verbal (n = 59)
Cultural Significance										
(1) not significant	239.3[a]	576.8	157.3[a]	163.6[a]	55.1	200.4	334.8	157.3[a]	163.6[a]	56.4
(2) minimal	643.6	781.6	221.3	246.9	51.2	227.4	323.0	221.3	246.9	46.3
(3) moderate	214.0	258.1	336.7	222.5	44.3	276.1	453.6	336.7	222.5	44.6
(4) major	605.4	508.3	–	–	–	130.2	–	–	–	30.0
Public Significance										
(1) not significant	272.1	564.2	118.4[a]	125.2[b]	38.5	161.8[c]	309.1	118.4[a]	125.2[b]	57.6
(2) minimal	195.3	411.2	206.2	199.9	54.2	249.1	388.6	206.2	199.9	35.9
(3) moderate	183.1	218.8	243.3	297.0	60.7	237.4	362.1	243.3	297.0	62.7
(4) major	292.3	1231.2	188.0	202.3	67.0	307.7	–	188.0	202.3	26.5

[a] $p < .05$; [b] $p < .01$; [c] $p < .001$.

significance was a predictor for verbal prominence for newspaper and television and in Bloemfontein for verbal television.

Of all the media, the prominence scores of the items in the Bloemfontein newspaper and on television news were most often related to social significance.

Deviance and Social Significance as Predictors of News Prominence

In addition to looking at the prominence scores for deviance and social significance, the influence of all the dimensions of deviance and social significance on news prominence was tested by calculating stepwise regression analyses. For each of the two cities, one analysis was calculated for the prominence of verbal newspaper content, the prominence of combined verbal and visual content in newspapers, as well as for television and radio prominence.

From Table 16.5 it can be concluded that for Johannesburg none of the independent variables explained a statistically significant amount of variance in news prominence. For Bloemfontein, the prominence of verbal content and combined visual and verbal content prominence were related to deviance and social significance, and the same was true of television. However, the amount of the variance explained was rather small (3% for *Die Volksblad*'s verbal content and 6% if combined with the visual content; 9% for OFM and 12% for television). The overall conclusion from this analysis is that deviance and social significance explain a small part of the prominence of news items in South Africa.

Looking at the data in greater detail, we find that statistical deviance alone affected the prominence in television in Johannesburg and actually had a negative effect on the prominence of verbal newspaper content in Bloemfontein. Also, for the Bloemfontein newspaper some significance can be observed regarding the constructs of social change deviance and public significance—looking at the verbal content and more than double for public significance when the visual and verbal content is combined. The only other significant results were obtained with the visual content of television for public significance, and the Bloemfontein radio's verbal content related to normative deviance.

People Defining News

Eight focus group discussions were conducted (four in each city). In Bloemfontein the four focus groups were composed of individuals from 18 to 60 years of age, and the majority of respondents were white, Afrikaans-speaking females (approximately 60%). The journalists' focus group consisted of individuals from *Die Volksblad*, OFM, and the SABC. The size of the groups varied from nine to eleven participants each. Blacks were best represented in the low socioeconomic status (SES) audience group. In Johannesburg, the focus groups comprised a representative majority of blacks (based on the readership of the Sowetan) in all groups (approximately 75%), with males and females

TABLE 16.5 Stepwise Regression Analyses of Intensity of Deviance and Social Significance on News Prominence

	Johannesberg								Bloemfontein							
	Newspaper Prominence Verbal only Total R^2 = ns (n = 581)		Newspaper Prominence Visual and Verbal Total R^2 = ns (n = 124)		Television Prominence Total R^2 = .12[c] (n = 130)		Radio Prominence Total R^2 = ns (n = 29)		Newspaper Prominence Verbal only Total R^2 = .03[c] (n = 918)		Newspaper Prominence Visual and Verbal Total R^2 = .06[b] (n = 114)		Television Prominence Total R^2 = .12[c] (n = 130)		Radio Prominence Total R^2 = .90[a] (n = 59)	
Independent Variables	r	Std. Beta	r	Std. Beta	r	Std. Beta	r	Std. Beta	r	Std. Beta	r	Std. Beta	r	Std. Beta	r	Std. Beta
Deviance																
– Statistical, verbal content	.04	ns	.00	ns	.25[b]	.21[a]	.19	ns	-.02	-.09[a]	.10	ns	.25[b]	.21[a]	.16	ns
– Statistical, visual content	–	–	.03	ns	.24[b]	ns	–	–	–	–	.20[a]	ns	.24[b]	ns	–	–
– Social change, verbal content	.01	ns	.02	ns	.17	ns	-.18	ns	-.11[b]	.09[b]	.15	ns	.17	ns	-.02	ns
– Social change, visual content	–	–	.03	ns	.15	ns	–	–	–	–	-.06	ns	.15	ns	–	–
– Normative, verbal content	.07	ns	-.02	ns	.10	ns	-.14	ns	-.04	ns	.15	ns	.10	ns	.29[a]	.29[a]
– Normative, visual content	–	–	-.01	ns	.12	ns	–	–	–	–	.05	ns	.12	ns	–	–

Social Significance

– Political, verbal content	-.00	ns	.02	ns	.15	ns	.16	ns	-.10[b]	ns	.14	ns	.15	ns	-.07	ns
– Political, visual content	—	—	.03	ns	.18[a]	ns	—	—	—	—	.03	ns	.18[a]	ns	—	—
– Economic, verbal content	-.00	ns	.06	ns	.25[b]	ns	.22	ns	-.07[a]	ns	.01	ns	.25[b]	ns	.06	ns
– Economic, visual content	—	—	.02	ns	.20[a]	ns	—	—	—	—	.08	ns	.20[a]	ns	—	—
– Cultural, verbal content	.07	ns	-.03	ns	-.24[b]	ns	-.07	ns	.07[a]	ns	.13	ns	24[b]	ns	-.13	ns
– Cultural, visual content	—	—	-.02	ns	.21[a]	ns	—	—	—	—	-.00	ns	.21[a]	ns	—	—
– Public, verbal content	-.02	ns	.02	ns	.21[a]	ns	.19	ns	.12[c]	.11[b]	.25[b]	ns	.21[b]	ns	.08	ns
– Public, visual Public	—	—	.01	ns	.27[b]	.24[b]	—	—	—	—	.09	ns	.27[b]	.24[b]	—	—

$^{a}p < .05$; $^{b}p < .01$; $^{c}p < .001$; ns = not part of final stepwise regression equation.

between 20 and 55 years of age almost equally distributed. The journalists' focus group consisted of journalists from the *Sowetan*, the *Star* (another daily newspaper in Johannesburg), the SABC, and one freelance journalist. The sizes of the groups varied between eight and twelve respondents. In Johannesburg, nine of the ten participants in the public relations focus group were employed by various organizations, and only one came from a consulting firm. In Bloemfontein the public relations group consisted of ten participants, with five from the private sector and five employed by institutions. All the focus groups were conducted in March 2001.

South Africa's crime statistics show that serious crimes such as murder, rape, carjacking, and aggravated robbery are common events in both communities. The media and their audiences are therefore very much preoccupied with the topic. Shoemaker (1996, 39) explains that people are interested in news and deviant events because of a genetically determined need to monitor the environment. The need for humans to execute surveillance of their environment was, for thousands of years, their key to survival. In modern life (and specifically in South Africa with its extremely high crime rate), the aforementioned serious crimes represent one form of an immediate physical threat similar to what humans faced in the ancient past. This was confirmed during the focus group interviews. Respondents felt as though the media perform surveillance on their behalf. Respondents indicated that they want to be informed about possible threats and that they expect the media to perform that function.

The focus groups confirmed that crime reports serve as a practical warning system that elicit specific responses from the audience, such as to be aware of personal safety, to be informed about possible threats, to improve security measures, to be careful of crime hot spots, and to question the character of the people they are dealing with.

Thus, respondents confirmed the basic notion of the theory that surveillance of the environment with the aim of survival is the dominant motive for obtaining news. Without explicitly referring to this function in news reporting, the focus group respondents spontaneously referred to this during the discussions. The impact of events on their lives—and proximity—were perceived as the critical factors in determining news value for them. Examples of comments illustrating this were:

- "Some things are crucial to survive; sport is interesting but not important, but what happens in politics and in the environment—the basic sources—that is going to touch me."
- "If it happened to her, it can happen to our kids ... plays directly on your sense of safety. If she wasn't safe, then we are not safe."
- "If it's happening to them it might even happen to us."

- "The (newspaper) told me there was someone murdered, for example in Sanlam Plaza. It is enough for me to know you cannot go there safely. You have to be careful."

In the same context as threatening events, respondents also made references to people. A participant in Bloemfontein said, "It took someone who you could always believe in and trust, and he now did something wrong. It makes you start to doubt a lot of things around you, because this guy who was so pure has now done something wrong. Now you start to think: but that guy I made that deal with last week—he could also be wrong."

The focus group results indicate that proximity to a news event has a direct impact on the way in which the respondents perceive a threat. One participant said, "If it is close, you are more threatened." Another stated, "So if something like that happens in Bloemfontein, then it catches your attention immediately, because now it's getting close to home. And now you want to know: what happened, why it happened—did the police do anything about it, did they catch anybody, that type of thing, Ok, stands out."

Deviance

The focus group discussions also confirmed that the unusual and unexpected (statistical deviance) contributes to the newsworthiness of news items. In almost every focus group the respondents referred to the frequency of events as determinants of their value, for example, by distinguishing between "normal" and "unusual" murders. In view of South Africa's high crime rate, a crime news event has to be exceptionally deviant to be seen as such. Although there was more or less consensus about extremely unusual news events, other serious events such as corruption or murder do not necessarily represent statistical deviance, depending on the frequency of these events in a specific society and the proximity of the event. To illustrate this point, on the 8 p.m. newscast of SABC3 on December 6, 2000, the bulletin was mainly devoted to reporting on the results of the local elections held in all of South Africa the previous day. In the second third of the news broadcast, it was reported that on the previous day six men arrived at the "wrong" voting station on the East Rand and were shot. In another context—such as in a country in Europe—this type of incident would have caused major headlines, whereas in South Africa it was only mentioned later in the newscast.

News items catching respondents' attention were described as "unexpected," "something different," "unusual," and so on. Some examples of quotes related to statistical deviance were:

- "It was the first one. It was the only one."
- "Because it happens for the first time."

- "When it is exceptional—for example, when it is an unusual murder. But the newspapers, a normal murder, when someone dies—it is only a small report, and the people will not buy the newspaper for that."
- "Some cases are exceptional. That person being shot … to us it's normal."
- "It was different from the ordinary. It was not only an issue of somebody murdering somebody to steal his wallet."
- "It's got to shock to the point where you have to read it."

Since the early 1990s, South Africa has gone through a period of extreme changes; therefore, references to issues related to potential for social change deviance were frequently encountered in the focus groups. Examples of these comments are:

- "And then Nelson Mandela's release changed the circumstances, everything in the country changed. … It had mass impact on everybody."
- "One is afraid of the unknown, when things change suddenly."
- "But essentially the white person who was living in a comfort zone, which was sort of official, all of a sudden saw that this is now."
- "And, as with the change of government, you do not know what is waiting for you. Until now you knew how things worked but now you have no idea. … You feel uncertain."
- "Yes perhaps—the black people felt that they obtained freedom, you know, and some of the whites were very much afraid. It is every time two opposites confront each other."
- "I think it affected us directly because it was the end of one era and the start of a new one."

Normative deviance was the one dimension of deviance that attracted the least attention as markers for newsworthiness among the respondents. They were not so much worried about the moral issues in a news event as they were concerned about the impact it may have on them. Typically they would refer to "unacceptable behavior" or express disgust about specific behavior such as child molesting. Examples of quotes are:

- "And everybody realizes that children may be affected."
- "Even if not everybody is a confessing Christian … many are believers and once it is threatened [they react]."
- "And now they are the new government, and they have a completely new viewpoint. They see the white people as invaders. This is the way they see it. History is a fact; you cannot change it. But it is the way it is conveyed—the wording—and that is what they want to change."
- "It [Satanism] is not acceptable in our society, and it is unacceptable for the community. This is one of the strongest weapons she could use,

because it is unacceptable in the community and now she labels it [the Harry Potter books] with this awful unacceptable thing."
- "That is totally unacceptable behavior."

Social Significance

The focus group respondents often referred to stories of social significance to illustrate their views on newsworthiness. Being in South Africa and in view of the radical political changes that have taken place since 1990 and that are still taking place on a daily basis, issues of political significance are referred to in historical terms. The beginning of the process (the release of Nelson Mandela) and the 1994 elections had a huge impact on almost every participant in the group discussions. The frequency and volume of political changes still going on led to complacency and political apathy when the current political impact was evaluated in the focus groups. Virtually all references to political significance of news events refer to the historical events of the first part of the 1990s. Current political events are overshadowed so greatly by the start of the political process in South Africa that they have lost much of their significance to the participants. Both black and white participants in the focus groups shared this response.

Economically significant events, on the other hand, were important for both cities' respondents—especially the high SES audience and public relations groups. Their responses were mostly related to the issues of survival and personal relevance. Examples of quotes referring to the economic significance of news events are the following:

- "I sometimes question if they [some people] ever come close to a newspaper, but then came the day before the [minister of finance's] budget speech—then they started talking about the budget and speculating, and the next day one of them immediately came and asked, 'May I have a quick look at the paper' and 'What happened with the budget?' And these are people who do not read the newspaper usually."
- "An announcement about the petrol price or interest rates which were lowered will affect people positively."
- "I think it [a hike in the fuel price] influences one's budget, because it is not only the petrol price which is higher. Sugar prices will rise—everything goes up."
- "Anything, as you said—if you see a headline, anything with monetary implications. Then I immediately get a newspaper. And then I will see this thing—if it will cost me something, then I immediately read it."
- "So my only interest is in economic issues because I think that affects my daily, my daily pocket, and everyone else's pockets so to say."

On one hand, cultural significance as a construct did not play a major role in the focus groups. The exception was where Afrikaans-speaking respondents in Bloemfontein expressed fears about their language being threatened. On the other hand, focus group respondents often referred to issues related to the construct of public significance. The first emotion expressed was that of empathy:

- "It's the number of people that are affected by it. If it ... affects a large number of other people—that makes it more intense."
- "The person whose name I often read about—here he is experiencing the same emotions I also had."
- "And he just sat there in an old chair, but all he wanted was for me to bring him medicine in order to become healthy again. I don't know—when I left I was really touched" [based on a story about the oldest man in the Bloemfontein region].
- "I think a lot of us have been touched by this—by the HIV AIDS problem."
- "Repeat pictures of how people die of poverty in southern African countries. I can't get that picture out of my head. To suddenly realize it is people who live like you in certain aspects. It is people with the same needs. It changed my life."

The other question was a "feel-good" emotion, which respondents illustrated as follows:

- "We are surrounded by issues such as murder, theft, robberies, and then suddenly there happens one good little thing. And then you will buy the newspaper to see this one good thing."
- "It affects—although we don't admit it—it affects every one of us."
- "After Mandela was released, now the country—all of us—we are now free, are going to be equal."
- "It gives me a reason why I can be positive at the moment in South Africa."
- "We were united as a nation, people from the Free State, Capetonions, blacks, whites, Zulus, everybody. We all had one vision—we had to win. For those few weeks we all spoke the same language; everybody agreed. That was absolutely the best two weeks of my life."

Personal Relevance

Another concept that came out of the focus group discussion was the issue of personal relevance. Loosely related to the concept of proximity, personal relevance also plays a role in bringing a news event closer to the members of the audience and makes it more newsworthy for them. Personal relevance would therefore refer to psychological distance rather than physical distance. In the focus groups the issue of personal relevance was expressed in the following examples of quotes:

- "If it does affect you, no matter how boring something might be if it affects you personally it makes it interesting."
- "Anything that affects me personally or anything that affects the world or my surrounding or humankind, personally."
- "Depends how it would affect you personally. If I don't see any personal involvement, I tend to ignore things like that happening in Bosnia and that type of thing. They don't really have a direct effect. But if you talk about something that would have a personal effect on the money in my pocket or my safety."
- "First of all I think it is something that can affect you or someone that's close to you. And then second of all, it's something that interests you, something that you personally are interested in, that ... could affect you."
- "If you know the people who are directly involved, then it will become much worse."

Comparing People's News Preference with What's in the Newspaper

A hypothesis in this study is that a relationship exists among the rankings of newsworthiness among journalists, public relations practitioners, and high and low SES audience members and that a relationship exists between the rankings of the four focus group members and the actual prominence of the newspaper items. This was tested by means of the gatekeeping task. Spearman rank-order correlation coefficients would determine the relationships between the actual rankings of the prominence of the newspaper items and the rankings of the four groups.

Table 16.6 presents the twenty Spearman rank-order correlation coefficients for the two cities. Of the twenty correlations, only eleven were statistically significant.

TABLE 16.6 Spearman Rank-Order Correlations between Newspaper Item Prominence and Focus Group Rankings

	Journalists	Public Relations	High SES Audeince	Low SES Audience	Newspaper
Journalists	−	.87[c]	.77[c]	.61[c]	.06
Public Relations	.88[c]	−	.88[c]	.79[c]	.06
High SES Audience	.60[c]	.70[c]	−	.83[c]	−.03
Low SES Audience	.47[b]	.58[b]	.31	−	−.03
Newspaper	.13	.01	.12	−.16	−

Note: Johannesburg coefficients are in the upper triangle; Bloemfontein coefficients are in the lower. SES = socioeconomic status.
[a]$p < .05$; [b]$p < .01$; [c]$p < .001$.

Two interesting phenomena emerge from the data. First, the intercorrelations among the focus group participants in Johannesburg were meaningfully higher than the counterparts in Bloemfontein. Second—and much more poignant—is the fact that in both cities there was virtually no relationship between the rankings of the focus group members and the actual line-ups of the newspapers in the two respective cities.

In Johannesburg, all six Spearman rank-order correlations among the four groups were significant, ranging from .61 to .88. In contrast, in Bloemfontein, the range was from a nonsignificant .31 (between the high and low SES participants) to .88, and on average the correlations were quite lower than in Johannesburg.

While the intercorrelations among the rankings of the different group members were relatively high, there was no relationship between the actual prominence of the newspaper items and the perceptions of group members. This total lack of relationship is quite intriguing as it includes not only the audience members but also the two groups of professionals—journalists and public relations practitioners—in both cities. How can these findings be explained? It is clear that the two newspapers seem to be out of touch with what their audiences see as important news. One can speculate that since the sweeping changes of the early 1990s the South African society transformed fundamentally and that newspapers' managements have not adapted to these changes to the same extent as society in general. Editors and associate editors may still feel responsible for shaping the agenda in society by focusing attention on issues that in their opinion are of importance instead of being guided by news values in congruence with their audiences. However, this warrants further investigation.

Discussion

The selection of the two cities and the different media was done with the aim of maximizing variance. The two communities and their media were supposed to represent a spectrum of the heterogeneous society in South Africa. The results of the study indicate this because the South African society as a whole shares similar experiences and is similarly affected by societal issues such as crime and political developments. All communities in the country share a common perception about their society and the issues related to it. There were no substantial differences between the two communities (and their media) in the study to indicate that their culture determines what is seen as deviant or socially significant. The focus group discussions confirmed that it is their common experience rather than cultural background that influences their perceptions of what news is.

The topical structure of the media was very similar according to the Spearman rank-order correlation coefficients between rankings of news topics in the various media. The only exception was the Johannesburg radio station, which was less intensely related to the other media. This may be ascribed to the low number of news items included in the study for the Johannesburg radio. The highest correlation was found between the two newspapers.

The main question the study attempted to answer is determining to what extent deviance and social significance would account for the prominence of news items in the different media. The results are somewhat ambivalent. Based on all the average prominence scores it can be concluded that the two constructs may have some influence on the prominence of news items, especially statistical deviance and normative deviance, as well as political, economic, and public significance. At the same time, the significant ANOVAs as well as the regression analyses indicate that the impact of the effect of deviance and social significance is limited.

However, the focus group data confirmed the basic concepts that form the foundation of this study. The news media assist people in doing surveillance of their environment in very practical ways. For the focus group participants, statistical deviance clearly affects the newsworthiness of news items with social change deviance and normative deviance playing a minor role. Three of the four constructs of social significance (political, economic, and public) also add to the newsworthiness of news items, but cultural significance as a construct did not figure prominently in the focus groups. From the focus group results it emerged that personal relevance also plays a major role in people's perceptions of the newsworthiness of news items.

Finally, the main results of the gatekeeping task indicate that although there was a basic consensus among the focus group participants about the newsworthiness of news items, their perceptions did not correlate at all with the actual prominence given to news items in the newspapers.

17
What's News in the United States?

ELIZABETH A. SKEWES AND HEATHER BLACK

The Media System in the United States

The United States is home to more than 291 million people of diverse backgrounds in race, ethnicity, religion, and sexual orientation (U.S. Census Bureau 2003). The mass media system in the United States is very large and well developed. Free speech and exchange of ideas are considered to be important fixtures in the democratic process and are protected by the United States Constitution. The system is driven primarily by private and commercial interests, operating under relatively limited federal government regulation. Historically, this structure of commercial competition has encouraged technological innovation and also has allowed for the relatively rapid construction of a comprehensive infrastructure. Very few areas of the country are not "wired," and even these are now accessible via satellite transmission. As a result, in the United States the mass media are just about everywhere. Where there is mass media, there is news.

Americans are connected to the mass media. In 2000, more than 98% of households had at least one television set, 70% of households had cable television, and 50% of households were connected to the Internet (Campbell, Martin, and Fabos 2002). In that same year there were more than 1,500 daily newspapers, more than 12,000 radio stations, more than 1,500 television stations, seven television networks (NBC, CBS, ABC, UPN, WB, Fox, and PBS), and more than 9,000 cable television systems in the United States (Campbell, Martin, and Fabos 2002). As the number of sources for information has increased, public attention has been increasingly more fragmented across all media sources.

Over the years, the advent of radio diluted newspaper readership, network television diluted the radio audience, and then cable television news further diluted the network television news audience. Newspaper readership has been declining for many decades. Research has shown that the number of newspaper readers has fallen off dramatically in the past thirty years. In 1970 a full 78% of adults in the United States read a newspaper once a week, but that declined to only 57% by 1999 (Campbell, Martin, and Fabos 2002, 295). Since

2000, research has shown that the number of Americans who reported reading the newspaper during the previous day had dropped by another 10%. Since the advent of 24-hour news programs on cable television, Americans are less reliant on the network news (ABC, CBS, and NBC). In the early 1990s, approximately 60% of Americans reported regularly watching nightly network news; by 2002, half of that network news audience had moved to watching cable news (Pew Research Center 2002). All in all, American news consumers divide their time among various news media. In 2002, they averaged fifteen minutes per day reading a newspaper, twenty-eight minutes watching daily television news, and sixteen minutes listening to radio news (Pew Research Center 2002).

The increase in the types and number of media outlets available to Americans over the past seventy years and the fragmentation of readership and viewing audiences have created a system of intense competition for audiences and advertising dollars. The downward trend in newspaper readership has caused some papers to cease operation and has driven an overall trend to consolidate ownership within the industry. Broadcast (television and radio) stations have been facing similar pressure and have seen dramatic increases in consolidations and sharp decreases in the numbers of independently owned stations.

Most of the media outlets in the United States are owned and operated by private companies or individuals and are funded by advertisements. The U.S. government does not own or operate mass media outlets available to the public. The federal government does exert some control over the media through regulation, which mainly governs ownership and seeks to limit defamation (libel), slander, and obscenity. Regulations also seek to limit monopolies, while maximizing both the diversity of content transmitted and the public's access to voice opinions in the public sphere.

In general, the regulations for broadcast media are much more stringent than for print media. The Federal Communications Commission (FCC) grants licenses to broadcast media and enforces ownership rules. For years the broadcast industry operated under the Federal Communications Act adopted in 1934. More recently ownership rules have been gradually relaxed with the Telecommunications Act in 1996 and then again in 2003. The trend toward consolidation in the media industry in recent years has caused great debate over whether public access and viewpoints will be too limited.

Overall, the news and news media serve important functions in the United States. Americans value their media and value the news in their lives. Increasingly, though, Americans are turning to a wider variety of sources to obtain the news and information that is important to them. Some get the majority of their news from alternative sources such as magazines, newsletters, and the Internet. Additionally, the line between news and entertainment is ever blurring, and a growing number of Americans get their news only from entertainment sources, such as late-night talk shows and other sources not considered to be traditional news media.

The Study Sample

The two cities selected for the United States study were New York City and Athens, Ohio. New York City is the largest city in the United States, with more than 8 million people. It is also the largest media market in the United States, with many media outlets. The leading newspaper among the four major dailies is the *New York Times*. The *Times* is one of the oldest and most prestigious newspapers in the country and historically serves as an informal newspaper of record for the United States. Though its readership is national, it is a local paper for New York City and includes a local news section. The Sunday *New York Times* contained 230 pages in this study sample. The radio newscast selected was the WCBS Newsradio 880 AM local evening broadcast from 6 to 7 p.m. This all-news format station is the leader of the CBS radio network, run by CBS News, and is owned and operated by Infinity Broadcasting Corporation. The television news broadcast selected was WNBC's *NewsChannel 4* local evening broadcast from 6 to 6:30 p.m. WNBC became the first commercial television broadcaster in the United States when it came on the air July 1, 1941.

Athens is a small city of just over 27,000 people located in southwestern rural Ohio. It is located in a very small media market and has few media outlets providing news. This small city is typical of many small cities in the United States. With only one daily newspaper, the *Athens Messenger* (owned by Brown Publishing Co.), is the leading newspaper in Athens. It is significantly smaller than the *New York Times*. The radio newscast selected was the WOUB-FM Athens local afternoon broadcast from 4 to 4:30 p.m. WOUB-FM began broadcasting in 1947 and became a founding member of the National Public Radio network in 1970. The television broadcast was WOUB's local evening *Newswatch* show, from 6:30 to 7 p.m.

Ohio University, located in Athens, owns the FCC licenses for both WOUB radio and television, which are public broadcast stations. Public broadcast stations occupy a relatively small portion of the U.S. media market. More than 80% of broadcast stations in the U.S. system are commercial. Many public radio stations are affiliated with universities because public radio does not generate profits as do commercial stations. They do not rely on advertising but are funded by a combination of grants from the federal government, corporate sponsorships, and contributions from the general public in the form of memberships.

In both New York and Athens, the sample of news cycles for the content analysis differed slightly from the prescribed dates for the study. WOUB-TV in Athens does not produce a weekend news show, so the newscasts and newspapers collected had to reflect that. In addition, the station took a vacation over the Thanksgiving holiday, so another news cycle was collected in January. In New York City there were a few taping problems, so other cycles were collected in January and February. Overall, the news cycles in both cities spanned from November 2000 to February 2001 and covered all the available days of the week (seven in New York City and five in Athens).

The focus groups for the U.S. study were conducted in March and April 2001. The sample of people for the focus groups stayed true to the overall study methodology, with the exception of the journalist group in Athens. The journalist focus group there included only three people, so three more in-depth interviews were conducted later in the spring. In both cities, the composition of the focus groups was fairly evenly distributed between males and females. The focus groups in New York City were much more racially diverse than those in Athens, although the majority of participants were white, with some black, Asian, and Hispanic participants. All of the participants in the Athens groups were white. In the New York groups there was also a fairly wide range of ages (25 to 67 years). The public relations professionals and journalists had from three to thirty years of work experience.

Topics in the News

While New York City and Athens are very different cities in population, pace, and lifestyle, some striking similarities exist in the topics covered by the news media in both communities. Most striking, perhaps, is the fact that only two topics managed to constitute 10% or more of the news coverage across all media in both communities—sports and internal politics, as seen in Table 17.1.

For most of the media in both New York City and Athens, sports coverage was dominant. In fact, in Athens sports received the most coverage in the *Athens Messenger* (22.7%); on the television program *Newswatch* (24.1%); and on the radio station WOUB-FM (21.8%). This is likely due to the fact that Athens is a small town that is heavily influenced by the events at Ohio University, a major employer and the community's cultural center.

In New York City, however, which offers an abundance of cultural opportunities in addition to sports, the amount of space devoted to sports coverage was high. On WNBC's television station, sports news took up 30.5% of the airtime—more than twice the time given to any other topic. On the CBS 880 radio station, sports coverage got 13.8% of the airtime and was second to coverage of internal politics; in the *New York Times*, sports received 13.5% of the news coverage and was second only to business/commerce/industry.

Internal politics—the issues of national, state, and local governance—received heavy radio coverage in New York (22.7%) and received substantial news coverage (14.2%) on both the television station analyzed for this study and in the *New York Times* (12.5%). The pattern of political coverage is similar in Athens, with the newspaper devoting 10.6% of its news time to coverage of internal political issues. By comparison, the television news program analyzed for this study devoted 10.7% of its time to political content and the radio broadcast gave 16.2% of its news airtime to internal political issues. While this seems counterintuitive—since newspapers generally offer more coverage of politics and are more relied upon for political news—newspapers typically have more "space" for all news than their broadcast counterparts, so the proportion of

TABLE 17.1 Distribution of General Topics of News Items by City and Medium

Topics	Newspaper New York	Newspaper Athens	Television New York	Television Athens	Radio New York	Radio Athens
Business/Commerce/ Industry	17.7	4.7	3.0	7.4	8.8	16.5
Sports	13.5	22.7	30.5	24.1	13.8	21.8
Internal Politics	12.5	10.6	14.2	10.7	22.7	16.2
Cultural Events	9.8	8.5	2.7	1.1	.6	.3
Human Interest Stories	7.8	11.1	8.5	6.3	2.5	3.5
International Politics	7.2	1.6	1.5	.0	3.6	2.9
Internal Order	5.7	7.6	10.3	7.6	6.8	7.6
Communication	3.3	.7	.9	2.7	5.0	2.4
Transportation	2.6	1.3	4.8	4.2	9.0	2.4
Economy	2.5	1.3	2.4	3.3	1.9	3.2
Health/Welfare/ Social Services	2.4	3.4	1.8	5.4	3.3	4.1
Education	2.0	4.4	1.5	6.0	1.5	5.6
Social Relations	1.7	1.4	2.4	.0	1.5	.3
Environment	1.4	1.8	.0	2.5	.5	2.6
Science/Technology	1.4	.3	1.5	.4	.8	.3
Military and Defense	1.1	.8	.0	.7	.6	.3
Energy	1.0	.6	.9	2.2	1.0	2.6
Entertainment	1.0	7.9	.0	.0	.0	.0
Other	1.0	.3	.0	.0	.0	.0
Ceremonies	.9	2.1	2.1	1.8	1.0	1.5
Disasters/Accidents/ Epidemics	.9	2.1	2.1	2.7	2.4	.6
Weather	.9	2.7	7.3	8.9	12.3	4.1
Housing	.7	2.0	.0	1.6	.1	.3
Fashion/Beauty	.6	.1	.0	.0	.0	.0
Labor Relations and Trade Unions	.4	.0	.9	.4	.1	.3
Population	.1	.1	.6	.0	.2	.6
Total[a]	100.0	100.0	100.0	100.0	100.0	100.0
	($n = 3006$)	($n = 709$)	($n = 331$)	($n = 448$)	($n = 960$)	($n = 340$)

Note: Distributions given in percent.

[a]Total percentage may not actually be 100.0 due to rounding error.

political news can be smaller in newspapers even while the volume of political coverage is greater.

Some differences in the coverage of news topics are worth noting. First, the *New York Times* gave its highest proportion of coverage to business/commerce/industry (17.7%). This is consistent with the fact that New York City is the seat of financial power in the United States and is home to both Wall Street and Madison Avenue, places where large sums of money change hands in business, service and commerce every day. The only other medium to give a substantial amount of coverage to business was the Athens radio station, but since it was a National Public Radio broadcast its news coverage was much broader than a standard local station. Also, the television stations and New York City's radio station gave a fair amount of news coverage to weather. For television, this is likely due to the fact that most stations have a staff member devoted to weather news, so the allocation of resources means that weather will get substantial airtime. For the New York radio station, the focus on weather (12.3% of air time) is likely due to the fact that so many New Yorkers commute to work and need to know how the weather may affect their trip.

An examination of the rank-order correlations between news topics (see Table 17.2) shows strong relationships between all of the media in both of the cities, with the strongest correlations among the broadcast media. There was a .87 correlation between topics of news coverage on Athens radio and television, and a .83 correlation between topics in the New York City broadcast media. The radio stations in the two cities had a .84 correlation in their distribution of news topics, and Athens television showed a high correlation (.82) with New York City radio. Essentially, the degree of similarity between what is considered news on broadcast stations ran quite high. For newspapers,

TABLE 17.2 Spearman Rank-Order Correlation Coefficients between Rankings of News Topics in Various Media

	New York Newspaper	Athens Newspaper	New York Television	Athens Television	New York Radio	Athens Radio
New York Newspaper		.62[b]	.65[c]	.55[b]	.68[c]	.63[b]
Athens Newspaper			.63[b]	.64[c]	.53[b]	.57[b]
New York Television				.74[c]	.83[c]	.70[c]
Athens Television					.82[c]	.87[c]
New York Radio						.84[c]
Athens Radio						

[a] $p < .05$; [b] $p < .01$; [c] $p < .001$.

comparing the *New York Times* to the *Athens Messenger*, the correlation was lower (.62), but still strong and statistically significant. The lowest correlations, although they were still statistically significant, were found between the *New York Times* and Athens television (.55) and between the *Athens Messenger* and the radio stations in New York City (.53) and Athens (.57). While there was agreement between the print and broadcast media in the two cities, that agreement was higher among the broadcast outlets than the newspapers.

Deviance in the News

Of all the one-way analyses of variance (ANOVAs) calculated, twenty-one of the thirty were statistically significant. Social change deviance—or the degree to which the information in the story is about some level of threat to the status quo—is linked to better play in U.S. media outlets. The greater the threat to the status quo, the more likely the story was to receive prominent coverage across all of the media in both New York City and Athens (see Table 17.3).

Stories involving intense social change deviance were far more likely to receive prominent play in the *New York Times*, although the newspaper was the only media outlet to have any stories about extreme threats to the status quo. This is not unusual since New York City is the largest city in the nation and since the *New York Times* is a national paper of record. It is more surprising for New York City that neither the television station nor the radio station broadcast any news items about extreme threats to the status quo. However, even among the New York broadcast media, there were strong indications that the more content in a story was about social change deviance, the more prominently it played in the news.

In the Athens media, there also were no stories that dealt with concerns that qualified as extremely deviant. But Athens, located in southeastern Ohio, is a quiet town, so it is unlikely that anything representing an extreme deviance from the social structure would occur. However, within the range of a common occurrence to one that is quite unusual, the more unusual a news item was, the more likely it was to be played up in the news. This trend held true for all of the Athens media.

Stories that represented statistical deviance— the indicator of how unlikely the event is—also tended to receive more prominent play, but the pattern across media was less consistent than it was for social change deviance. For instance, a more statistically deviant visual item typically was placed prominently—measured by the amount of time or space, and the relative position in the newscast or page assignment in a newspaper—in newspapers and television in both Athens and New York City. However, stories that represented greater statistical deviance did not always get more prominent play. While the *Athens Messenger* did give more prominent play to text about statistically deviant events, the *New York Times* did not. The findings also differ for television: both verbal and visual statistical deviance mattered to New York City's station, but visual statistical deviance was the only factor in Athens—and for radio,

TABLE 17.3 Mean Verbal and Visual Prominence Scores for Deviance Intensity

Intensity of Deviance	New York City					Athens				
	Newspaper		Television		Radio	Newspaper		Television		Radio
	Verbal only ($n = 2215$)	Verbal plus Visual ($n = 744$)	Verbal ($n = 200$)	Visual ($n = 200$)	Verbal ($n = 514$)	Verbal only ($n = 618$)	Verbal plus Visual ($n = 144$)	Verbal ($n = 246$)	Visual ($n = 246$)	Verbal ($n = 206$)
Statistical Deviance										
(1) common	326.1	610.0[c]	56.8[c]	51.2[c]	50.9	175.5[c]	381.4[c]	63.9	61.5[c]	51.8[c]
(2) somewhat unusual	368.9	852.5	87.0	135.6	61.6	244.4	545.6	95.4	120.0	148.1
(3) quite unusual	400.4	1047.3	152.9	166.1	62.6	315.0	607.4	68.1	113.7	187.4
(4) extremely unusual	496.8	2338.5	205.4	115.4	65.7	366.8	1819.1	125.6	–	185.2
Social Change Deviance										
(1) not threatening to status quo	341.9[b]	665.7[b]	65.8[c]	84.5[c]	53.1[c]	186.2[c]	425.3[a]	65.7[b]	72.9[c]	88.0[c]
(2) minimal threat	319.3	669.8	174.7	257.4	78.3	399.0	399.0	121.7	283.0	162.9
(3) moderate threat	238.2	733.0	224.8	248.3	96.1	828.0	1536.0	119.2	342.0	307.3
(4) major threat	3759.0	3759.0	–	–	–	–	–	–	–	–

Normative Deviance

(1) does not violate any norms	346.6	654.3[b]	82.7[b]	89.7[a]	55.8	200.8	427.4[b]	78.4	76.5	113.6
(2) minimal violation	227.7	806.0	210.8	254.1	79.5	199.5	324.6	58.4	93.0	167.5
(3) moderate violation	345.3	1236.9	262.0	292.0	59.3	346.8	1819.1	64.7	–	103.9
(4) major violation	606.3	2320.7	103.2	81.0	66.5	129.8	169.0	77.9	–	111.5

[a] $p < .05$; [b] $p < .01$; [c] $p < .001$.

where New York City's station did not use statistical deviance as an indicator of newsworthiness but Athens' station did.

Issues of normative deviance—a measure of the degree to which a news item relates to an event that violates laws or social norms in the community—presents an even less cohesive picture. New York City's television station used normative deviance as a significant indicator of newsworthiness, but then all three measures of deviance—social change, statistical, and normative—were related to how prominently a story was played on New York City television. Normative deviance also was related to the play that an item received in both the *New York Times* and the *Athens Messenger*, but only for verbal plus visual content. For verbal content only—with the exception of New York City television—there was no statistically significant relationship between the degree to which the news item related information about violated laws or norms and the amount of prominence it received.

Across all media, the perceived threat to the social system was a strong indicator of newsworthiness. Given what we know about the role of hegemony as an influence on news content, this is not a surprise. What is more surprising is the relatively limited role of normative deviance in relation to a story's prominence. While an event that threatens the social system, or the status quo in a community, will make top headlines, an event that involves breaking that society's laws or conventions is treated as more of a statistical blip. This could be an artifact of higher crime rates in the United States and the sense that breaking the law is not that unusual, or it could be the result of a culture of independence in the United States. Crimes typically are viewed as individual acts: It is rare that anyone considers crime to be the result of broader social conditions, like poverty and a lack of education.

Social Significance in the News

News organizations in the United States are designed to cover politics and political figures. Even small newspapers are certain to have reporters designated to cover the "essential" beats—cops, courts, city hall, and the school board. But many events that occur on these beats are not especially important. A traffic accident with minor injuries, for instance, typically merits only a brief mention in a smaller newspaper and may not be covered at all in a larger newspaper. But other events, such as a shooting involving a police officer, may dominate the news for several days. The findings bear this out, showing that across all media, in both New York City and Athens, as events increase in political significance, they are more likely to be covered more prominently (see Table 17.4). They will be at the top of the news or on the front page, will receive more time or space, and will have more visual content to draw attention to them. And the relationship is a strong one. In fact, of the forty ANOVAs conducted, twenty-six were significant. For instance, in the *New York Times*, the mean verbal prominence ranking more than quadruples from 334.5 for items not really politically significant to 1,439.3 for items of major political

TABLE 17.4 Mean Verbal and Visual Prominence Scores for Intensity of Social Significance

| Intensity of Social Significance | New York | | | | | Athens | | | | |
| | Newspaper | | Television | | Radio | Newspaper | | Television | | Radio |
	Verbal only ($n = 2196$)	Verbal plus Visual ($n = 727$)	Verbal ($n = 200$)	Visual ($n = 200$)	Verbal ($n = 514$)	Verbal only ($n = 616$)	Verbal plus Visual ($n = 143$)	Verbal ($n = 246$)	Visual ($n = 246$)	Verbal ($n = 246$)
Political Significance										
(1) not significant	334.5[c]	636.7[b]	56.9[c]	67.2[c]	53.6[a]	180.5[c]	407.8[b]	72.4[b]	73.0[b]	92.7[a]
(2) minimal	290.5	760.3	165.6	197.8	60.3	271.3	624.7	69.9	120.0	142.1
(3) moderate	569.5	1134.6	150.8	133.8	48.0	459.0	562.3	152.9	208.0	202.8
(4) major	1439.3	1587.0	273.6	287.4	79.2	1167.0	1536.0	226.0	307.5	206.8
Economic Significance										
(1) not significant	302.4[b]	665.3	88.9	91.7[b]	58.4	200.7	434.9	71.1	77.6	107.1
(2) minimal	428.4	702.2	161.6	340.0	47.9	222.9	537.5	103.6	46.0	120.6
(3) moderate	533.0	856.4	117.5	134.0	67.8	288.4	196.0	73.3	60.0	186.7
(4) major	–	966.0	–	–	82.5	–	–	66.0	–	139.5

[a] $p < .05$, [b] $p < .01$, [c] $p < .001$.

(Continued)

TABLE 17.4 Mean Verbal and Visual Prominence Scores for Intensity of Social Significance (Continued)

Intensity of Social Significance	New York										Athens									
	Newspaper		Television		Radio						Newspaper		Television		Radio					
	Verbal only ($n = 2196$)	Verbal plus Visual ($n = 727$)	Verbal ($n = 200$)	Visual ($n = 200$)	Verbal ($n = 514$)						Verbal only ($n = 616$)	Verbal plus Visual ($n = 143$)	Verbal ($n = 246$)	Visual ($n = 246$)	Verbal ($n = 246$)					
Cultural significance																				
(1) not significant	312.9[c]	613.0[c]	131.6[c]	131.5[c]	61.6[a]						192.5	437.6	88.7[b]	86.6[a]	133.6					
(2) minimal	361.5	765.3	34.4	35.2	43.8						220.3	421.9	48.2	50.7	77.8					
(3) moderate	786.9	1037.1	159.6	166.1	62.5						137.5	–	202.3	100.0	–					
(4) major	910.0	3759.0	216.0	138.0	–						–	–	–	–	–					
Public significance																				
(1) not significant	331.5[c]	637.3[c]	57.7[c]	67.4[c]	49.2[c]						192.2[a]	424.8[b]	61.3[b]	73.74[a]	67.3[c]					
(2) minimal	368.0	920.9	137.2	202.3	78.5						238.6	461.6	98.7	192.3	262.4					
(3) moderate	431.7	2631.4	271.3	363.0	78.2						469.5	1536.0	163.1	186.0	250.8					
(4) major	1535.3	3759.0	471.0	474.0	–						–	–	–	–	111.0					

[a] $p < .05$, [b] $p < .01$, [c] $p < .001$.

significance. The pattern repeats itself for television and radio, but interestingly the difference between the scores at the ends of the scale is not as great. Most likely, this is a result of the constraints of broadcast news, which have less "space" to work with—typically about twenty-two minutes for a television news show and often only a five-minute news update on radio.

While political significance is the strongest and most consistent predictor of the play that a story might get, other types of significance also matter. Particularly in New York City, which offers a much wider variety of cultural activities than Athens, increased cultural significance was related to greater prominence. Again, this is particularly the case for New York City's newspaper, which showed a stronger and more linear pattern, than it is for television or radio. And the relationship is particularly pronounced for verbal plus visual content in the *New York Times.* Cultural events—a concert or play, celebrity news, food, fashion—often are covered with large visual elements, like a photograph or photo illustration. Contemporary U.S. newspaper design, particularly for feature sections, calls for visual elements to dominate a page, which is reflected in the finding that the visual prominence score for events of major cultural significance is more than six times greater than the score for events that are not culturally significant.

Increased public significance—a measure of the degree to which a story might affect the public— is also related to increased prominence in its coverage, but here the picture is a little less consistent. For instance, the *New York Times* gives greater verbal plus visual prominence to an item that has greater public significance, but there is not a statistically significant difference in the verbal prominence the story will merit. Broadcast media in New York City and Athens give greater prominence to publicly significant stories. Athens television does so with its verbal content.

The relationship of economically significant items to news coverage is even more puzzling. Despite the fact that capitalism is at the heart of the economic system in the United States and that wealth is a key measure of success in society, stories of economic significance were weakly related to prominence, and then only for verbal content in the *New York Times* and visual content on New York City's television station. Economic significance was not important in the placement of stories in any of the Athens media. While this seems somewhat counterintuitive—particularly for newspapers, which often devote a section to business—several factors may account for it. First, visuals for business stories often are not compelling: A story on a shake-up at a major corporation may be accompanied by a portrait of the company's chief executive officer or a photo of the company itself. It is not as dramatic as other kinds of visual content, so is less likely to be given a lot of prominence. Additionally, economic stories often are more complex. A story about the Federal Reserve cutting overnight interest rates by a quarter of a percent has to go into some depth to explain what this might mean, and the explanation often is not as strong in its narrative structure as other news stories. Since narrative structure is an important element for a story, an economic story that cannot be told in terms of protagonists and antagonists is likely to get weaker play.

Deviance and Social Significance as Predictors of News Prominence

While the analysis so far has shown that elements of deviance (statistical, normative, and social change), as well as elements of significance (political, cultural, public, and economic), all can influence the play a story gets in a newspaper or on television or radio, it is important to examine how these various factors interact with each other in the production of news. Putting the elements together into a single stepwise regression model yields some interesting results (see Table 17.5).

First, deviance and social significance are the strongest predictors of news prominence for television in New York, but they are the weakest predictors for the *New York Times*, where deviance and significance account for only 2% of the variance in prominence. The difference, as mentioned previously, is likely to be at least partly an artifact of the large number of stories that the *New York Times* covers, in comparison to the much smaller number of stories that can appear in a television or radio news broadcast. In fact, as others have noted, a Sunday edition of the *New York Times* contains more factual information than the average person living in the 1800s encountered in his or her lifetime (Van Winkle 1998). Clearly, deviance and social significance are important elements in how the *Times* plays stories. Looking at the combined visual and verbal prominence scores in the *New York Times*, the prominence of visuals was affected by social change deviance (−.11), public significance (.23), and normative deviance (.11). The prominence of verbal content was most affected by political significance (.21), economic significance (.09), and cultural significance (.23). For New York City's television station, the two strongest predictors in the model were the political significance of verbal content and the public significance of visuals.

For Athens media, the regression model is strongest for the *Messenger*, explaining 20% of the variance in prominence (when visual and verbal scores are combined), and for the radio station, where it accounts for 19% of the variance. However, it also has good predictive power for prominence on the television station, where it explains 15% of the variance. Across the media in Athens, public significance is an important element in explaining the prominence of verbal content, with the standardized beta ranging from .19 to .33. For newspaper and radio verbal content, statistical deviance emerged as an important factor, and for the Athens television station social change deviance and political significance of visuals were the other key predictors of story prominence.

Looking at all of the independent variables, the two most consistently included in the models, both in New York City and Athens and across the various media, are measures of public significance (a predictor in half of the regression models) and political significance (a predictor in five of the eight

TABLE 17.5 Stepwise Regression Analyses of Intensity of Deviance and Social Significance on News Prominence

	New York City								Athens							
	Newspaper Prominence Verbal only Total R^2 = .02[c] (n = 2191)		Newspaper Prominence Visual and Verbal Total R^2 = .12[c] (n = 700)		Television Prominence Total R^2 = .30[c] (n = 200)		Radio Prominence Total R^2 = .07[c] (n = 511)		Newspaper Prominence Verbal only Total R^2 = .10[c] (n = 616)		Newspaper Prominence Visual and Verbal Total R^2 = .20[c] (n = 128)		Television Prominence Total R^2 = .15[c] (n = 246)		Radio Prominence Total R^2 = .19[c] (n = 206)	
Independent Variables	r	Std. Beta	r	Std. Beta	r	Std. Beta	r	Std. Beta	r	Std. Beta	r	Std. Beta	r	Std. Beta	r	Std. Beta
Deviance																
– Statistical, verbal content	.03	ns	.16[c]	ns	.30[c]	ns	.09	ns	.18[c]	ns	.41[c]	.30[b]	.06	ns	.31[c]	.18[a]
– Statistical, visual content	–	–	.17[c]	ns	.29[c]	ns	–	–	–	–	.36[c]	ns	.24[c]	ns	–	ns
– Social change, verbal content	.00	ns	.05	ns	.34[c]	ns	.18[c]	ns	.24[c]	.15[b]	.32[c]	ns	.21[b]	ns	.29[c]	ns
– Social change, visual content	–	–	.07	-.11[b]	.28[c]	ns	–	–	–	–	.20[a]	ns	.28[c]	.20[b]	–	–
– Normative, verbal content	-.00	ns	.06	ns	.19[b]	ns	.06	ns	.04	ns	.20[a]	ns	-.03	ns	.02	ns
– Normative, visual content	–	–	.14[c]	.11[b]	.17[a]	ns	–	–	–	–	.10	ns	.01	ns	–	–

[a]p < .05; [b]p < .01; [c]p < .001; ns = not part of final stepwise regression equation.

(Continued)

TABLE 17.5 Stepwise Regression Analyses of Deviance and Social Significance Intensity on News Prominence (*Continued*)

	New York City								Athens							
	Verbal Newspaper Prominence Total $R^2 = .02^c$ (n = 2191)		Visual and Verbal Newspaper Prominence Total $R^2 = .12^c$ (n = 700)		Television Prominence Total $R^2 = .30^c$ (n = 200)		Radio Prominence Total $R^2 = .07^c$ (n = 511)		Verbal Newspaper Prominence Total $R^2 = .10^c$ (n = 616)		Visual and Verbal Newspaper Prominence Total $R^2 = .20^c$ (n = 128)		Television Prominence Total $R^2 = .15^c$ (n = 246)		Radio Prominence Total $R^2 = .19^c$ (n = 206)	
Independent variables	r	Std. Beta	r	Std. Beta	r	Std. Beta	r	Std. Beta	r	Std. Beta	r	Std. Beta	r	Std. Beta	r	Std. Beta
Social Significance																
– Political, verbal content	.04	.08[c]	.15[c]	.21[c]	.46[c]	.37[c]	.13[b]	.14[b]	.26[c]	.22[c]	.30[b]	ns	.18[b]	ns	.21[b]	ns
– Political, visual content	–	–	.14[c]	ns	.44[c]	ns	–	–	–	–	.27[b]	ns	.24[c]	.20[b]	–	–
– Economic, verbal content	.07[b]	.11[c]	.02	.09[a]	.11	ns	–.02	ns	.05	ns	.08	ns	.08	ns	.10	ns
– Economic, visual content	–	–	.05	ns	.14[a]	ns	–	–	–	–	–.02	ns	–.05	ns	–	–
– Cultural, verbal content	.06[b]	.11[c]	.13[b]	.23[c]	–.15[a]	ns	–.12[b]	ns	.05	.13[b]	.04	ns	–.13[a]	ns	–.14[a]	ns
– Cultural, visual content	–	–	.13[c]	ns	–.19[b]	ns	–	–	–	–	.02	ns	–.15[a]	ns	–	–
– Public, verbal content	.03	ns	.16[c]	ns	.37[c]	ns	.21[c]	.22[c]	.10[a]	ns	.36[c]	.22[a]	.24[c]	.19[b]	.40[c]	.33[c]
– Public, visual content	–	–	.22[c]	.23[c]	.42[c]	.31[c]	–	–	–	–	.18[a]	ns	.17[b]	ns	–	–

[a] $p < .05$; [b] $p < .01$; [c] $p < .001$; ns = not part of final stepwise regression equation.

models). While deviance measures were a factor in many cases, they were not as consistently a predictor of news prominence as the social significance measures.

People Defining the News

The focus groups in Athens were held in March 2001, and the New York City focus groups were conducted in April 2001, just a few months after the U.S. Supreme Court issued its ruling in December 2000 that finally ended thirty-six days of legal and political wrangling over the 2000 presidential election.

Deviance

Across the focus groups conducted in the United States, the concept of deviance emerged as a key element in their definitions of news. This was true for the stories that they thought were the most important ones, which is discussed later in this chapter, as well as for stories they encountered in their day-to-day lives. In this sense, deviant information was information that caught them by surprise or that they had not expected to hear. One example mentioned in the low socioeconomic status (SES) audience focus group in New York was the plane crash that killed John F. Kennedy Jr., his wife, and his sister-in-law in July 1999.

Man #1:   Like a death of a 91-year-old, it's expected, you know.

Woman #1:   Right, but like the death of John-John ... It didn't make a difference what race you was, he was everybody's son and there's thousands of people who die in a plane crash, but he seemed to have caught the nation's attention because it was the history of the Kennedys ....

Woman #2:   And it hurt ... because he was a really nice guy ....

Man #2:   I think, you know, you feel it more because, just because of the shock of it. You're not expecting it.

For those in the news business, either as journalists or public relations professionals, the idea of news as being the unexpected is critical. Several journalists and public relations people said simply that news is anything they do not know, but it is particularly so when it is something they did not anticipate and when it represents a threat. One woman journalist in New York City said that the story about the school shootings at Columbine High School in Littleton, Colorado, in April 1999 was something that touched people across the nation because it occurred in an upper middle-class area where people normally feel insulated from violence. So, as an Athens journalist noted, the fact that violence came to the neighborhood violated the normal expectations and, to a degree, the status quo. Talking about the stories on Walt Disney World in Florida and the fact that some theme-park employees were pedophiles, this journalist said that the real news value of the story was that "you think that Disney is the icon of a family, you know, and what makes it unusual is that if the family is not safe in Disney World, that is in opposition to the status quo."

Another person, who is a public relations professional in Athens, said that the importance of an event or story is a function of its intensity. "You think of the intensity of the Kennedy assassination and the sorrow, and you think about the intensity of the Berlin Wall coming down." Both of these events, in her mind, were intense because of their unexpected nature.

Social Significance

The woman in the New York City focus group who said that John F. Kennedy Jr.'s death "hurt … because he was a really nice guy" raised a second dimension that makes a story newsworthy to U.S. audiences—impact, which is measured by the social significance variables in the study. The impact can be emotional, as in the case of the plane crash, or it can be more direct, as in the case of an increase in the price of gas at the pump. And it can be either short term or long term. One woman in the high SES audience focus group in New York City said that information about the economy and job losses gets her attention because "it has an impact, but it may not be immediate." However, she also said that stories with immediate impact are more likely to grab her attention than those with long-term impact. "I go to the local news because there are things that affect me. Like revitalization [of run-down neighborhoods] was a big, big factor, and now it's going to affect the whole town. Those things that immediately affect me are important to me."

Even stories that have less apparent impact, such as a school shooting in Colorado or California, were important for another woman in the high SES audience focus group in New York City because it raises the potential for impact. "You say, 'Could this happen to my child? Could this happen to a family member or to my neighbor's child?' I listen to the news; I watch the news with a very discerning spirit. I see the facts come in, but then I sift them and say, 'Okay, what does that mean for me?'"

The scope of the information's impact also surfaced as a criterion that individuals use to assess the importance of information. Stories that affect a lot of people, even indirectly, can be newsworthy, but so can stories that have less breadth of impact but more immediate consequences. For audience members in Athens, there was a similar pattern. Immediacy and impact were critical factors in determining the news value of a story. One woman said that she pays more attention to stories that impact her directly, whether it is something about coal miners because her family worked in coal mines, or lottery numbers because she buys lottery tickets. A man in the high SES audience group in Athens said that while he sometimes pays attention to information that does not impact him, "I'm more concerned about having food to eat today than I am about whether we're going to run out of food in twenty years or not. And if I'm hungry and haven't eaten for two days, I'm not the least bit interested in these projections of national issues. I'm concerned with the immediate future and how it affects me personally."

For news professionals—both journalists and public relations practitioners—impact also is an important factor, but they often apply two standards for this criterion. On the one hand, much like the audience members, news is information that has a personal impact on some level; economic or public impact were the two categories most often mentioned. But news professionals, and particularly the journalists in the groups, also have a more distanced view of the news, one that causes them to pay attention to information even if it does not seem personally relevant. A male journalist in the New York focus group said that he was personally interested in the idea of a commuter tax, "but in terms of work, it's ... zero interest, so it's down a peg even though I have a personal stake in it. ... A work story, which might have been what happened in the Middle East today, with Israel hitting Syrian targets that they haven't hit for twenty years, is more interesting to me ... because of what I'm going to be dealing with tomorrow." What's worth noting, however, is that even though the Middle East story is not personally relevant to the journalist in his private live, it does have personal relevance for his professional life since it will affect what he does the next day at work. Perhaps, then, just as all politics is local, all new is—by some measure—personal.

The personal nature of news is also evidenced through the consensus among all the focus groups that news often has a utilitarian function as well. So while the impact may not be as great from a regional or national scope, important news for most people is the weather report, so that they know whether to bring an umbrella, or the traffic report (especially in New York), so that they know what the commute to and from work will be like. Or it could be the movie reviews, so that they have a better idea about whether to go to a show, and the movie listings, so that they know what time to be at the theater. For one woman in the New York City low SES group, information about child-rearing always catches her eye because she's the single mother of a 7-year-old boy: "Sometimes he's just out of control, and I try to read different things ... I try to find out as much information as I can, like what I'm doing wrong." Another woman in the same group, who has a 15-year-old son, echoed the same view, saying that she follows news to know what trouble her son might be getting into and what she, as a parent, can do to keep him out of harm's way.

And sometimes, even when the relevance may not be immediate, people will attend to news they believe could have value in the future. A person living in an apartment might pay attention to changing rates on home loans because she is thinking about buying a home in the future. One woman in an Athens focus group said she "collects" bits of information and stores them "because they almost invariably come in handy sometimes."

Other Determinants of News

While the concepts of deviance and social significance play a large role in most people's ideas about what makes something interesting and newsworthy, the two concepts do not fully explain the level of interest many people have in news about celebrities and sports. While entertainment news does serve a cultural function, another reason that many people follow sports or celebrities is for their value as a social lubricant. Knowing how the local sports team fared over the weekend or whom the actress Julia Roberts is dating gives many people a common ground from which to launch discussions. As one Athens journalist noted, news, from this perspective, is "something you can share, um, 'share-ability.'" An Athens public relations professional said it is the information she hears and then wishes she had someone she could tell it to. Even within her family, where conversations typically are easier, "the common, everyday routine announcements tended to be at the bottom of my list with more broad subjects that I would never go home and tell my husband."

To serve this function, the news has to be more neutral, like the weather, in order to reduce the risk of offending someone or of getting into a heated debate with a friend or colleague. One man in an Athens focus group said that while he does not really support the policies of the current administration of U.S. President George W. Bush, he listens to news about the president—particularly stories that praise his accomplishments—because it helps him to better understand and relate to his coworkers who are Bush supporters.

However, one public relations professional in the New York City focus groups said that while the definition of news is broad, people also are becoming more jaded by a media system that now spits news at them twenty-four hours a day. "I think a lot of what's presented to us as news is always presented as, 'This is the biggest, greatest, most momentous thing that ever happened.' And it isn't," he said. Since news is being pitched to viewers and readers in more strident tones these days, news consumers are increasingly using utilitarian filters to determine what they pay attention to and what they process. If it does not have a use—personally, professionally, or socially—it is much less likely to get past an individual's filtering system.

News Valence

While focus group members agreed that the news they read, see, and hear is largely negative in its tone, there seemed to be little understanding about why that is the case. Journalists argued that they report on car accidents and shootings because that is what they will get the most calls about. One Athens journalist said that if traffic were tied up in town due to an accident, people in town would call him and expect to find out what happened in the next day's paper. But audience members in both the low and high SES audience groups said that news is negative because crime and conflict sell papers or bring viewers to the screens. They echoed the popular sentiment that "if it bleeds, it leads."

"Every newspaper is competing. Every TV station is competing, and they're going to have what's going to catch people's ears," said one high SES audience member in New York City. Another member of that focus group said that violence seems to intrigue people—that it's "just human nature" to take an interest in negative news. A journalist in Athens said that negative news simply has more shock value and oftentimes, therefore, has more emotional impact: "Our negative emotional reactions—fear, shock, sorrow—tend to be stronger, you know, than our positive reactions which, you know, if the strongest positive emotion is love, that's pretty much a personal thing." But another journalist in that focus group said that, in part, the focus on negative news is a byproduct of news beats that focus on police and courts and government. "I don't have a 'good news' reporter," he said.

Most Important News Events in Life

For focus group members in New York City and Athens, the stories considered to be the most important typically were also those that were the most deviant—statistically, normatively, or in terms of their potential social change impact. In some cases, the most important stories were those that embodied more than one kind of deviance. Among them were the assassinations of John F. Kennedy in 1963, his brother Robert Kennedy in 1968, and Martin Luther King Jr., also in 1968. The Apollo moon landing in 1969 and the explosion of the Challenger space shuttle in 1986 also turned up frequently, as did the death of Britain's Princess Diana in 1997 and the need, for the first time in U.S. history, to have a presidential election adjudicated by the U.S. Supreme Court in 2000. Said one journalist in New York City, "You go, you vote, you hear the outcome the next day. This is something that, I don't know, I guess we haven't seen, I certainly haven't seen in my lifetime—the fact that we went, we voted, and we didn't have a clue who won."

These key events in a person's lifetime certainly share statistical deviance as a common thread, and several of them, particularly the assassination of a sitting president and the shuttle explosion, had the potential to significantly alter the political world. These stories also share high degrees of political, economic, cultural, and public significance. But for the focus group members, these events are memorable often because they alter how they see the world. For one male journalist in New York City, the Challenger explosion was important because "it was sort of a big setback to the space program in my lifetime, you know. In a time where, you know, the space program had gotten to be sort of just taken for granted. It was just … you know, they always went up, they always came down."

The idea that the shuttle might not take off safely and complete its flight was something that simply had not occurred to him until he saw the plumes of smoke heading in different directions. And they are periods of national—or sometimes global—mourning or celebration. One Athens journalist said that although he was just a child when Kennedy was assassinated, the memory has stayed with him not because he was a fan or because he followed politics at that point in his life but instead because "I just remember how stunned everybody was by it. I remember how jubilant everybody was at the moon landing .... Both events were like national obsessions and it turned out that both were significant in history."

Another journalist in Athens said that the important events were ones that everyone was talking about, like Operation Desert Storm, the first war the United States waged in the Persian Gulf. Technology allowed people to have a front-row seat to the war, with its Patriot missiles and night-vision photography, so people watched it "like a sporting event," the journalist said. A New York City journalist said much the same thing: "It was the first major war that I ever experienced. Everyone was talking about it."

Comparing People's News Preferences with What's in the Newspaper

One of the more interesting things that came out of the focus group discussions about news valence was an apparent disconnect between what people say they want, which is generally more positive news, and what they say they see, which is largely negative. Focus group members argued that people read negative news because they are fed a steady diet of it. But an examination of their "news judgment"—of how they think a story should be played—shows some strong relationships between how audience members view news and how their local news professionals view it (see Table 17.6). However, there is little correlation between how individuals—whether audience members, public relations

TABLE 17.6 Spearman Rank-Order Correlation Coefficients between Newspaper Item Prominence and Focus Group Rankings

	Journalists	Public Relations	High SES Audience	Low SES Audience	Newspaper
Journalists	–	.76[c]	.54[b]	.55[b]	.28
Public Relations	.82[c]	–	.55[b]	.65[c]	.21
High SES Audience	.70[c]	.74[c]	–	.56[c]	.04
Low SES Audience	.79[c]	.88[c]	.73[c]	–	.10
Newspaper	.42[a]	.35	.14	.25	–

Note: New York City coefficients are in the upper triangle, Athens, Ohio, coefficients in the lower. SES = socioeconomic status.
[a] $p < .05$; [b] $p < .01$; [c] $p < .001$.

professionals, or working journalists—would have played the stories and how they actually were placed in the newspaper.

For this exercise, focus group members were given three sets of cards with stories from their local newspapers—either the *New York Times* or the *Athens Messenger*—and were asked to rank the stories from 1 to 10 based on their newsworthiness. For both high and low SES audiences in New York City, the rank-order correlations with journalists or public relations practitioners ranged from .54 to .65. The agreement between journalists and public relations workers was even higher, at .76. But none of the focus groups ranked the stories in the same way that they appeared in the *Times*. The highest degree of correlation with the actual newspaper coverage was only .28, and it was not statistically significant.

In Athens, journalists and their newspapers agreed somewhat (.42) in how they ranked the stories and how the *Athens Messenger* played them. But no other focus group in Athens shared the newspaper's vision of the relative newsworthiness of the stories. However, as in New York City, all four Athens focus groups showed strong agreement among themselves in terms of how they would have played the stories for each of the three news days. The higher degree of agreement between the Athens audiences and news professionals is not surprising. The *Messenger* is a much smaller paper than the *New York Times*, and the region itself is not nearly as diverse as the New York area.

Still, the striking level of disagreement between focus group rankings and the prominence of the stories in the newspapers is evidence that what is news to each of us as individuals is different than an aggregate sense of what is most interesting on any given day. As the New York City journalist who discussed the Middle East mentioned, his judgment about what is newsworthy for the paper might not mesh with his personal interests directly. But what runs in a newspaper or on television is the product of many news judgments, from the reporter on the street to the editors and producers in the newsroom who discuss the story line-up several times a day to the night editors who have to decide whether to scrap one story to fit another in as deadlines loom. The lack of this kind of consensus building in each of the focus groups could be one factor in the disconnect between their news judgments and the way the stories were played in the *Times* and the *Messenger*. Still, it is likely that some of the lack of correlation to the actual newspapers found here is real evidence of a difference between news values and individual values.

Discussion

The clear message from both the focus groups and the content analysis of print, radio, and television media in New York City and Athens is that stories about events or people who threaten the normal order and usual expectations are the ones that will grab the big headlines and the attention of readers. Deviance of all kinds is a factor in determining a story's newsworthiness, but social change deviance is the critical determinant. Likewise, social significance, particularly political significance, factors into how and where a story is played in the news.

However, the truly memorable stories most often combine elements of deviance and social significance. Kennedy's assassination was statistically deviant, since political leaders are not killed on a regular basis, and it was normatively deviant, since it violated social law. And certainly his assassination represented a threat, on some level, to the status quo. While the political system in the United States did not change as a result of his death, certainly the key players in that system did. The story also had elements of political significance, as just noted, and likely had cultural and economic fallout as well.

But the big stories are easy to identify as news. What this study also shows is that after the big stories—the assassinations, the moon landing, the odd outcome in the 2000 presidential election or even the World Trade Center attacks of September 11, 2001—most people value news for its utilitarian functions, its role as a social bridge to other people, and its unique personal value in their lives.

Part 4
Conclusions and Appendices

18
What's News? Theory Revised

The purpose of our project was to study *newsworthiness* around the world. Our theory posits that news is defined by two constructs, deviance and social significance, and that this is true of news in all countries. Following Shoemaker's (1996) theory, our work is based on two main notions: (1) biological evolution has led people to instinctively survey the environment for *deviance* (generally "bad" news); and (2) cultural evolution has resulted in people paying attention to different topics among countries, but that these topics should be high in *social significance*, as well as in deviance. In other words, we believe that these two constructs can predict the newsworthiness of events, ideas, and people around the world.

We gathered a wealth of data—a content analysis involving more than 32,000 news items from sixty news organizations (newspapers, television news, and radio news in twenty cities in ten countries). We also interviewed audience members, journalists, and public relations practitioners in eighty focus groups (four groups in each of two cities in the ten countries). In addition, the focus groups yielded quantitative data from a gatekeeping exercise that show how journalists, public relations practitioners, and consumers of news perceive newsworthiness, compared to their local newspapers. Two hundred and forty sets of ten newspaper stories were rank ordered and analyzed.

As a result of studying so much media content and talking with so many people, we have made two important revisions to our theory. First, we now propose that the constructs *news* and *newsworthiness* are not synonyms, but rather are two distinct theoretical concepts and must be explicated as such. Second, we have added a new construct to our theory, the *complexity* of an event. Complexity is related to the construct *intensity*, which we have measured in the content analyses, but complexity offers a different way of looking at the cognitive processes that determine whether an event, idea, or person is considered newsworthy. The concept of complexity in the news is not new. Cohen, Adoni and Bantz (1990) studied social conflicts in television news, and described intensity and complexity in their portrayal of conflict in the news, as well as in terms of how people perceive social conflicts.

We will discuss complexity in some detail, but first we offer some thoughts about the difference between *news* and *newsworthiness* and how these help us interpret the results of our study.

News and Newsworthiness

At the beginning of our study, we made an assumption that events, people and ideas that are prominently displayed in the news media are also newsworthy: the more prominent, the more newsworthy. Therefore in our content analyses, we used the prominence of the news item as a surrogate for newsworthiness. Our goal has been throughout this study to derive theoretical definitions of newsworthiness, and therefore we first looked at what actually had become news in our ten countries during one period of time, and then asked people to talk about what makes an event, idea or person newsworthy to them. Finally, our gatekeeping exercise asked people to use their own judgments about how newsworthy stories from their local newspapers were, and then we compared these judgments to how prominently the stories had actually been displayed in the newspapers.

As we have seen in the country-by-country analyses, however, there is a disconnect between the results of the content analyses and of the focus groups. Whether asked generally about their information needs or more specifically about what ought to be news, people in all countries volunteered their own versions of our deviance and social significance dimensions. They acknowledged that information about deviant people, ideas, and events need to be news, even if the news is "bad." Although the focus group participants wished for more positive news, they understood both that people need to pay attention to deviance in order to protect themselves and their societies, and that human beings intrinsically are interested in things that are unusual, that break laws or norms, or that threaten to change their societies. In addition, they were interested in information about things that have political, economic, cultural, or public significance. Some of this interest was connected to their desire for more positive news, but they clearly were interested in events, people, and ideas of social significance that are also deviant. This finding is consistent across the ten countries, even considering that the countries are very different from one another. The only other theme from the focus groups was an interest in information of personal salience to the participants. We think it is safe to assume therefore that people would rate people, events, or ideas that are deviant, socially significant, and personally salient as the most newsworthy. So why do our content analyses show consistently across countries that deviance and social significance are related in a modest way to the prominence of news items?

Our content analyses measured in detail the intensity of deviance and social significance for both verbal and visual forms of information included in newspapers, television, and radio news programs. Yet these indicators could not explain much of the variability in news item prominence. On the other hand, conversations with people in the focus groups do support our theory. Participants said they want news that is both deviant and socially significant, but the content analysis of the news media shows that both constructs are present in only modest proportions. Why?

Part of the answer can be found in our quantitative data from people in the focus groups—our gatekeeping exercise with journalists, public relations practitioners and audience members. Their rank ordering of actual stories (represented by headlines) from their local newspapers revealed a remarkable level of agreement among the people in the focus groups, both within and between the groups. People are people when it comes to assessing newsworthiness. Regardless of their country, city, or socioeconomic status, people tended to rate stories as similarly newsworthy. But, when what the people wanted was compared to how prominently the stories were covered in their own newspapers, we found that agreement between people and the newspaper was much lower or there was no agreement at all. In a few instances, people wanted the opposite of what their newspapers emphasized. Therefore, from this study we can conclude that:

- People say they want to know about things that are deviant and/or socially significant.
- People disagree with their newspapers about what should be covered most prominently.
- This is because newspapers don't give sufficiently prominent coverage to deviance and social significance.
- Hence, we must conclude that we are looking at two constructs and not just one. *News* and *newsworthiness* are theoretically distinct.

Our assumption that what is covered most prominently by the news media is also the most newsworthy is in question. *News* is a social artifact: the product, the output of journalistic routines that is made available to the audience, and therefore news is what we studied in our content analyses, not newsworthiness as we had assumed. In contrast, *newsworthiness* is a cognitive concept, a mental judgment made by individual people, and therefore our focus group data—both quantitative and qualitative—measure newsworthiness. *Therefore a major finding of our study is that what people—even journalists—think is newsworthy is not necessarily what becomes news.* This is consistent with what Shoemaker and Reese (1996) have argued, that there are many influences that shape mass media content. Individual people's assessments of an event's newsworthiness exist on the micro, individual level of analysis. Factors that affect the news also include routine practices of media organizations, organizational characteristics, social institutions, and macro variables.

From this perspective, we can conclude that variance in the prominence of news items which is unexplained by the deviance and social significance dimensions is an estimate of these other, unmeasured influences. From government control in China to the volatility of the Middle East region for Israel and Jordan, from a sports-loving culture in Australia to tumultuous social change in Russia, a myriad of factors shape what finally—after a lengthy gatekeeping process—becomes the news sent out to the audience.

That what the audience wants to receive may not be what the news media give it is not a new idea. There has long been a tension between those who believe the news should give the audience what it needs to know (social responsibility) or what it wants to know (marketing). A third variation (censorship) exists in countries such as China, where the *People's Daily* is written by the Communist Party and government with the specific purpose of communicating information that the *party wants* people to know. Conversations with people from China have revealed that another of our assumptions may be in jeopardy—that the most important news is inside the newspaper, rather than on the front page. This possibly explains the negative relationship between people's assessments of newsworthiness and how prominently the stories were presented by the newspapers.

In other parts of the world, such as the United States, it seems that the globalization of business and its convergence into a smaller and smaller set of mega-corporations has given more power to market forces. Still, our study suggests a fourth relationship between the audience and news content—that the attempt of businesses to maximize profits by giving the audiences what they want is not working. What audiences, journalists, and public relations practitioners want in the news is much the same, and they want something different than they are getting from the news media. So if journalists are trying to give the audience what it needs, they're not doing a very good job of it. If marketers are trying to give the audience what it wants, they are also failing. Neither the social responsibility nor marketing approaches can be supported.

Instead, we are left with a disconnect between what people think is newsworthy and what ultimately becomes news, and this disconnect is probably greatest with newspapers, since they consistently offer audiences less deviance and less socially significant news than television and radio do. Perhaps this accounts to some extent for, in the United States, a decades-long decline in newspaper readership and an increase in the number of people who say that they rely on television and the Internet for news. The difference between *newsworthiness* and the *news* may be increasing.

Having done the content analyses before the focus groups, we did not conceptualize or measure the personal salience of the news items to the audience. We did not collect any survey data about the audience's wants and needs, but if such data were available, it would be possible to add personal salience to the content analyses involving deviance and social significance. Possible, but problematic. It would be difficult to create a measure of personal salience against which all news items could be compared. The idea of personal salience seems so specific and idiosyncratic to individuals that a general measure that could be applied to all news items might be useless, with some people's intense interest in a specific event being neutralized by other people's disinterest. Still, the idea of personal salience relating to newsworthiness deserves more attention in future studies. Unlike our content analyses, which used news item as the unit of analysis, such research would most likely use the individual person as the unit being studied.

Next we return to our analysis of the news from the ten countries. We have come to realize that the extent to which deviance and social significance predict newsworthiness may be enhanced by the bifurcation of these constructs into not only their intensity, but also by their complexity.

Complexity: A New Construct for Studying the News

Although we predicted in chapter 2 that deviance and social significance combine in an important way when defining newsworthiness, the data we gathered and analyzed in this project have shown us that there are actually two ways in which deviance and social significance define newsworthiness: (1) the *intensity* of the relationship between the event and each of the seven dimensions of deviance (statistical, social change, and normative) and social significance (political, economic, cultural, and public); and (2) the *complexity* of those relationships. In previous chapters, intensity was measured as seven four-point scales. In this chapter, we show that the *number* of dimensions of deviance and social significance that an event involves is also important. Our theory suggests that evaluating an event, person, or idea as being both deviant and socially significant will make it more newsworthy than if it relates to only one or the other concept. In this chapter, we argue that when deviance and social significance combine in one way they measure the intensity of the deviance and social significance, but if combined in another way, they create a new construct, which we call *complexity*. We show in this chapter that complexity is a predictor of how prominently events are covered; in other words, complexity is an indicator of newsworthiness.

We conceive of complexity as a theoretical continuum that describes the extent to which a potentially newsworthy event, person, or idea[13] affects people's construction of social reality, both by assessing the parts of the social world impacted and the extent to which the event, person, or idea is of innate interest to people. Although we assume that people individually construct their own social reality, we believe that, because the mass media present news of the world as a sort of objective reality, news content can form connections to each person's view of the world (Adoni and Mane 1984). Information about an event may change an individual's social reality or be deflected as "not newsworthy." Most of what people know about the larger world (from outside of one's home to what is happening anywhere around the globe) comes from the news media; however, not everything in a newspaper or television or radio news program is considered as being newsworthy by everyone in the audience.

As we have seen in previous chapters, newsworthiness is partly predicted by the intensity of an event's deviance and social significance. Complexity deals with the multidimensional nature of these concepts and suggests that the more dimensions an event has, the more likely it is to be considered newsworthy. To illustrate this point, we have only to think about our long-distant ancestors.

An innate interest in deviance is probably a result of human evolution over millennia, especially when attention to threats gave a family group a reproductive advantage and the chance to pass their genetic structure to subsequent generations. The more types of deviance our ancestors identified and dealt with, the greater their chances of surviving and reproducing. Surveillance of the environment was key to survival, but sometimes the environment was simpler to deal with than at other times. Saving one's family by frightening a tiger with a burning stick is threatening enough, but if in the process the family's bedding is set afire, then the threat is more complicated. Tigers present a simple, but intense, threat—kill or be killed. When the process of fighting the tiger causes the environment to grow more complex (fire in the cave), surveillance is necessary in multiple directions, and the situation becomes more serious and more difficult to handle successfully. So, although environmental threats are innately newsworthy, if one event presents multiple threats, then it is more likely to become news.

The presence and process of environmental threats affects each person's social reality—an understanding of what the world is like. By definition, people's social realities differ from each other, because each person has a unique set of life experiences, knowledge, and attitudes. Our use of the term presumes that no individual person can be an objective observer of the world—that each person has a unique and, in the context of the whole world, a subjective social reality. Variance in how people perceive newsworthiness can be explained by the fact that their social realities differ.

But the environment is made of more than tigers and fires. In fact, the environment is largely organized and interpreted by cultural evolution. Maybe tigers are considered sacred. Or maybe they are not sacred when the moon is full. Maybe the number of stripes a tiger has affects its holiness. Perhaps tigers are fair game when other food sources are scarce. The rules and practices of a culture are often highly complex and sometimes contradictory. If seeing a tiger causes a need to consider the phase of the moon, its stripes, whether food is needed, and its general level of holiness, then a complex situation arises. Mistakes can be made; some people may benefit at the expense of others; and the stakes are high. An event that is culturally complex connects in many ways to people's social reality and therefore is also evaluated as newsworthy.

In fact, the construct *complexity* is rather complicated, being the result of factors shaped by both biological and cultural evolution and, on a more concrete level, the combination of the concepts *deviance* and *social significance* and their seven dimensions. The more dimensions of deviance and social significance an event taps into, the more complex it is and the more newsworthy people think it is.

We deliberately use the word *people* here instead of *journalists* or *reporters*. As we demonstrated in chapter 7, people tend to agree on what is newsworthy, regardless of their socioeconomc status. When asked to rank-order stories that appeared in their local newspaper a few months earlier, it made little difference whether the person was a reporter, a housewife, or a public relations account executive. There is a tendency for people in all walks of life to agree about the relative newsworthiness of events, people, or ideas. Therefore, we can talk in general about how a *person's* social reality is constructed and about how the complexity of an event, person, or idea helps define that person's social reality.

The Relationship of Events to Social Reality

In essence, *newsworthiness* is the extent to which information about an event, person, or idea touches various parts of a person's social reality, and this is true whether that person sends the information or receives it. All of us continually evaluate the newsworthiness of things in our worlds. Whether a woman tells her coworkers about her dental appointment will depend in large part on what happened during the visit. A routine cleaning is less likely to be communicated to others than a painful and expensive procedure. If the woman is a journalist, however, she is unlikely to rate her painful and expensive dental experience as newsworthy enough to include in that day's newspaper. In contrast, if this journalist is involved in a fifteen-car crash on the way to work, her personal experience will be rated as newsworthy not only by her but also by her coworkers and a large part of the audience.

The difference between these events is the extent to which they connect to a person's social reality—the person's understanding of the world. Social reality includes not only what people know about the world but also how important they rate its various aspects. Each person's social reality is continually changing. An Australian man's social reality may in general include people, events, and ideas that are relevant to his local community; he is aware of the larger world, but he has only vague mental connections to it. Things that happen in Chile are not within the scope of his social reality until his daughter gets a new job and moves there. Or a woman may have a sense of the world as a peaceful place, but if her country goes to war, the war becomes part of her social reality. The makeup of her psychological and physical worlds has changed and become more complex.

Figure 18.1 illustrates how the seven dimensions of deviance and social significance may combine to determine the complexity of the event's connections to a person's social reality—in other words, how newsworthy it is evaluated to be. Imagine two events, one of which involves the routine promotion of the assistant police chief to be the new chief (lower circle on left). This event has only one connection to the person's social reality, through political significance (because of an administrative change in a politically appointed job). In contrast, the second event involves a drunken police chief, his mistress, and a bicyclist who was hit by the police chief's car. This event connects to the person's social reality in four ways: statistical deviance (police chiefs not normally arrested and bicyclists not usually hit by cars), normative deviance (laws broken), political significance (subsequent administrative change), and public

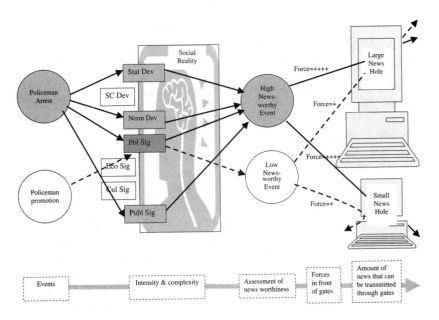

Fig. 18.1 Stages of news transmission.

significance (safety of bicyclists). Some events have many connections to a person's social reality, whereas others are of no relevance. Of the billions of events that occur around the world every day, only a small number relate to an individual's social reality. Events that have complex relationships to a news source's social reality and then to a journalist's social reality are more likely to pass through individual and organizational gates and eventually be communicated as news.

As gatekeeping theory predicts, the intricacy of these connections to the journalist's mental world creates a force in front of a gate. Forces can be positive, neutral, and negative and can vary in strength. Therefore, the more an event connects to the journalist's social reality (the more complex it is), the stronger its force will be. The figure shows that the stronger force an event has, the greater its likelihood to pass through the gate. In other words, the more complex an event is, the more likely it is to be considered newsworthy. There are many gates in the process that determine whether an event becomes news, such as the kind of information sources give reporters, which events reporters decide to pass along, which details the reporter selects, how the news is shaped by multiple editors and producers, and so on (Shoemaker 1991).

This perspective views newsworthiness not as a characteristic of an event but rather the outcome of a person's mental judgment. This explains why not all people judge events as equally newsworthy. If newsworthiness were a simple characteristic of the event, then people would be able to perceive its newsworthiness level and all would understand whether it should be communicated to others. This is obviously a much too simple explanation. Although we have shown that people agree in general about what is newsworthy, we also know that there is variation in their judgments based on differences in their social realities. Therefore, newsworthiness is a cognitive variable. When we write about "very newsworthy" events, such as Tuchman's (1978) news category *what a story!*, we must not assume that the event makes itself newsworthy but rather that there is a consensus among people that the event is complex and intense and therefore touches many aspects of their social realities.

Measuring Complexity

Although the idea of complexity is highly theoretical, it can be easily operationalized. As we noted in chapter 5, about two-thirds of the 32,000+ news items we studied have some degree of intensity (from weak to strong) on one or more of the three verbal (text) deviance dimensions (statistical, social change, and normative); for visual news items, the amount is about half. Chapter 6 shows that almost 90% of news items were coded as having from weak to high intensity on one or more of the four verbal social significance dimensions (political, economic, cultural, and public). The same is true for just under half of visuals. The scales shown in Table 5.1 and Table 6.1 measure the number of deviance dimensions and the number of social significance dimensions, respectively, that are related to news items.

To put it another way, these scales measure the complexity of the news items. To illustrate, let us consider the events represented in Figure 18.1 in more detail. A news item that is about the routine promotion of a town's assistant police chief as the top police official has no elements of deviance but does have minimal political significance because of the impending change in administration at the police department. This is a simple event and will probably be determined as of minimal newsworthiness, except perhaps by the new chief's family. In contrast, consider this event: The police chief resigned after he was arrested late at night for hitting a bicyclist with his car because he had drunk too much alcohol, his passenger turned out to be his secret mistress, *and* he admitted he was picked up for drunk driving three times before but was not punished. Now that's a story! It has political significance not only because of the administrative change but also because he committed two crimes—hitting the bicyclist and driving while under the influence of alcohol—and we find out that he should have been arrested on earlier occasions. Because a bicyclist was hit by a car, there is an element of public significance, such as questions about the bicyclist's health and the safety of bicycling on that road. Thus we have a score of 2 on the social significance complexity scale (in two of four dimensions—economic and cultural—there was no social significance at all). But this event is also deviant. It should be coded as normatively deviant, first because of the crimes; second because a public official committed the crimes; and third because the presence of a secret mistress suggests he has been cheating on his wife. There is also a hint of statistical deviance, because police officers are rarely arrested, and bicyclists are not generally hit by cars. That's a score of 2 on the deviance complexity scale (two of three dimensions). We do not need these scales to tell us that the second event is more complex than the first, but they do help us operationally define complexity.

Table 18.1 shows a seven-point, additive scale, which measures the combined number of deviance and/or social significance dimensions on which the news items in our study received some score. A news item coded as having no deviance and no social significance is given the value of 0. A value of 7 indicates the news item was coded somewhere between low and high on all three types of deviance and all four types of social significance. A value of 5 indicates that there were some types of deviance and some types of social significance but that the exact combination is unknown. Still, a news item with a score of 5 can be interpreted as being less complex than one with a score of 7 but much more complex than a news item with a score of 1.

We see in Table 18.1 that deviance and/or social significance are present in 88.8% of the verbal content of the 32,000+ news items. The remaining news items, 11.2%, were coded as not deviant and not socially significant on all seven dimensions—in other words, as having no complexity. For visual news content, nearly two-thirds of items were coded as having deviance and/or social significance, with 37.2% of visuals having neither. It is worth noting that, as the project's primary investigators, we wrote coding instructions that defined (according to our own social realities) some topics and content as inherently not deviant or not significant (see chapter 3). These include content such as comics (but not political cartoons), weather maps, listings of

TABLE 18.1 Distributions of Verbal and Visual News Items Coded as either Deviant or Socially Significant

Number of Deviant and/or Socially Significant Dimensions in News Items	Verbal News Items	Visual News Items
0	11.2	37.2
1	23.0	22.6
2	19.4	15.4
3	16.5	9.8
4	13.7	7.2
5	10.0	4.8
6	4.7	2.3
7	1.4	.7
	100.0	100.0

Note: The top row shows how many news items were coded as having no deviance and no social significance. The following rows note the percentage of items coded on 1 to 7 of the deviance and social significance dimensions. Percentages may not add up to 100.0 because of rounding error.

stock market quotations, and horoscopes. Therefore, before the coding began, our instructions guaranteed that some news items would not be considered deviant and some would not be considered socially significant.

When we began the study, we expected that a news item might be coded as having more than one type of deviance or even deviance and social significance. We did not, however, anticipate that news items would convey many dimensions of these concepts: Nearly half of the verbal content is coded on four or more dimensions, but the corresponding percentage for visuals is only 7.8%. Apparently news is composed of text that is more complex than its images, and this is consistent with our findings in the previous chapters that the intensity of deviance and social significance in verbal content is higher than in visual content.

Complexity and News Topics

As we have seen in previous chapters, some news topics are covered more prominently than others. Here we look at the extent to which the aggregation of news items into topics can predict how prominently the topics are covered. Whereas all previous analyses in this book have used "news item" as the unit of analysis, with $n = 32,000+$, the analyses for the remainder of this chapter use "topic" as the unit of analysis, $n = 26$. Earlier we looked at differences among news items; now we are looking at differences among topics.

Radio and Television News

In Figures 18.2 to 18.4, the four quadrants represent combinations of high and low values of prominence and complexity. Roughly, our theory predicts that news should fall either in the upper-right or lower-left quadrants; that is, news of high complexity should have high prominence, and news of low complexity should have low prominence. In all three figures, the majority of topics fall in these two quadrants, as predicted: twenty-one radio topics, seventeen television verbal topics, and sixteen television visual topics. For a more precise estimate of the relationship between complexity and prominence, we ran regression equations to test the hypothesis that the more complex items in a topic are, the more prominently they are covered. For radio news, complexity explains 62% of variance in prominence ($p < .05$). As for television news, complexity explains 48% and 38% of verbals and visuals, respectively ($p < .05$). Most topics falling outside of the predicted quadrants are not far from the regression lines and therefore are close to the prediction.

The most complex and most prominently covered topics in radio and television news programs are labor, internal politics, internal order, international politics, and population. The least complex and least prominently covered topics were weather, sports, and fashion. Also among the least complex topics were entertainment, culture, and human interest.

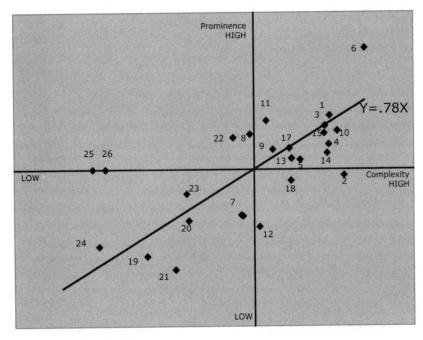

R² = .621, p<.05, n=26

1=internal politics, 2=international politics, 3=military and defense, 4=internal order, 5=economy, 6=labor relations and union, 7=business/commerce/industry, 8=transportation, 9=health/welfare/social service, 10=population, 11=education, 12=communication, 13=housing, 14=environment, 15=energy, 16=science and technology, 17=social relations, 18=disasters/accidents/epidemics, 19=sports, 20=cultural events, 21=fashion/beauty, 22=ceremonies, 23=human interest, 24=weather, 25=entertainment, 26=other

Fig. 18.2 Relationship between complexity and prominence in radio news.

Newspaper News

As we have seen all along in this project, newspaper prominence is related quite differently to deviance and social significance than is the prominence of television and radio news. Figures 18.5 and 18.6 show this difference: For verbal newspaper content, half of the topics fall within the predicted quadrants, and for verbal content, only eleven topics appear as predicted. Neither regression analysis shows a statistically significant relationship between complexity and prominence. Also, in contrast to radio and television news, the most prominent newspaper topics are fashion, ceremonies, the environment, the military, and housing. The very complex topics of population, international politics, internal order, and labor receive below-average prominence in newspapers.

Why the difference? Perhaps because in most cases television and radio news programs have a fixed and small amount of time to communicate the news of the day. News producers must make difficult decisions about the few news items that pass through the television and radio news gates in a given day. In contrast, the number of pages in a newspaper may vary from day to

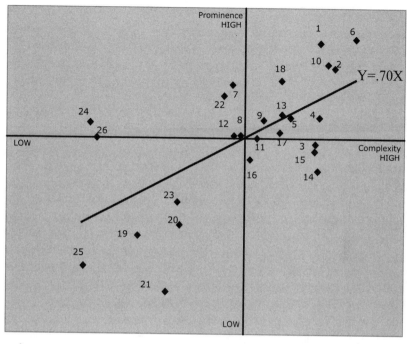

* $R^2 = .484$, p<.05, n=26

1=internal politics, 2=international politics, 3=military and defense, 4=internal order, 5=economy, 6=labor relations and union, 7=business/commerce/industry, 8=transportation, 9=health/welfare/social service, 10=population, 11=education, 12=communication, 13=housing, 14=environment, 15=energy, 16=science and technology, 17=social relations, 18=disasters/accidents/epidemics, 19=sports, 20=cultural events, 21=fashion/beauty, 22=ceremonies, 23=human interest, 24=weather, 25=entertainment, 26=other

Fig. 18.3 Relationship between complexity and prominence in verbal content of television news.

day. If the number of pages is small, then it is possible that newspaper editors will work under the same gatekeeping pressures as television and radio news producers—make the best news decisions and still meet the deadline. If, on the other hand, the number of newspaper pages is larger on some days, editors can allow more news items to flow through the newspaper gates. Having said this, however, there are occasions—such as when major terror events take place—that television and radio news holes are expanded and regular programming is pre-empted.

Variability in the amount of news that newspapers can include from day to day influences editors' gatekeeping decisions. For example, the size of many newspapers' news holes is determined by policy of the publisher and is based on the amount of advertising space sold for a particular day. If more advertising is sold, the news hole is larger; if fewer ads are sold, the news hole is smaller. This suggests that, when the news hole is small, only the most newsworthy events will be covered; however, if the news hole is larger, then there is also room for less newsworthy content, features, and other types of soft news.

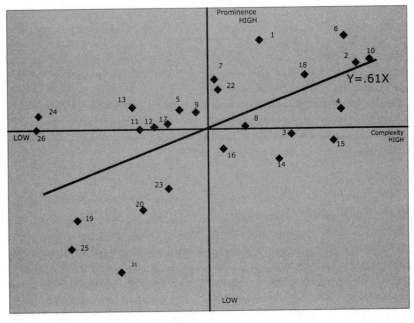

*R^2 = .377, p<.05, n=26

1=internal politics, 2=international politics, 3=military and defense, 4=internal order, 5=economy, 6=labor relations and union, 7=business/commerce/industry, 8=transportation, 9=health/welfare/social service, 10=population, 11=education, 12=communication, 13=housing, 14=environment, 15=energy, 16=science and technology, 17=social relations, 18=disasters/accidents/epidemics, 19=sports, 20=cultural events, 21=fashion/beauty, 22=ceremonies, 23=human interest, 24=weather, 25=entertainment, 26=other

Fig. 18.4 Relationship between complexity and prominence in visual content of television news.

Of course, many newspapers routinely allocate some space for soft news about people's lifestyles, sports, and weather regardless of the size of the news hole presumably because they believe their readers expect sections on topics such as sports.

Some newspapers use a fixed advertising-to-news ratio to determine the size of the news hole and the amount of space for news may vary by day of the week. Some newspapers have a fixed size that rarely varies. Others may increase the size of the news hole according to the editor's assessment of how newsworthy the day's events are. Whatever the determinant of the news hole size, however, we propose that when news holes are smaller, news selection decisions will be more similar to that of television and radio news producers. As Figure 18.1 illustrates, a smaller news hole means the gate is "smaller" and more difficult to enter. Therefore, the intensity and complexity of potential news items (according to their deviance and social significance) predict which items pass through the gate. When either or both intensity and complexity are high, the force in front of the gate is strong, indicating that such items are likely to pass through. If the news hole is larger and therefore the gate is larger, not only will items with a strong, positive force attached to them enter the

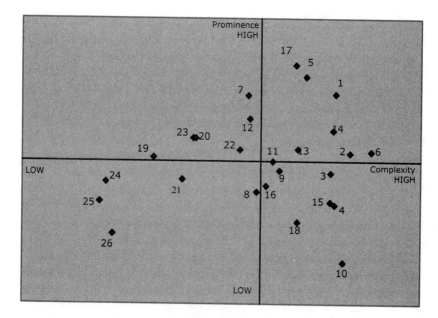

1=internal politics, 2=international politics, 3=military and defense, 4=internal order, 5=economy, 6=labor relations and union, 7=business/commerce/industry, 8=transportation, 9=health/welfare/social service, 10=population, 11=education, 12=communication, 13=housing, 14=environment, 15=energy, 16=science and technology, 17=social relations, 18=disasters/accidents/epidemics, 19=sports, 20=cultural events, 21=fashion/beauty, 22=ceremonies, 23=human interest, 24=weather, 25=entertainment, 26=other

Fig. 18.5 Relationship between complexity and prominence in verbal content of newspaper news.

gate, but also items with neutral or even negative forces may go through the gate. That items with negative forces might pass through the gate sounds counterintuitive, but when an editor must fill a large news hole, items of almost any sort may become news. We sometimes laugh about "slow news days," when the mundane nature of the newspaper makes it seem that nothing of importance has happened in the world, but in fact that is sometimes true—at least in the judgment of our local newspapers' editors.

Universals and Differences

Our theory predicts similarities among the ten countries' news, but as we indicated, the nature of statistical tests requires that we test hypotheses of difference. In this section, we refer back to the research questions and hypotheses in chapter 2.

The first set of research questions asks about the intensity of deviance and social significance indicators in the news we studied in sixty media from ten countries. We find deviance and social significance in the news from all countries (chapters 5 and 6), and it is in general of low-to-moderate intensity. The intensity of deviance or social significance varies minimally between countries, between news media, and between large and small cities. However, it is important to emphasize that although statistically significant (due to our large

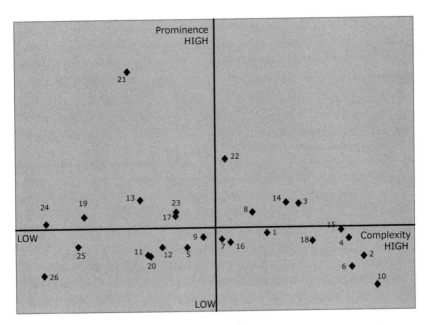

1=internal politics, 2=international politics, 3=military and defense, 4=internal order, 5=economy, 6=labor relations and union, 7=business/commerce/industry, 8=transportation, 9=health/welfare/social service, 10=population, 11=education, 12=communication, 13=housing, 14=environment, 15=energy, 16=science and technology, 17=social relations, 18=disasters/accidents/epidemics, 19=sports, 20=cultural events, 21=fashion/beauty, 22=ceremonies, 23=human interest, 24=weather, 25=entertainment, 26=other.

Fig. 18.6 Relationship between complexity and prominence in verbal content of visual content of newspaper news.

sample size), the differences are usually minor. In other words, there is a universal tendency for the news media in the ten countries to include news items that on average are of low-to-moderate intensity, with extremely deviant and extremely socially significant news appearing only from time to time.

Of the 32,000+ news items we studied, sports was coded as the most common or one of the most common news topics in every country; however, we find that sports has low deviance and social significance, suggesting that although there are many sports stories, they are of lower prominence. Other common news topics in all countries include internal order and politics (both domestic and international).

The fact that few stories have the highest intensity on all of the dimensions of deviance and social significance is not surprising, since many news items deal with one important point—such as the political significance of a flawed election—rather than being more complex. Yet the presence of news items that have high intensity on most or all of the deviance and social significance dimensions points out topics that are almost always newsworthy, such as international or domestic (internal) politics and the violation or maintenance of internal order.

There is also a universal tendency for newspapers to have less intense deviance and social significance. If we had studied only front pages, or just the international or national news sections, we would almost certainly have measured more deviance in newspaper articles. The blandness of most newspaper content dilutes the more sensational parts of the newspaper and yields overall low deviance and social significance scores. This, too, appears to be true across countries.

A universal tendency also exists for visuals to be less deviant and especially less socially significant than verbal text. This says much about the visual gatekeeping process: If a news medium's goal was sensationalism, it could easily show sensational photographs of bodies, body parts, and other gruesome sights. In contrast, there is a tendency for information about the more gruesome side of events to be carried in the text and not to be shown visually. Perhaps the audience is more tolerant of unpleasant information in verbal form than in visual, or maybe journalists believe (incorrectly?) that the audience has no appetite for such fare. For television journalists this creates a problem, since videographers capture many minutes of images that must be viewed before gatekeeping decisions can be made to choose the few seconds that will be aired and become news. Which images can accurately portray the horror of terrorism without making the audience turn away?

Our last universal conclusion comes from chapter 7, where we show strong agreement among journalists, public relations practitioners, and the audience in the ten countries regarding the newsworthiness of events from their local newspapers. Across the board, there is a positive relationship between the newsworthiness ratings of audience and journalists, audience and public relations practitioners, and journalists and public relations practitioners. However, the relationships between the various people's ratings of newsworthiness and how their own newspapers covered the events were generally far weaker, and at times even negative. In all ten countries, people (including journalists) disagreed at least partially with how the news was selected and presented in their newspapers. Our eighty focus groups also revealed that people want similar types of information, such as about their families' health or the state of their nations. When asked about the most newsworthy events in their lifetimes, the majority of people from all countries selected disasters and political events that changed their nation or the world.

This suggests that are forces that shape news content in the ten countries other than what the audience or journalists want. Shoemaker and Reese (1996) offer a hierarchical model for organizing what has become a vast literature that explains how news content comes to be. In addition to what individuals want, news is shaped by routine practices of news work (such as the beat system), policies and characteristics of the news organization, pressures from social institutions (including government), and characteristics of the social system (including ideology and the type of economic system).

Although we did not initially hypothesize about complexity as a predictor of newsworthiness, we did find, as we presented earlier in this chapter, that complexity may be a stronger predictor of prominence than intensity. Additional studies will determine their relative contributions to newsworthiness.

And Finally ...

Despite our large-scale project, we were limited to only ten countries. The next step ought to be an attempt to study more countries and to replicate some of our work in another time and in other places. Scholars should consider the ecological validity of such research—that is, the extent to which we can generalize these findings to other time periods. Because we based our analysis of the news contents in the ten countries on only one composite week, consisting of only seven days, it is reasonable to ask whether this particular time period was normal or atypical. There is much variability across the globe in terms of what is news on any particular day, as Malik and Anderson (1992) ably showed in their study of television news around the world on a single "normal" day. It is only when an extraordinary event occurs that news around the world would be fairly similar (e.g., the assassination of a world leader, a major terrorist event such as the attacks on the World Trade Center or the London underground system, a pope's funeral, the signing of a peace treaty between warring countries, or even a major sports event). In this respect, our sample was "normal," and no special events were recorded.

We have proposed a new construct, *complexity*, which addresses the extent to which an event connects with a person's social reality. This is operationalized as the number of deviance and social significance dimensions to which an event relates, in contrast to *intensity* scores which are measured on a four-point scale for each of the seven deviance and social significant dimensions. The relationship between the intensity with which deviance and social significance connect to an event and the event's complexity has not been fully explored.

As it turns out, complexity may be the more important construct. The mean intensity of deviance and social significance in chapters 5 and 6 as well as the "country reports" (chapters 8–17) are on the low-to-moderate side—certainly lower than we expected based on the theory. When results are not as expected, there are usually two options to explore. First, while our measures are reliable and valid, there may actually be less intensity conveyed by deviance and social significance in the news media than we expected. There is no way to know for sure, but anecdotal evidence suggests that news is often deviant and occasionally socially significant. It is possible that our expectations were too high—that the most prominent articles are deviant and socially significant but that a large quantity of other content does not have these characteristics and therefore drags the means down. Second, there may be methodological problems in the study, such as low intercoder reliability. Our intercoder reliability is acceptable within each country, and since the results are similar from country to country, this suggests that there is probably not a problem across countries. Certainly deviance and social significance could have been operationalized differently; however, Shoemaker, Danielian, and Brendlinger (1991) used the same scales with satisfactory results, and that study of U.S. news media was one impetus for our current international project. Since our theory predicts that countries would have similar deviance and social significance scores, maybe the similar but low means are a true reflection of news content.

Or perhaps a second construct is needed. The discussion of means and how high they are is about the intensity of the deviance and social significance concepts and not about their presence or absence. Therefore, in post hoc analysis we demonstrate that complexity—that is, counting the number of dimensions (three for deviance; four for social significance) to which a news items relates—is a good predictor of how prominently that news item is conveyed by the news media. For example, instead of looking at the intensity of normative deviance in a story, we look at whether the story is also statistically deviant and perhaps politically and economically significant. Such a story connects with people's lives in three important ways and therefore measures newsworthiness as the complexity or intricacies with which the story can meaningfully touch someone else. If a story is both complex and intense then it should be very newsworthy.

That is just one direction more scholarship could take. Even after such a large project, we cannot completely define *newsworthiness*, but we have some good ideas about where to continue our search and we now realize that newsworthiness is not synonymous with prominence. There is no doubt that deviance and social significance are important, and now we have two ways of using them to create constructs—intensity and complexity. We have also confirmed that some topics are more newsworthy than others, although there seems to be some disagreement between newspapers and the electronic media about that. In addition, our project has applied gatekeeping theory to our current theoretical base of biological and cultural evolution. We hope that others will tackle some of the ideas that this project has suggested.

As technology develops and expands and interactions with people across the globe take place in real time (for ordinary people, not just the wealthy), there is an increasing need for information to help make sense of the world. Therefore, people have a greater need for information that relates to their social realities today than at any time in the past. Our cave-dwelling ancestor may have been stressed about lurking tigers, but when the "fight or flight" moment came, a decision was easy. People today are confronted with stress from many more sources, and while fighting is not socially acceptable, fleeing is often impossible. Unfortunately, there are hundreds of lurking tigers—bosses, governments, family, friends, terrorists, landlords, and natural disasters such as hurricanes—that must be dealt with every day, some intensely important and others trivial. Events that involve these "tigers" capture one's attention so as to determine whether the information is useful or threatening. Such events stand a good chance of becoming news, whether communication is between two people or between a journalist and the world.

Appendix A
Content Analysis Codebook

News item number

Names of media studied

1 *Sydney Morning Herald,* newspaper, Sydney, Australia
2 BL, radio, Sydney, Australia
3 TCN-9, television, Sydney, Australia
4 *Courier Mail,* newspaper, Brisbane, Australia
5 4QR, radio, Brisbane, Australia
6 QTQ-9, television, Brisbane, Australia
7 *El Mercurio,* newspaper, Santiago, Chile
8 Diario de Cooperativa, radio, Santiago, Chile
9 Teletrece, television, Santiago, Chile
10 *El Sur,* newspaper, Concepción, Chile
11 Radio Bio Bio, radio, Concepción, Chile
12 Canal 9 Regional de Concepcion, television, Concepción, Chile
13 *People's Daily,* newspaper, Beijing, China
14 CNR1, radio, Beijing, China
15 CCTV1, television, Beijing, China
16 *Jinhua Evening Post,* newspaper, Jinhua, China
17 JHPR1, radio, Jinhua, China
18 JHTV2, television, Jinhua, China
19 *Berliner Zeitung,* newspaper, Berlin, Germany
20 Berliner Rundfunk, radio, Berlin, Germany
21 Abendschau, television, Berlin, Germany
22 *Allgemeine Zeitung Mainz,* newspaper, Mainz, Germany
23 Radio RPR morning newscast, radio, Mainz, Germany
24 Rheinland-Pfalz aktuell, television, Mainz, Germany
25 *Hindustan Times,* newspaper, New Delhi, India
26 All India Radio, New Delhi, India
27 Doordarshan News, television, New Delhi, India
28 *Eenadu,* newspaper, Hyderabad, India
29 All India Radio, Hyderabad, India
30 Telugu, television, Hyderabad, India
31 *Ha'ir,* newspaper, Tel Aviv, Israel
32 Mish'al Al Ha'Boker, radio, Tel Aviv, Israel
33 Tahama Merkazit, television, Tel Aviv, Israel

34 *Kol Hanegev,* newspaper, Be'er Sheba, Israel
35 Hayom Ba'Darom, radio, Be'er Sheba, Israel
36 Sihat Ha'ir, television, Be'er Sheba, Israel
37 *Al-Rai,* newspaper, Amman, Jordan
38 Radio Jordan, newscast, Amman, Jordan
39 Channel One, television, Amman, Jordan
40 *Shihan,* newspaper, Irbid, Jordan
41 Radio Jordan, newscast, Irbad, Jordan
42 Regional bulletin from Channel One, television, Irbid, Jordan
43 *Izvestia,* newspaper, Moscow, Russia
44 Radio Rossiya, Moscow, Russia
45 ORT, television, Moscow, Russia
46 *Tula Izvestia,* newspaper, Tula, Russia
47 The Reporter from Moja Tula, radio, Tula, Russia
48 The Reporter from State Television, television, Tula, Russia
49 *The Sowetan,* newspaper, Johannesburg, South Africa
50 Metro FM, radio, Johannesburg, South Africa
51 SABC3, television, Johannesburg, South Africa
52 *Die Volksblad,* newspaper, Bloemfontein, South Africa
53 OFM, Bloemfontein, South Africa
54 SABC3, television, Bloemfontein, South Africa
55 *New York Times,* newspaper, New York City, United States
56 WCBS 880, radio, New York City, United States
57 WNBC4, television, New York City, United States
58 *Athens Messenger,* newspaper, Athens, United States
59 WOUB FM, radio, Athens, United States
60 Newswatch from WOUB, television, Athens, United States

Date the news item was published

Weekday the news item was published

1 Monday
2 Tuesday
3 Wednesday
4 Thursday
5 Friday
6 Saturday
7 Sunday

Topic of news item

Internal politics

101 Legislative activities (e.g., discussion of a new law)
102 Executive activities (e.g., announcement by the president)

103 Judicial decisions
104 Constitutional issues
105 Elections
106 Political fundraisers and donations
107 Political appointments
108 Statements and activities of individual politicians
109 Interparty relations
110 Internal party relations
111 Activities of interest groups
112 Referendum
113 Public opinion/polling
114 Abuse of political power, corruption
199 Other

International politics
201 Activities of international political organizations
202 Activities of individual politicians
203 Activities of political parties
204 Diplomatic visits
205 Diplomatic negotiations and agreements
206 Promises of aid or cooperation
207 Policy statements
208 Wars between countries
209 International tensions and disagreements
210 International terrorism
211 Embargo
299 Other

Military and defense
301 Military activities
302 Appointments and firings in the military
303 Government defense policy and action
304 Protest at government defense policy
399 Other

Internal order
401 Civil war
402 Peaceful demonstrations
401 Violent demonstrations
401 Terrorism
401 Crime
401 Petit-crimes
401 Police
401 Fire brigade

401 Prisons
401 Corruption (not political)
401 Espionage
499 Other

Economy

501 State of economy
502 Economic indexes (e.g., domestic production numbers)
503 Job market
504 Appointments
505 Fiscal measures
506 Budget issues
507 Natural resources
508 Monopolies
509 Tariffs
510 Economic legal issues
511 Donations
599 Other

Labor relations and union

601 Union activities (e.g., lobbying)
602 Disputes
603 Strikes
604 Legal measures and policy
605 Foreign/guest workers
699 Other

Business/commerce/industry

701 Business activities
702 Legal measures and policy
703 International business
704 Globalization
705 Stock market
706 Mergers and acquisitions
707 E-commerce
708 Technology
709 Tourism
710 Agriculture
711 Trade with foreign countries
712 Appointments and firings
799 Other

Transportation

801   Transportation infrastructure/transportation systems
802   Public transportation issues
803   Automobiles
804   Driving behavior
805   Driving conditions
806   Parking issues
807   Aviation
808   Railway/trains/subway
809   Transportation-related construction
899   Other

Health/welfare/social service

901   State of health system
902   Health policies and legal measures
903   Health insurance issues
904   State of social services
905   Nonprofit organizations
906   Benefit events for a good cause
907   Developments in medical practice
908   Malpractice suits
909   Poverty
999   Other

Population

1001   General population statistics
1002   Immigration
1003   Emigration
1004   Visa issues
1099   Other

Education

1101   General educational facilities
1102   Higher education (colleges and universities)
1103   Teachers and faculty
1104   Students
1105   Parental issues
1106   Level of teaching and teaching standards
1107   School curricula
1108   Relations between teachers and parents
1109   Relations between teachers and students

1110 Registration
1111 Opening and closing of schools
1112 Preschool education
1113 Sectorial education (e.g., religious versus secular)
1199 Other

Communication

1201 Industry-wide issues and statistics
1202 Journalism and media in general
1203 Newspapers
1204 Network television
1205 Cable television
1206 Radio
1207 Magazines
1208 Internet
1209 Phones/cell phones/mobile phones
1210 Media regulation
1211 Technical aspects of communication
1299 Other

Housing

1301 Housing supply
1302 Living conditions
1303 Construction
1304 Mortgages
1305 Building permits
1306 City planning
1307 Housing demolition
1399 Other

Environment

1401 Threats to environment (e.g., pollution)
1402 Activities of environmental organizations
1403 Garbage collection
1404 Conservation
1499 Other

Energy

1501 Energy supply
1502 Energy costs
1599 Other

Science/Technology

1601 Standards
1602 Inventions
1603 Individual scientists
1604 Scientific organizations
1605 Computer issues
1606 Multimedia issues
1607 Space exploration
1608 Problems related to science/technology
1699 Other

Social relations

1701 Gender relations
1702 Sexual orientation issues
1703 Ethnic relations
1704 Class relations
1705 Age differences
1706 Religious groups
1707 Family relations
1708 Minority–majority relations
1799 Other

Disasters/accidents/epidemics

1801 Natural disasters
1802 Traffic accidents (includes airplane crashes)
1803 Work accidents
1804 Military-related accidents
1805 Home accidents
1806 Crowd accidents
1807 Health epidemics (e.g., AIDS)
1899 Other

Sports

1901 Competition/results
1902 Training
1903 Records
1904 Individual athletes/coaches/teams
1905 Leagues
1906 Fans/supporters
1907 Legal measures
1908 Appointments and firings
1999 Other

Cultural events

2001 Music (including musicals)
2002 Theatre, opera, ballet, and cabaret
2003 Film and photography
2004 Literature and poetry
2005 Painting and sculpturing
2006 Television shows (e.g., soap operas)
2007 Radio shows
2008 Museums
2009 General exhibits

Fashion/beauty

2010 Festivals and competitions
2011 Prizes and awards
2099 Other
2101 Fashion shows
2102 Beauty contests
2103 Models
2104 Fashion products
2105 Fashion trends (e.g., trend colors, body piercing)
2199 Other

Ceremony

2201 Official government/political ceremonies
2202 Holidays
2203 Ethnic ceremonies/commemorations
2204 Anniversaries
2299 Other

Human interest

2301 Celebrities
2302 Noncelebrities
2303 Animal stories
2304 Travel stories
2305 Record attempts
2306 Supernatural or mystical stories
2307 Mystery
2308 Food
2309 Advice (e.g., on love, insurance, stock)
2399 Other

Weather

2401 Weather maps and statistics
2402 Weather forecasting

2403   General weather stories (e.g., coldest winter)
2499   Other

Entertainment

2501   Horoscopes
2502   Comics
2503   Crossword puzzles and other games
2504   Gossip
2505   Lottery, betting, raffles, contests
2506   Literature, short stories
2507   Poetry
2599   Other

Other

2600   Specify

Is item story or teaser?
1   Story
2   Teaser

Nature of news item
1   Hard news
2   Soft news
3   Editorials
4   Reviews
5   Columns
6   Letters to the editor
7   Other

Newspaper: Square cm of verbal content

Newspaper: Square cm of visual content

TV/Radio: Item length in seconds

Newspaper: Item placement
1   Any other position
2   Section front
3   Front page

TV/Radio: Item placement
1   In last third of news program
2   In middle third of news program
3   In first third of news program

Statistical deviance intensity, verbal content
1   Common information
2   Somewhat unusual information

3   Unusual information
4   Extremely unusual information

Statistical deviance intensity, visual content
1   Common information
2   Somewhat unusual information
3   Unusual information
4   Extremely unusual information

Social change deviance intensity, verbal content
1   Not threatening to status quo
2   Minimal threat to status quo
3   Moderate threat to status quo
4   Major threat to status quo

Social change deviance intensity, visual content
1   Not threatening to status quo
2   Minimal threat to status quo
3   Moderate threat to status quo
4   Major threat to status quo

Normative deviance intensity, verbal content
1   Does not violate any norms
2   Minimal violation of one or more norms
3   Moderate violation of one or more norms
4   Major violation of one or more norms

Normative deviance intensity, visual content
1   Does not violate any norms
2   Minimal violation of one or more norms
3   Moderate violation of one or more norms
4   Major violation of one or more norms

Political significance intensity, verbal content
1   Not at all politically significant
2   Minimal political significance
3   Moderate political significance
4   Major political significance

Political significance intensity, visual content
1   Not at all politically significant
2   Minimal political significance
3   Moderate political significance
4   Major political significance

Economic significance intensity, verbal content
1   Not at all economically significant
2   Minimal economic significance
3   Moderate economic significance
4   Major economic significance

Economic significance intensity, visual content
1   Not at all economically significant
2   Minimal economic significance
3   Moderate economic significance
4   Major economic significance

Cultural significance intensity, verbal content
1   Not at all culturally significant
2   Minimal cultural significance
3   Moderate cultural significance
4   Major cultural significance

Cultural significance intensity, visual content
1   Not at all culturally significant
2   Minimal cultural significance
3   Moderate cultural significance
4   Major cultural significance

Public significance intensity, verbal content
1   Not at all significant to public's well being
2   Minimal significance to public's well being
3   Moderate significance to public's well being
4   Major significance to public's well being

Public significance intensity, visual content
1   Not at all significant to public's well being
2   Minimal significance to public's well being
3   Moderate significance to public's well being
4   Major significance to public's well being

Size of city
1   Major city
2   Peripheral city

Type of medium
1   Newspaper
2   Television
3   Radio

Newspaper: prominence of verbal content =
size of news item multiplied by its placement

Newspaper: prominence of visual content =
size of news item multiplied by its placement

TV: prominence of verbal content =
length of news item multiplied by its placement

TV: prominence of visual content =
length of news item multiplied by its placement

Radio: prominence of verbal content =
length of news item multiplied by its placement

Country
1   Australia
2   Chile
3   China
4   Germany
5   India
6   Israel
7   Jordan
8   Russia
9   South Africa
10   United States

Appendix B
Distributions of Deviance Intensity Scores
for the Ten Countries

Statistical deviance/verbal

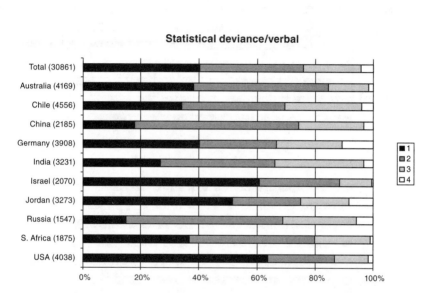

Statistical deviance/visual

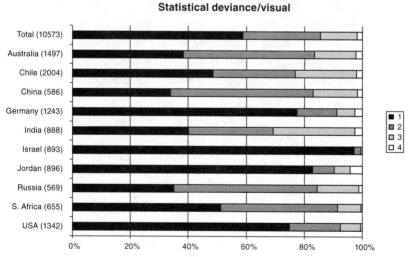

Note: The number in parentheses indicates the total number of stories analyzed in each country. Differently shaded blocks in the bars indicate the intensity of statistical deviance: 1 = common information; 2 = somewhat unusual information; 3 = unusual information; 4 = extremely unusual information.

Social change deviance/verbal

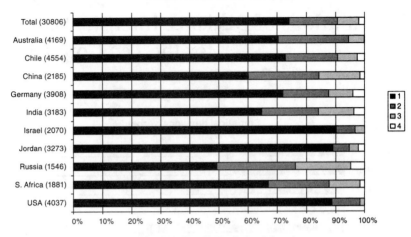

Social change deviance/visual

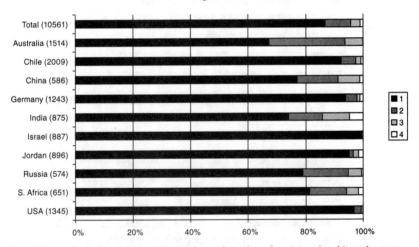

Note: The number in parentheses indicates the total number of stories analyzed in each country. Differently shaded blocks in the bars indicate the intensity of social change deviance: 1 = not threatening to status quo; 2 = minimal threat to status quo; 3 = moderate threat to status quo; 4 = major threat to status quo.

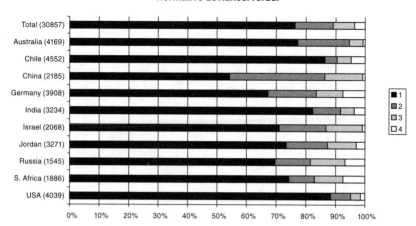

Normative deviance/verbal

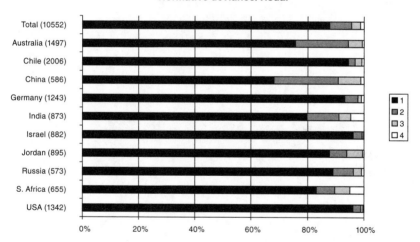

Normative deviance/visual

Note: The number in parentheses indicates the total number of stories analyzed in each country. Differently shaded blocks in the bars indicate the intensity of normative deviance: 1 = does not violate any norms; 2 = minimal violation of one or more norms; 3 = moderate violation of one or more norms; 4 = major violation of one or more norms.

Appendix C
Distributions of Social Significance
Intensity Scores for the Ten Countries

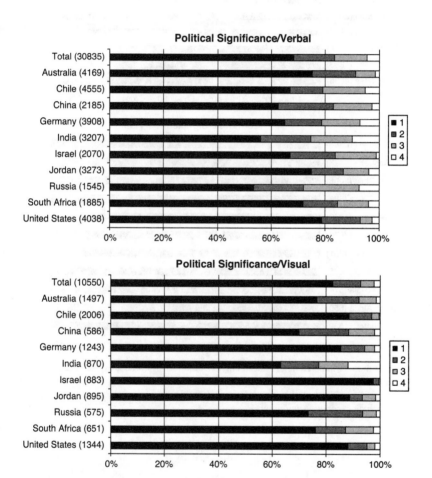

Note: The number in parentheses indicates the total number of stories analyzed in each country. Differently shaded blocks in the bars indicate the intensity degree of political significance: 1 = not at all politically significant; 2 = minimal political significance; 3 = moderate political significance; 4 = major political significance.

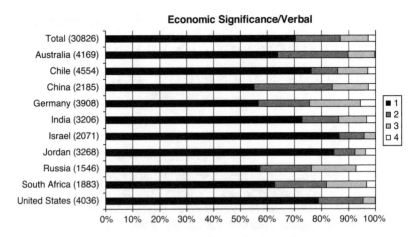

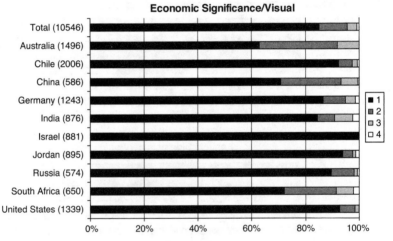

Note: The number in parentheses indicates the total number of stories analyzed in each country. Differently shaded blocks in the bars indicate the intensity of economic significance: 1 = not at all economically significant; 2 = minimal economic significance; 3 = moderate economic significance; 4 = major economic significance.

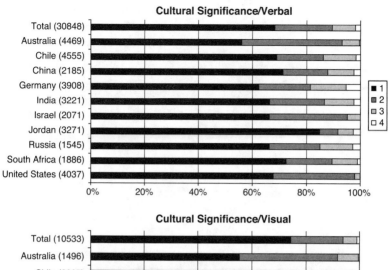

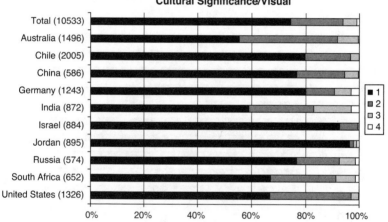

Note: The number in parentheses indicates the total number of stories analyzed in each country. Differently shaded blocks in the bars indicate the intensity of cultural significance: 1 = not at all culturally significant; 2 = minimal cultural significance; 3 = moderate cultural significance; 4 = major cultural significance.

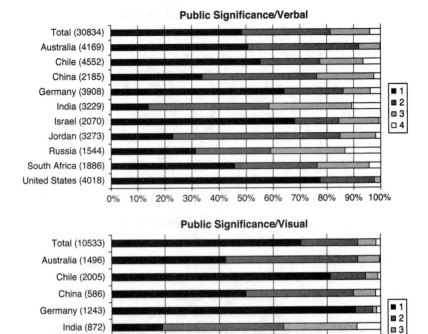

Public Significance/Verbal

Total (30834)
Australia (4169)
Chile (4552)
China (2185)
Germany (3908)
India (3229)
Israel (2070)
Jordan (3273)
Russia (1544)
South Africa (1886)
United States (4018)

0% 10% 20% 30% 40% 50% 60% 70% 80% 90% 100%

■ 1
■ 2
▣ 3
□ 4

Public Significance/Visual

Total (10533)
Australia (1496)
Chile (2005)
China (586)
Germany (1243)
India (872)
Israel (884)
Jordan (895)
Russia (574)
South Africa (652)
United States (1326)

0% 20% 40% 60% 80% 100%

■ 1
■ 2
▣ 3
□ 4

Note: The number in parentheses indicates the total number of stories analyzed in each country. Differently shaded blocks in the bars indicate the intensity of public significance: 1 = not at all significant to public's well being; 2 = minimal significance to public's well being; 3 = moderate significance to public's well being; 4 = major significance to public's well being.

Endnotes

1. There are two other definitions of deviance in the literature, but these do not lend themselves easily to the study of news definitions around the world. *Labeling* deviance describes situations in which labels or words are used by others to describe a person (Becker 1963; Erikson 1966). A teacher who describes a student as a slow learner may affect that student's life thereafter when subsequent teachers hear of and adopt the slow-learner label. It is clear that a teacher's expectations of a student influence the student's success.

   Related to labeling is *self-conception* deviance (Wells 1978). If students adopt a teacher's or parent's opinion as self-descriptive, their self-conception has been changed. Students think what others say about them must be true and then believe it to represent their inner self.

   While labeling and self-conception are valid and important types of deviance, they do not apply to this study, in which the content of news is being studied and what people think about the news is analyzed. Mass media content has no self-conception; while people do, our emphasis is not on what people think of themselves but on what they think of the news.

2. The Scott's pi coefficient ranges from .00 to 1.00; the closer to 1.00, the more reliable the coding. Scott's pi was used to measure intercoder agreement on all variables, with the exception of *topic* and *size/length*. The topic variable has twenty-six main categories and well over 100 subcategories, so the Scott's pi coefficient was inappropriate. Measuring the size of a newspaper article in square centimeters and the length of television and radio stories in seconds was a more objective task than assessing its normative deviance, so no reliability check was performed. The training process included explicit rules for making the size and time measurements.

3. *Socioeconomic status* (SES) is defined here as a person's education, income, and occupational prestige.

4. It should be emphasized that the newspapers used for the gatekeeping exercise appeared at least several months prior to, and in some case about a year prior to, conducting the focus groups. This time gap was inevitable given the design of the study, but it also served to decrease the chances that the focus group participants would recall the specific events used and in turn would remember the newspaper's treatment of them.

5. As stated in chapter 3, in the Israeli case the only local newspapers available in both cities were weeklies, hence the emphasis on softer news stories.

6. Some newspaper articles include visuals (e.g., photographs, maps, figures), but most do not. A very small number include visuals only. In our study, of the 25,886 newspaper items analyzed only 6,688 include both text and visuals, and 1,049 items consist of visuals only. Therefore, to compare the deviance in verbal newspaper text and visuals, we are limited to the subset of items that includes both types of information. Television news items always include both visuals and text. Thus, we look at a total of 9,458 news items: the 6,688 items from newspaper news and 2,770 items from television news.

7. There are exceptions, however. For example, during the 1970s, very graphic and deviant videos from the Vietnam War were shown on U.S. television networks in early-evening newscasts. Ultimately, so much criticism of this occurred that U.S. television news today rarely shows very deviant images. Following the events of September 11, 2001, the same dilemma came up in the United States and in some other countries.

8. The two deviance intensity indexes (deviance of visuals and deviance of verbal text) are the averages of the three deviance (statistical, social change, normative) scores for each country.

9. The two social significance intensity indexes (social significance of verbal content and social significance of visual content) are the averages of the four social significance (political, economic, cultural, and public) scores for each country, both visual and verbal.

10. As items with major significance are rare, cell sizes in "major significance" categories are extremely small, and some cells had no items at all. This leads to instability in the analysis of political, economic, cultural, and public significance at this level. For example, Brisbane newspaper verbal ratings had only three items in this cell. The mean prominence score appears to be very low (87.7 compared to over 300.0 at other levels of public significance); however, this may be interpreted as a result of the small cell size.

11. The newspaper shut down in April 2004 due to financial difficulties.

12. It should be noted that in September 2002 it was replaced by a daily newscast *Vesti-Tula* (News of Tula).

13. The construction *event, person, or idea* conveys the variability in what news is about, but is inelegant. In the remainder of this chapter, we use the term *event* to represent news about events as well as news about people and ideas. This is reasonable, because most news is about events. People and ideas are most often included in news that centers on an event, such as the president (person) of a country giving a news conference talking about a certain policy (idea).

References

Adoni, H., and S. Mane. 1984. Media and the social construction of reality: Toward an integration of theory and research. *Communication Research* 11: 323–340.

Australian Broadcast Authority. 2002. *Annual report, 2001–2001.* Sydney: Australian Broadcast Authority.

Asociación Nacional de la Prensa (ANP). 2003. Empresas asociadas (available at www.anp.cl/site/pags/empresas/index.html).

Avraham, E. 2000. Cities and their news media image. *Cities* 17(5): 363–370.

———. 2002. Sociopolitical environment, journalism practice, and coverage of minorities: The case of the marginal cities in Israel. *Media, Culture and Society* 24(1): 69–86.

Ball-Rokeach, S. J. 1985. The origins of individual media-system dependency: A sociological framework. *Communication Research* 12(4): 485–510.

Bandura, A. 1971. *Social learning theory.* Morristown, NJ: General Learning Press.

Barkow, J. H. 1989. The elastic between genes and culture. *Ethology and Sociobiology* 10: 111–129.

Becker, H. S. 1963. *Outsiders: Studies in the sociology of deviance.* New York: Free Press.

Breed, W. 1955. Social control in the newsroom: A functional analysis. *Social Forces* 33: 326–335.

Brosius, H. B., and A. Fahr. 1996. *Die Berichterstattung des SAT.1 Regional reports Rheinland-Pfalz/Hessen: Informationsleistung und Regionalisierung* [Regional news in SAT.1 Rhineland-Palatinate/Hesse]. Ludwigshafen, Germany: LPR.

Buss, D. M. 1991. Evolutionary personality psychology. *Annual Review of Psychology* 42: 459–491.

Campbell, R., C. R. Martin, and B. Fabos. 2002. *Media and culture: An introduction to mass communication.* 3rd ed. Boston, MA: Bedford/St. Martin's.

Carey, J. W. 1987. The dark continent of American journalism. In *Reading the news,* edited by R. K. Manoff and M. Schudson, 146–196. New York: Pantheon.

Carter, S., F. Fico, and J. A. McCabe. 2002. Partisan and structural balance in local television election coverage. *Journalism and Mass Communication Quarterly* 79(1), 41–53.

Caspi, D. 1986. *Media decentralization: The case of Israel's local newspapers.* New Brunswick, NJ: Transaction Books.

Caspi, D., and Y. Limor. 1999. *The in/outsiders: Mass media in Israel.* Cresskill, NJ: The Hampton Press.

Central Bureau of Statistics. 2002. *Statistical yearbook of Israel, 2001.* Jerusalem, Israel: Israel Government Printing Office.

Chen, C. S., J. H. Zhu, and W. Wu. 1997. The Chinese journalist. In *The global journalist: Studies of journalists around the world,* edited by D. H. Weaver, 9–30. Cresskill, NJ: Hampton Press.

Chomsky, N. 1968. *Language and mind.* New York: Harcourt, Brace.

CNTV. 2003. *Consejo nacional de television.* (Retrieved from http://www.cntv.cl/link.cgi/Consejo).

Cohen, A., H. Adoni, and C. R. Bantz. 1990. *Social conflict and television news: Sage library of social research.* Thousand Oaks, CA: Sage Publications.

Cohen, A. A., M. R. Levy, I. Roeh, and M. Gurevitch. 1996. *Global newsroom, local audiences: A study of the Eurovision News Exchange.* London: John Libbey.

Cole, M., and S. Scribner. 1974. *Culture and thought: A psychological introduction.* New York: Wiley.

Cosmides. L., and J. Tooby. 1987. From evolution to behavior: Evolutionary psychology as the missing link. In *The latest and the best: Essays on evolution and optimality,* edited by J. Dupré, 277–306. Cambridge MA: MIT Press.

Dal, V. I. 1989. *Tolkovij slovar velikorusskogoy yazika* [Dictionary of Russian language]. Moscow, Russia: Russkij Yazik [Russian Language Publishing House].

Darschin, W. and S. Kayser. 2001. Fernsehgewohnheiten und Programmbewertungen im Jahr 2000 [Habits of media use and program images in the year 2000]. *Media Perspektiven:* 162–175.

Darwin, C. 1936a [1860]. *The origin of species.* New York: Random House.

———. 1936b [1871]. *The descent of man.* New York: Random House.

Dermota, K. 2002. *Chile Inédito, el periodismo bajo democracia.*

Donsbach, W. 1999. Journalism research. In *German communication yearbook*, edited by H.-B. Brosius and C. Holtz-Bacha. Cresskill, 159–180: Hampton Press.

Eckland, B. 1982. Theories of mate selection. *Social Biology* 29: 7–21.

Eilders, C. 1997. *Nachrichtenfaktoren und rezeption: Eine empirische analyse zur auswahl und verarbeitung politischer information* [News factors and news reception. An empirical analysis of the selection and processing of political information]. Opladen, Germany: Westdeutscher Verlag.

Eimeren, B. V., and H. Gerhard. 2000. ARD/ZDF-Online-Studie 2000. *Media Perspektiven* 8: 338–349.

Erikson, K. T. 1966. *Wayward puritans: A study in the sociology of deviance*. New York: John Wiley and Sons.

Fishman, M. 1980. *Manufacturing the news*. Austin: University of Texas Press.

Galtung, J., and M. H. Ruge. 1965. The structure of foreign news: The presentation of the Congo, Cuba, and Cyprus crises in four Norwegian newspapers. *Journal of Peace Research* 1: 64–91.

Gans, H. J. 1979. *Deciding what's news*. New York: Vintage Books.

Geiger, G. 1990. *Evolutionary instability: Logical and material aspects of a unified theory of biosocial evolution*. Berlin, Germany: Springer-Verlag.

Gerbner, G., and G. Marvanyi. 1977. The many worlds of the world's press. *Journal of Communication* 27(1): 52–66.

Godoy, S. 1999. *Gestión de Radio y TV*. Ediciones Universidad Católica de Chile.

Gove, W. R. 1987. Sociobiology misses the mark: An essay on why biology but not sociobiology is very relevant to sociology. *American Sociologist* 18: 257–277.

Government Communication and Information System. 2002. *South Africa yearbook, 2002–2003*. Pretoria, South Africa: Government Communication and Information System.

Gronemeyer, M. 2002. El reto de formar periodistas autónomos e independientes. *Cuadernos de Información* 15: 53–70.

Hall, S. 1981. The determination of news photographs. In *The manufacture of news: A reader*, edited by S. Cohen and J. Young, 226–243. Beverly Hills, CA: Sage.

Handwerker, W. P. 1989. The origins and evolution of culture. *American Anthropologist* 91, 313–326.

Hans-Bredow-Institut. 2000. Internationales handbuch für hörfunk und fernsehen, 2000–2001 [International handbook of radio and television, 2000–2001]. Baden-Baden, Germany: Nomos.

Harcup, T., and D. O'Neill. 2001. What is news? Galtung and Ruge revisited. *Journalism Studies* 2(2): 261–280.

Hester, A. 1973. Theoretical considerations in predicting volume and direction of international information flow. *Gazette* 19(4): 239–247.

Hilgartner, S., and C. L. Bosk. 1988. The rise and fall of social problems: A public arena model. *American Journal of Sociology* 94: 53–78.

Instituto Nacional de Estadísticas (INE). 2003. *Censo Nacional de Población y VI de Vivienda*. Santiago: Instituto Nacional de Estadísticas.

Iyer, V. 2000. Mass media laws and regulations in Asia (India). *Series of monographs on mass media laws and regulations in Asia*. Singapore: Asian Media Information and Communication Centre.

Jensen, K. B. (Ed.) 1998. *News of the world: World cultures look at television news*. London: Routledge.

Karan, K., and R. Mathur. 2003. India. In *Asian Communication Handbook*, edited by A. Goonasekara, L. C. Wah, and S.Venkatraman, 93–122. Singapore: Asian Media Information and Communication Centre.

Kemper, T. D. 1987. How many emotions are there? Wedding the social and the autonomic components. *American Journal of Sociology* 93: 263–289.

Kenez, P. 1985. *The birth of the propaganda state: Soviet methods of mobilization, 1917–1929*. Cambridge, UK: Cambridge University Press.

Kim, K., and G. A. Barnett. 1996. The determinants of international news flow: A network analysis. *Communication Research* 23(3): 323–352.

Kim, Y. Y. 1977. Communication patterns of foreign immigrants in the process of acculturation. *Human Communication Research* 4: 66–77.

Kleinsteuber, H. J. 1997. Federal Republic of Germany (FRG). In *The media in Western Europe: The euromedia handbook*, edited by B. S. Østergaard, 75–97. London: Sage.

Klingler, W. and D. K. Mueller, Hoerfunknutzung in Deutschland [Radio use in Germany]. *Media Perspektiven* 9: 434–449.

Koch, T. 1990. *The news as myth: Fact and context in journalism*. New York: Greenwood Press.

Krohne, W. 2002. *La libertad de Expresión en Chile bajo la atenta mirada de la crítica.* Santiago, Chile: Fundación Konrad Adenauer.

Lang, P. 1985. The cognitive psychopsysiology of emotion: Fear and anxiety. In *Anxiety and the anxiety disorder,* edited by A. Tuma and J. Maser, 117–130. Urbana: University of Illinois Press.

Lasswell, H. D. 1960. The structure and function of communication in society. In *Mass communications,* edited by W. Schramm, 117–130. Urbana: University of Illinois Press.

Lazarsfeld, P. F., and R. K. Merton. 1948. Mass communication, popular taste, and organized social action. In *The communication of ideas,* edited by L. Bryson, 95–118. New York: Harper and Brothers.

Lopreato, J. 1984. *Human nature and biocultural evolution.* Boston, MA: Allen and Unwin.

Lumsden, C. J., and R. O. Wilson. 1981. *Genes, mind, and culture: The coevolutionary process.* Cambridge, MA: Harvard University Press.

Malamuth, N. M., C. L. Heavey, and D. Linz. 1993. Predicting men's antisocial behavior against women: The interaction model of sexual aggression. In *Sexual aggression: Issues in etiology, assessment, and treatment,* edited by G. N. Hall, R. Hirschman, J. Graham, and M. Zaragoza, 63–97. Washington, DC: Hemisphere.

Malek, A., and A. P. Kavoori (Eds.). 2000. *The global dynamics of news: Studies in international news coverage and news agenda.* Stamford, CT: Ablex.

Malik, R. and K. Anderson. 1992. The global news agenda survey. *InterMedia* 20(1): 8–70.

Mankekar, D. R. 1978. *Whose news? Whose freedom?* New Delhi, India: Clarion.

Matza, D. 1969. *Becoming deviant.* Englewood Cliffs, NJ: Prentice Hall.

McCombs, M. E., and J. H. Zhu. 1995. Capacity, diversity and volatility of the public agenda: Trends from 1954 to 1994. *Public Opinion Quarterly* 59: 495–525.

McQueen, H. 1978. *Australia's media monopolies.* Melbourne, Australia: Visa Books.

Molotch, H., and M. Lester. 1974. News as purposive behavior: On the strategic use of routine events, accidents, and scandals. *American Sociological Review* 39: 101–112.

Mosciatti, M. 2003. *En Entrevista virtual efectuada a Mauro Mosciatti* (available at: http://www.telefonicamundo.cl/clientes/noticia11.htm).

Nataranjan, K., and H. Xiaoming. 2003. An Asian voice? A comparative study of Channel News Asia and CNN. *Journal of Communication* 53(2): 300–314.

Nelson, K. 1989. Remembering: A functional developmental perspective. In *Memory: Interdisciplinary approaches,* edited by P. R. Soloman, G. R. Goethals, C. M. Kelley, and B. R. Stephens, 127–150. New York: Springer-Verlag.

Neuberger, C. 2000. Massenmedien im Internet. *Media Perspektiven* 3: 102–109.

Newell, A. 1990. *Unified theories of cognition.* Cambridge, MA: Harvard University Press.

Newhagen, J. E., and B. Reeves. 1992. The evening's bad news: Effects of compelling negative television news images on memory. *Journal of Communication* 42(2): 25–41.

Nisbett, R., and L. Ross. 1980. *Human inference: Strategies and shortcomings of social judgment.* New York: Prentice-Hall.

NRS. 2002. Title (available at: http://www2.cddc.vt.edu.pipermail/icernet/2002).

Osenberg, R. J. 1964. The social integration and adjustment of postwar immigrants in Montreal and Toronto. *Canadian Review of Sociology and Anthropology* 1: 202–214.

Pasadeos, Y., E. E. Hoff, Y. Stuart, and L. Ralstin. 1998. *Guiding lights of international news-flow research: A temporal comparison of influential authors and published works.* Paper presented at the annual meeting of the Association for Education in Journalism and Mass Communication, in Baltimore, Maryland, August 1998.

Patterson, T. 1998. Political roles of the journalist. In *The politics of news—the news politics,* edited by D. Graber, D. McQuail, and P. Norris, 17–32. Washington, DC: CQ Press.

Pew Research Center for the People and the Press. June 9, 2002. Public's news habits little changed by Sept. 11: Americans lack background to follow international news. Washington, DC: The Pew Research Center for the People and the Press (Available at: http://people-press.org/reports/display.php3?ReportID=156).

Pfetsch, B. 2001. Political communication culture in the United States and Germany. *Harvard International Journal of Press/Politics* 6(1): 46–67.

Phillips, E. B. 1976. Novelty without change. *Journal of Communication* 26(4): 87–92.

Piaget, J. 1972. Intellectual evolution from adolescence to adulthood. *Human Development* 15: 1–12.

Press in India. 2001. *45th annual report of the registrar of newspapers for India.* New Delhi, India: Ministry of information and Broadcasting, Government of India.

Reinemann, C. 2003. *Medienmacher als mediennutzer: Kommunikations und einflußstrukturen im politischen journalismus der gegenwart* [Media makers as media users: Structures of communication and influence in contemporary political journalism]. Cologne, Germany: Boehlau.

Remington, T. F. 1988. *The truth of authority: Ideology and communication in the Soviet Union.* Pittsburgh, PA: University of Pittsburgh Press.

Rindos, D. 1986. The evolution of the capacity for culture: Sociobiology, structuralism, and cultural selectionism. *Current Anthropology* 27: 315–326.

Roelofse, K. 1996. *Unit 3: The history of the South African press.* In *Introduction to Communication: Journalism, press and radio studies,* edited by L. M. Oosthuizen, 66–118. Cape Town, South Africa: Juta.

Roeper, H. 2000. Zeitungsmarkt 2000: Konsolidierungsphase beendet [Newspaper market 2000: The end of a phase of consolidation] *Media Perspektiven* 7: 297–309.

Rogers, A. R. 1988. Does biology constrain culture? *American Anthropology* 90: 819–831.

Rovee-Collier, C. 1989. The joy of kicking: Memories, motives, and mobiles. In *Memory: Interdisciplinary approaches,* edited by P. R. Solomon, G. R. Goethals, C. M. Kelley, and B. R. Stephens, 151–180. New York: Springer-Verlag.

Schejter, A. 1999. From a tool for national cohesion to a manifestation of national conflict: The evolution of cable television policy in Israel, 1986–1998. *Communication Law and Policy* 4(2): 177–200.

Schoenbach, K. 1997. *Zeitungen in den neunzigern: Faktoren ihres erfolgs* [Newspapers in the nineties: Factors of their success]. Bonn, Germany: ZV Service.

Schoenbach, K., and L. Goertz. 1995. *Radio-nachrichten: Bunt und fluechtig* [Radio news: Colorful and superficial]? Berlin, Germany: Vistas.

Schuetz, W. J. 2000. Deutsche tagespresse 1999 [German daily newspapers 1999]. *Media Perspektiven* 1: 8–29.

Schulz, R. 1999. Nutzung von Zeitungen und Zeitschriften [Usage of newspapers and magazines]. In *Mediengeschichte der Bundesrepublik Deutschland* [Media history of the Federal Republic of Germany], edited by J. Wilke. Cologne, Germany: Boehlau.

Schuring, G. K. 1993. *Sensusdata oor die tale van Suid-Afrika in 1991* [Census data on the languages of South Africa in 1991]. Unpublished work document. Pretoria: HSRC.

Shils, E. 1988. Center and periphery: An idea and its career, 1935–1987. In *Center: Ideas and institutions,* edited by L. Greenfeld and M. Martin, 250–282. Chicago: University of Chicago Press.

Shlapentokh, V. 1986. *Soviet public opinion and ideology: Mythology and pragmatism in interaction.* New York: Praeger.

———. 1989. *Public and private life of the Soviet people.* New York: Oxford University Press.

Shoemaker, P. J. 1991. *Gatekeeping.* Newbury Park, CA: Sage.

———. 1996. Hardwired for news: Using biological and cultural evolution to explain the surveillance function. *Journal of Communications* 46: 32–47.

Shoemaker, P. J., T. Chang, and N. Brendlinger. 1987. Deviance as a predictor of newsworthiness: Coverage of international events in the U.S. media. In *Communication Yearbook 10,* edited by M. L. McLaughlin, 348–365. Newbury Park, CA: Sage.

Shoemaker, P. J., L. H. Danielian, and N. Brendlinger. 1991. Deviant acts, risky business, and U.S. interests: The newsworthiness of world events. *Journalism Quarterly* 68(4): 781–795.

Shoemaker, P. J., and S. D. Reese. 1996. *Mediating the message: Theories of influences on mass media content,* 2nd ed. New York: Longman.

Shoemaker, P. J., S. D. Reese, and W. Danielson. 1985. Spanish-language print media use as an indicator of acculturation. *Journalism Quarterly* 62: 734–740.

Singer, J. L. 1980. The power and limitations of television: A cognitive-affective analysis. In *The entertainment functions of television,* edited by P. H. Tannenbaum, 31–65. Hillsdale, NJ: Erlbaum.

Sreberny-Mohammadi, A. 1984. Results of international cooperation. *Journal of Communication* 34(1): 121–134.

Statistics South Africa. 2000. *Census 1996.* Pretoria, South Africa: Government Printer.

Tuchman, G. 1978. Making news: A study in the construction of reality. New York: Free Press.

Union of Journalists of Russia. 1997. *Atlas of Russian TV.* Moscow, Russia: Union of Journalists of Russia.

———. 1998. *Mass Media 1997: Analysis, tendencies, forecasts.* Moscow, Russia: Union of Journalists of Russia.

U.S. Census Bureau, Population Division. July 13, 2003. U.S. and world population clocks (available at: http://www.census.gov/main/www/popclock.html).

Van Dijk, T. A. 1988. *News as discourse*. Hillsdale, NJ: Erlbaum.

Van Winkle, W. February 1998. Information overload: Fighting data asphyxiation is difficult but possible. *Computer Bits* 8(2) (available at: http://www.computerbits.com/archive/1998/0200/infoload.html).

Venkateshwaran, K. S. 1993. *Mass media laws and regulations in India* [compilation]. Singapore: Asian Mass Communication Research and Information Centre.

Wallis, R., and S. Baran. 1990. *The known world of broadcast news: International news and the electronic media*. London: Routledge.

Weaver, D. H. (Ed.). 1998. *The global journalist: News people around the world*. Cresskill, NJ: Hampton Press.

Wells, L. E. 1978. Theories of deviance and the self-concept. *Social Psychology* 41: 189–204.

White, D. M. 1950. The "gate keeper": A case study in the selection of news. *Journalism Quarterly* 27: 383–390.

Wigston, D. 1996. *Unit 12: A historical overview of radio*. In *Introduction to Communication: Journalism, press and radio studies*, edited by L. M. Oosthuizen, 283–326. Cape Town, South Africa: Juta.

Wilkins, L., and P. Patterson. 1987. Risk analysis and the construction of news. *Journal of Communication* 37: 80–92.

Wilson, E. O. 1978. *On human nature*. Cambridge, MA: Harvard University Press.

Wozniak, P. R. 1984. Making sociobiological sense out of sociology. *Sociological Quarterly* 25: 191–204.

Wu, H. D. 2000. Systematic determinants of international news coverage: A comparison of 38 countries. *Journal of Communication* 50(2): 110–130.

Zassoursky, Y., E. Vartanova, I. Zassoursky, A. Raskin, and A. Richter. 2002. *Mass media in post-Soviet Russia*. Moscow, Russia: Spekt Press.

Zhao, Y. 1998. *Media, market, and democracy in China: Between the party line and the bottom line*. Urbana: University of Illinois Press.

Zhu, J. H., and Z. He. 2002. Perceived characteristics, perceived needs, and perceived popularity: Adoption and use of the Internet in China. *Communication Research* 29(4): 466–495.

About the Contributors

Mohammed Issa Taha Ali Assistant professor of economics, Princess Sumaya University of Technology, Amman, Jordan. He was previously the head of industrial studies at the Computer Technology Centre of the Royal Scientific Society. He also served as the head of the industrial economics division and as an economic researcher at the Royal Scientific Society. He received his Ph.D. from the University of Leeds in the United Kingdom and is the author of four books, thirty-two articles, and six papers that have been presented at international conferences. Dr. Ali's research interests include regional economic cooperation, econometric models, and opinion polls.

Heather L. Black Ph.D. candidate in mass communications, S.I. Newhouse School of Public Communications, Syracuse University, New York, United States; and senior project director (2003–present), Berrier Associates, Narberth, Pennsylvania, United States. From 2000–2002 she was research assistant to Dr. Pamela J. Shoemaker, the John Ben Snow Professor, S.I. Newhouse School of Public Communications. She earned her M.A. in public relations from the S.I. Newhouse School of Public Communications in 2000. From 1998–1999 she was communications associate for the Philadelphia Corporation for Aging. From 1996–1998 she was communications specialist for the Pennsylvania Association of Non-Profit Homes for the Aging.

Natalia Bolotina A journalist and independent researcher and graduate of the School of Journalism, Moscow State University. She defended her Ph.D. thesis in the history of Russian journalism at Moscow State in 1999. She also worked as a researcher at the Russian–Finnish Centre for Research in Journalism, Mass Media, and Culture, Moscow State University. Recently, Bolotina has been contributing to various leading Canadian media outlets like Canadian Broadcasting Corporation (CBC) radio and television, Discovery Channel, and Women's Television Network.

Akiba A. Cohen Professor of communication (1996–present), Tel Aviv University, Israel. Cohen is the founding chair of the Department of Communication at Tel Aviv University. He was previously at the Hebrew University of Jerusalem, where he served as chair of the Department of Communication and Journalism (1990–1993) and as director of the Smart Family Communications Research Institute (1986–1990). His books include *The Holocaust and the Press: Nazi War-Crimes Trials in Germany and Israel* (with Tamar Zemach-Marom, Jürgen Wilke, and Birgit Schenk; Hampton Press, 2002), *Global Newsrooms, Local Audiences: A Study of the Eurovision News Exchange* (with Mark R. Levy, Michael Gurevitch, and Itzhak Roeh; John Libbey, 1996), *Social Conflict and Television News* (with Hanna Adoni and Charles Bantz; Sage, 1990), and *The Television News Interview* (Sage, 1987). Cohen has served as president of the International Communication Association and is an elected fellow of the Association.

Danie F. du Plessis Chair of the Department of Communication Science (acting chair since 2001 and chair from 2004), University of South Africa (UNISA). He joined the Department of Communication Science at UNISA in 1988. His books include editing *Introduction to Public Relations and Advertising* (Juta Academic, 2001), *Multilingualism in the Workplace* (with Gerhard Schuring), an adaptation of the public relations classic *Effective Public Relations* (Cutlip, Center, and Broom), and chapter contributions to other books. He served for a term on the executive board of the South African Communication Association.

Martin Eichholz Director of research at Frank N. Magid Associates, New York, NY (2000–present). His consumer and business-to-business research activities focus on branding and positioning, new product development, audience segmentation, and customer satisfaction. Eichholz serves clients in a variety of media-related industries such as television (USA Networks, ABC Family, Sesame Workshop, National Geographic television), publishing (Tribune, Reed Business Information), entertainment (Disney, Gaylord Entertainment), technology (Microsoft, British Telecom), gaming (Electronic Arts, Maxis), and Internet (CBS's MarketWatch.com, AskJeeves.com, Cars.com). He is a former Fulbright scholar who received his Ph.D. in mass communication from S.I. Newhouse School of Public Communications, Syracuse University. He is a member of the American Association for Public Opinion Research and has published articles in the *International Journal of Public Opinion Research* (2002) and *Journalism & Mass Communication Quarterly* (2001).

Kavita Karan Ph.D. (LSE), M.Phil, M.C.J., B.C.J. from Osmania University, Hyderabad, India. She is on the faculty of the School of Communication and Information, Nanyang Technological University, Singapore (2001–present). From 1999–2001, she was associate professor and head, Department of

Communication and Journalism, Osmania University. From 1996–1998 she was chair of the Board of Studies in the same department and served on the faculty from 1984–1998. She obtained her Ph.D. from the London School of Economics and Political Science, United Kingdom, with specialization in political communication. She received the prestigious Nehru Centenary British Scholarship for pursuing the doctoral program in England. She has published chapters in books and journals. Her interests include political communication, advertising and market research, international communication, media, as well as and children and health communication.

Chris Lawe-Davies Has taught in the School of Journalism and Communication at the University of Queensland, Australia, for the past twelve years. During that time he also was director of studies for three years in the Faculty of Social Sciences and acting head of school for eight months. He has held academic appointments at Curtin University of Technology, Australia, and the University of Southern Queensland, Australia. He has served as president, vice president, and treasurer of the Australasian Journalism Education Association. His writing and research have been published in the fields of multicultural broadcasting; indigenous music; and popular culture, news, and audiences. He has also worked as a journalist in Australia and Scandinavia and has worked for periods of time in advertising and consumer research.

Robyne M. Le Brocque Research officer at the University of Queensland, Australia. She has coordinated a large longitudinal study looking at the transition to adulthood in a high-risk sample. Her interest is in methodological issues including issues arising from longitudinal data. In addition, she has published in the area of maternal mental health and child behavior problems. Recent papers include "Methodological Issues in the Effects of Attrition: Simple Solutions for Social Scientists" (*Field Methods*, 2004), "Maternal Depression, Parent–Child Relationships, and Resilient Outcomes in Adolescence" (*Journal of the American Academy of Child and Adolescent Psychiatry*, 2003), and "Maternal Depression, Paternal Psychopathology, and Adolescent Diagnostic Outcomes" (*Journal of Consulting and Clinical Psychology*, 2002).

Noa Loffler-Elefant M.A. in political science, Tel Aviv University. She is head of the Research and Information Department of Israel's Second Television and Radio Authority (the regulator of commercial television and radio). In this capacity she was part of a research team studying the representation of minorities in commercial broadcasting in Israel. She was previously a member of the founding team of the Research and Information Department of the Knesset (the Israeli parliament) and served as the coordinator of the research assistants of several disciplines. Her research interests include comparative research on the regulation of television and radio broadcasting as well as multiculturalism in commercial media.

Constanza Mujica Instructor, School of Journalism, Pontificia Universidad Católica de Chile (2001–present). Her teaching and research areas are journalistic writing and audiovisual analysis of television contents. With a group of researchers (Francisca Alessandri, Silvia Pellegrini, Soledad Puente, and William Porath) from Universidad Católica she developed a way of measuring quality of the press in the Spanish-speaking media through the concept of journalistic added value (JAV), which she won a government grant for in 2003. Currently she is working on her doctoral thesis on the analysis of the contrasts between narrative–verbal and audiovisual contents in Chilean soap operas, and together with Professors William Porath and Francisco Fernández she has applied for a government grant to develop an instrument to measure the quality of nonjournalistic television contents.

Soledad Puente Associate professor, School of Journalism, Pontificia Universidad Católica de Chile (1993–present). From 1995–2003 she was director of the School of Journalism. Her teaching and research areas are television journalism and speech communication. Her most well-known book is the *La noticia se cuenta*, part of her doctoral thesis, which shows how television journalism can learn to tell stories as in fiction. The book has been reprinted three times in Latin America. Other books include *Producción y Productores de la Televisión Chilena* (with Hugo Miller) and *Periodismo Audiovisual* (with Hugo Miller). With a group of researchers (Francisca Alessandri, Constanza Mujica, Silvia Pellegrini, and William Porath) from Universidad Católica, she developed a way of measuring quality of the press in the Spanish-speaking media through the concept of journalistic added value (JAV), which she won a government grant for in 2003.

Carsten Reinemann Assistant professor, Institut fuer Publizistik, Johannes Gutenberg-Universität, Mainz, Germany (2003–present). His research interests include journalism studies and political communication. He has published three books, including *Medienmacher als Mediennutzer. Kommunikations—und Einflussstrukturen im politischen Journalismus der Gegenwart* [Media Makers as Media Users—Structures of Communication and Influence in Today's Political Journalism] and *Schroeder gegen Stoiber—Nutzung, Wahrnehmung und Wirkung der TV-Duelle* [Schroeder versus Stoiber—Uses, Perceptions, and Effects of the Televised Debates] (with Marcus Maurer). He has published articles in *Press/Politics* and the *European Journal of Communication* and presented a Top Paper in Political Communication at the 2004 annual meeting of the International Communication Association in New Orleans (with Marcus Maurer).

Pamela J. Shoemaker John Ben Snow Professor, S.I. Newhouse School of Public Communications, Syracuse University (1994–present). From 1991–1994 she was director of the School of Journalism at the Ohio State University and was on the faculty of the Department of Journalism at the University of Texas at Austin from 1982–1991. Her books include *How to Build Social Science Theories* (with James Tankard and Dominic Lasorsa; Sage, 2003), *Mediating the Message: Theories of Influences on Mass Media Content* (with Stephen Reese; Allyn & Bacon, 1996), *Gatekeeping (Communication Concepts)* (Sage, 1991), and an edited volume, *Communication Campaigns about Drugs: Government, Media, Public* (LEA, 1989). She is coeditor of *Communication Research* (1997–present) and was assistant editor of *Journalism Quarterly* (1990–1992). Shoemaker has served as president of the Association for Education in Journalism and Mass Communication and received its Kreighbaum Under-40 Award for Achievement in Research, Teaching, and Public Service.

Elizabeth A. Skewes Assistant professor, School of Journalism and Mass Communication, University of Colorado at Boulder, United States (2001–present). She has a Ph.D. in mass communications from Syracuse University, a master's degree in journalism from the Ohio State University, and a bachelor's degree in political science from University of California, Los Angeles, United States. Before returning to graduate school, she was the alumni magazine editor at Dickinson College, Carlisle, Pennsylvania, and a newspaper reporter at the *Tampa Tribune* in Tampa, Florida, and the *Herald-Dispatch*, Huntington, West Virginia. She is the author or coauthor of several journal articles and papers that have been presented at national and international communications conferences. She was named the outstanding faculty member at the University of Colorado's journalism school in 2004.

Guo-liang Zhang Professor, Journalism School (1996–present), and director, Center for Information and Communication Studies (CICS), Fudan University, Shanghai, China. He has a Ph.D. in history and currently is president of the Chinese Association of Communication (CAC). From 1991–1994 he was vice dean of Fudan University. He is coeditor of *Asian Journal of Communication* (2004–present). Zhang's representative work includes contemporary mass communication, news media and society, and the history of mass communication in Japan. His main research interest is Chinese audience and communication effects.

Jonathan J. H. Zhu Professor of communication and new media, Department of English and Communication, City University of Hong Kong. He taught at Fudan University from 1984 to 1986 and the University of Connecticut, Hartford, United States, from 1990 to 1999. He has published articles in such journals as *Public Opinion Quarterly, Journal of Communication, Communication Research, Human Communication Research, Political Communication, Journalism Quarterly,* and *International Journal of Public Opinion Research* and has received research awards from the International Communication Association, American Association for Public Opinion Research, World Association for

Public Opinion Research, and Association for Education in Journalism and Mass Communication. He was president of the U.S.-based Chinese Communication Association (2002–2004), is a member of the editorial board of such journals as *Communication Research, International Journal of Public Opinion Research,* and *Asian Journal of Communication,* and is a guest professor of Shanghai Jiaotong University, Shenzhen University, and others in China.

Index